Combinatorial Optimization Under Uncertainty

This book discusses the basic ideas, underlying principles, mathematical formulations, analysis and applications of the different combinatorial problems under uncertainty and attempts to provide solutions for the same. Uncertainty influences the behaviour of the market to a great extent. Global pandemics and calamities are other factors which affect and augment unpredictability in the market. The intent of this book is to develop mathematical structures for different aspects of allocation problems depicting real-life scenarios. The novel methods, which are incorporated in practical scenarios under uncertain circumstances, include the STAR heuristic approach, Matrix geometric method, Ranking function and Pythagorean fuzzy numbers, to name a few. Distinct problems which are considered in this book under uncertainty include, scheduling, the cyclic bottleneck assignment problem, the bilevel transportation problem, the multi-index transportation problem, retrial queuing, uncertain matrix games, optimal production evaluation of cotton in different soil and water conditions, the healthcare sector, the intuitionistic fuzzy quadratic programming problem and the multi-objective optimization problem. This book may serve as a valuable reference for researchers working in the domain of optimization for solving combinatorial problems under uncertainty. The contributions of this book may further help to explore new avenues leading toward multidisciplinary research discussions.

Advances in Metaheuristics

Series Editors: Patrick Siarry, *Universite Paris-Est Creteil, France*
Anand J. Kulkarni, *Dr. Vishwanath Karad MIT World Peace University, Pune, India*

Handbook of AI-based Metaheuristics
Edited by Patrick Siarry and Anand J. Kulkarni

Metaheuristic Algorithms in Industry 4.0
Edited by Pritesh Shah, Ravi Sekhar, Anand J. Kulkarni, Patrick Siarry

Constraint Handling in Cohort Intelligence Algorithm
Ishaan R. Kale, Anand J. Kulkarni

Hybrid Genetic Optimization for IC Chip Thermal Control: with MATLAB® applications
Mathew V K, Tapano Kumar Hotta

Handbook of Moth-Flame Optimization Algorithm: Variants, Hybrids, Improvements, and Applications
Edited by Seyedali Mirjalili

Combinatorial Optimization Under Uncertainty: Real-Life Scenarios in Allocation Problems
Edited by Ritu Arora, Prof. Shalini Arora, Anand J. Kulkarni, Patrick Siarry

For more information about this series please visit:
https://www.routledge.com/Advances-in-Metaheuristics/book-series/AIM

Combinatorial Optimization Under Uncertainty

Real-Life Scenarios in Allocation Problems

Edited by

Dr. Ritu Arora
University of Delhi, Delhi

Prof. Shalini Arora
Indira Gandhi Delhi Technical University for Women, Delhi

Dr. Anand J. Kulkarni
Dr. Vishwanath Karad MIT World Peace University, Pune, India

Dr. Patrick Siarry
Universite Paris-Est Creteil, France

CRC Press
Taylor & Francis Group
Boca Raton London New York

CRC Press is an imprint of the
Taylor & Francis Group, an **informa** business

First Edition published 2023
by CRC Press
6000 Broken Sound Parkway NW, Suite 300, Boca Raton, FL 33487-2742

and by CRC Press
4 Park Square, Milton Park, Abingdon, Oxon, OX14 4RN

CRC Press is an imprint of Taylor & Francis Group, LLC

Library of Congress Cataloging-in-Publication Data
Names: Arora, Ritu (Mathematician), editor. | Arora, Shalini (Mathematician), editor. | Kulkarni, Anand Jayant, editor. | Siarry, Patrick, editor.
Title: Combinatorial optimization under uncertainty : real-life scenarios in allocation problems / edited by Dr. Ritu Arora (University of Delhi, Delhi), Prof. Shalini Arora (Indira Gandhi Delhi Technical University for Women, Delhi), Dr. Anand J. Kulkarni (Nanyang Technological University, Singapore), Dr. Patrick Siarry (Paris-Sud University, France).
Description: First edition. | Boca Raton : CRC Press, 2023. |
Series: Advances in metaheuristics series | Includes bibliographical references and index. |
Identifiers: LCCN 2022047843 (print) | LCCN 2022047844 (ebook) | ISBN 9781032316581 (hardback) | ISBN 9781032358550 (paperback) | ISBN 9781003329039 (ebook)
Subjects: LCSH: Combinatorial optimization. | Uncertainty. | Assignment problems (Programming)
Classification: LCC QA402.5 .C5468 2023 (print) | LCC QA402.5 (ebook) |
DDC 519.6/4–dc23/ eng20230210
LC record available at https://lccn.loc.gov/2022047843
LC ebook record available at https://lccn.loc.gov/2022047844

ISBN: 978-1-032-31658-1 (hbk)
ISBN: 978-1-032-35855-0 (pbk)
ISBN: 978-1-003-32903-9 (ebk)

DOI: 10.1201/9781003329039

Typeset in Times
by Newgen Publishing UK

Contents

Preface

Combinatorial problems play a significant role in the field of optimization. Distinct mathematical models are developed for different aspects of allocation problems depicting real-life situations. Uncertainty can be seen in every sphere of life. It is an important factor which affects the market in each and every aspect. The global pandemic is another factor due to which uncertainty in the market increase manyfold and thus the market behaviour needs to be examined under such situations. Uncertain parameters can be dealt with via different approaches such as fuzzy programming, grey programming, robust optimization and queues etc. The objective of this edited book is to develop mathematical structures for different aspects of allocation problems depicting real-life situations. The intent is to find optimal, compromise optimal or satisfactory solutions for various allocation problems not only with the existing optimization techniques but also with other novel and modified approaches along with the usage of the software. The novel methods which are incorporated in practical scenarios under uncertain circumstances include the STAR heuristic approach, Matrix geometric method and Ranking function, to name a few. Distinct problems which are incorporated in this book under uncertainty include scheduling, the cyclic bottleneck assignment problem, the bilevel transportation problem, the multi-index transportation problem, retrial queuing, uncertain matrix games, the optimal production evaluation of cotton in different soil and water conditions, the healthcare sector, the intuitionistic fuzzy quadratic programming problem and the multi-objective optimization problem. The book intends to discuss the basic ideas, underlying principles, mathematical formulations, solutions and analysis and applications of the different combinatorial problems. It may provide guidelines to potential researchers about the choice of such methods for solving a particular class of problems at hand. The contributions of the book may further help to explore new avenues leading toward multidisciplinary research discussions. Every chapter submitted to the book has been critically evaluated by at least two expert reviewers. The critical suggestions by the reviewers helped and influenced the authors of the individual chapters to enrich the quality in terms of experimentation, performance evaluation, representation etc. The book may serve as a valuable reference for researchers working in the domain of optimization for solving combinatorial problems.

Estimation of Uncertainties for Multiserver Queuing Systems with Bernoulli Feedback presents a retrial queuing model with feedback and geometric loss. If a group of customer enters the system and only one out of them finds the free server then he accepts the service instantly and the remaining members will join the retrial orbit. But if every server is busy then they can either depart the system or enter the retrial orbit. Also, after the service is completed one can come and join the queue again if he/she is dissatisfied with the service or the service was unsuccessful. The system state governing equations are obtained and then performance measures under steady state conditions are examined. The impact of various parameters like arrival rate, service rate and retrial rate on performance measures like queue length and waiting time have been demonstrated through graphs depicting the significance of the defined problem. This model has been constructed as a quasi birth and death process. The steady state governing equations are solved numerically using the Matrix Geometric Method. Performance measures have been derived using MATLAB®. The application of the defined model can be seen in the banking services system, control service channels, computer job processing and various other applications. This could help in decision-making to improve the availability and quality of the system.

Optimality for Fuzzy Transportation Problem under Ranking Method shows the potential of a ranking approach to a transportation problem. The optimal solution for the transportation problem is obtained by applying the north west corner rule, Vogel's approximation approach and the uv–method. The resilient ranking approach is applied to solve the fuzzy transportation problem and to obtain an optimal solution by making continuous improvements in the original basic feasible solution. The transportation costs are represented as generalised trapezoidal fuzzy numbers. It gives the analyst a basic and straightforward mathematical programme and a preferred optimal solution. The robust fuzzy transportation problem is tackled in this chapter under particular conditions in order to obtain the membership function of the fuzzy total transportation costs. The concept is founded on the principle of extension. In robust optimizations, certain decisions must be made before knowing the real values of the uncertain parameters, while others, referred to as recourse decisions, must be made once the information is known. The methodology has been explained by applying the concept to the least time fuzzy transportation problem.

Solution of Bilevel Linear Fractional Transportation Problem with Pythagorean Fuzzy Numbers presents the bilevel linear fractional transportation problem with uncertain parameters. Due to the unpredictable behaviour of the market, especially during the time of the pandemic, the cost, demand and supply of the goods are ambiguous. The uncertainty in the cost coefficients of the objective functions, supply and demand parameters at both levels are Pythagorean fuzzy numbers. The score function transforms these fuzzy numbers into their respective crisp parameters. The satisfactory solution for the objective functions at both levels are obtained by two distinct procedures: fuzzy programming and goal programming using LINGO 17.0. A comparative analysis of the solutions obtained from the two methods is also given. The objective of the defined problem is to minimize the transportation cost at two levels. At the same time, the maximum number of units of goods should be transported so as to serve the maximum people. As an application, the methodology has been applied to explain the situation where the essential commodities are delivered from provider states to the states in need and then on to the people at large through the nodal zones.

Optimal Production Evaluation of Cotton in Different Soil and Water Conditions in Sundarban of West Bengal under Hesitant Interval Fuzzy Environment Using Projection Measure considers an optimal production evaluation of cotton in different land and water conditions. Sundarban of West Bengal (India) has been taken as the area of study. Different crops in Sundarban are harvested, mainly food-grains like paddy with other non-food crops like oilseeds, cotton, sunflower and so forth alongside vegetables. In Sundarban, the agriculture system depends on the quality of the soil to a large extent. The main characteristics includes soil-salinity, flooding, environmental changes such as cyclonic weather, inadequate supply of fresh water irrigation etc. The proposed methodology has been applied to a case study evaluating the production of cotton in Sundarban. The linguistic-based hesitant interval fuzzy real-life data are analysed to explain the problem, and a projection measure is considered to attain the optimal solution. This study may help researchers in the agriculture sector and might be useful for better policy support to farmers and administrators.

A Novel Approach for Feature Detection in Vector Graphics proposes a novel algorithm for corner detection from vector graphics. Although almost all computer vision applications are done using raster images, many techniques require high computational power and usually include complex processes. This algorithm provides an advantage by reducing the computational power by utilizing the advantages of vector graphics. The proposed algorithm processes the Extensible Markup Language (XML) code in SVG to output coordinates of corners in the image. Conventional algorithms work on raster images and require resizing of the image on multiple scales, resulting in significant computational complexity. In contrast to the standard algorithms, vector graphics can be scaled indefinitely without any information loss. They use sequential commands or mathematical statements to represent an image which makes the image file comparatively smaller in size. The advantages and disadvantages of Scalable Vector Graphics (SVG) over Portable Network Graphics (PNG) or Joint Photographic Expert Group (JPEG) image formats are carefully weighed. The algorithm has been implemented and tested on an image dataset and the results were compared with the performance of Harris and Feature from Accelerated Segment Test (FAST) corner detectors on PNG images. This implementation was successfully tested for the following test cases: an image with a single element, a combination of different elements and complex images taken from external sources. The results indicated that the proposed algorithm worked exceptionally well on most of the simple and complex images.

On Uncertain Matrix Games Involving Linguistic Pythagorean Fuzzy Sets describes matrix games using payoffs with uncertain qualitative variables by employing Linguistic Pythagorean Fuzzy Variables. A mathematical formulation and solution concepts for the aforementioned matrix game under uncertainty is presented by converting bi-objective models into linear or non-linear programming problems and thus finding solution(s) for them. The concept of Pareto optimal solutions is used to solve matrix games incorporating Linguistic Pythagorean Fuzzy Variables. The expected value of the game is computed via implementing the linguistic Pythagorean fuzzy weighted average operator. The effectiveness of the proposed strategy has been demonstrated numerically in real-life decision-making scenarios. The significance of the proposed technique is that it precisely defines belongingness and non-belongingness of an entity by using Linguistic Pythagorean Fuzzy Variables. Due to the reduced probability of data loss or distortion during the resolution of these uncertain matrix games employing Linguistic Pythagorean Fuzzy Variables, players can ultimately choose the optimum course of action.

Cyclic Surgery Scheduling using Variations of Cohort Intelligence describes the problems in the healthcare domain. Variations of their Cohort Intelligence (CI) algorithm, namely, follow best, follow better, follow itself, alienation and random selection approach, follow worst and follow median, are applied for healthcare scheduling problems. Healthcare scheduling problems include the scheduling of the admisson of patients problem, the nurse scheduling problem, the operation room scheduling problem and the surgery scheduling problem. Healthcare scheduling is a demanding job with several constraints like accommodating the patient's post-surgery in the recovery unit along with scheduling the surgery, the availability of nurses, hospital ward members and particular doctors, the availability of equipment etc. The defined problem is a new variant of the assignment problem that aims to minimize the maximum number of patients in the recovery unit for a cyclic surgery schedule. This kind of a problem is referred to as a Cyclic Bottleneck Assignment Problem (CBAP). The developed algorithm is substantiated by solving 16 problems of sizes ranging from 5×5 up to 50×50, with 10 instances for each case. The six variations of CI have been investigated to solve CBAP and it is observed that the Follow itself and alienation variations gave the best results and also showed improvements compared to the roulette wheel approach.

Cone Method for Uncertain Multiobjective Optimization Problems with Minmax Robustness represents a robust optimization to find feasible solutions for every possible scenario of any uncertain input data. Robust optimization is a well-developed method to deal with single objective decision-making problems in the presence of uncertainty. However, various real-world decision-making problems have multiple decisions or goals. Thus, this work exhibits the use of the cone method for robust multi-objective optimization problems. The cone method for deterministic multi-objective programming problems and the method of minmax robustness for single-objective uncertain optimization problems are also described. The robust counterpart of an uncertain multi-objective programming problem and the concept of robust efficiency for the same are also discussed. The formulation of a parametric single objective programming problem for an uncertain optimization problem using the cone method is described. Interpretation of a robust counterpart of an uncertain multi-objective optimization problem using the idea of objective-wise worst case is also discussed.

Solving Multi-Index Transportation Problem with Axial Constraints Having Impaired Flow presents a multi-index transportation problem with axial constraints in which heterogeneous commodities are transported from sources to destinations through different types of conveyances. This defined problem addresses these issues and the case of emergency situations when the total flow in the market needs to be impaired. A solution method is developed to find an optimal basic feasible solution of the given problem. The given problem is transformed into a related transportation problem by adding a source, a destination and a commodity, which then converts the defined problem into an equivalent standard axial sum problem. Thus, the defined problem is solved by transforming the specified problem into an equivalent multi-index transportation problem. It has been shown that the optimal basic feasible solution obtained by the proposed procedure is better than other contemporary algorithms/techniques. Numerical examples are illustrated and their computational efficiencies are also exhibited by comparing their results with examples from the literature.

STAR Heuristic Method: A Novel Approach and Its Comparative Analysis with CI Algorithm to solve CBAP in Healthcare developes the STAR heuristic method for the healthcare system and implements this method for solving the Cyclic Bottleneck Assignment Problem (CBAP) in Hospitals. The STAR Heuristic is a novel algorithm and it aims to reduce hospital overcrowding and improve cyclic scheduling. The proposed algorithm is compared to more traditional methods like the Cohort Intelligence (CI) algorithm. A comparison has been made against other algorithms using parameters like average CPU time and percentage of lower results (max column Sum). Our algorithms outperform the existing methods, such as the CBAP CI algorithm. With the non-cyclic approach to scheduling patients for each week, the STAR Heuristic method gives better efficiency than CI in terms of time taken. The algorithm was assessed on a variety of 5×5 to 50×50 matrices which have a weekly schedule for the doctor and the number of patients to be operated on/treated on that particular day, each with ten test cases. The results were compared them to existing CI outcomes. On the basis of the computational results, the method presented here shows outstanding performance in the majority of problem situations. The STAR Heuristic Method provides an optimized weekly schedule that benefits doctors and patients by using all the resources to their fullest.

Development and Optimization of Quadratic Programming Problems with Intuitionistic Fuzzy Parameters scrutinizes the construction and solution methodology of quadratic programming problems under an intuitionistic fuzzy environment. Quadratic programming problems with all the parameters as intuitionistic fuzzy numbers have been examined and two distinct techniques are proposed to obtain the optimal solutions. Due to the presence of fuzziness in the input parameters of the defined problem, the optimal objective value obtained is an interval rather than a crisp single value. In the proposed methodology, α and β cuts are applied on the objective and the constraints. Thus, the problem IF-QPP is subdivided into max-min and min-min type optimization problems. Furthermore, the upper bound subproblem, which involves max-min type optimization, is managed by introducing a duality and a computationally efficient direct solution technique. The presented mathematical explanation supports the hypothesis that the proposed direct approach results in the same optimal value for the upper bound problem as the duality theory, along with significantly reducing the number of variables and constraints in the model, thus yielding a more efficient algorithm. A numerical example is solved at the end of the chapter to establish the efficacy of the introduced technique. The presented direct approach requires substantially less computing time which permits it to be used to handle big-data problems in various different sectors.

About the Editors

Dr. Ritu Arora received her PhD degree from the University of Delhi, India. She has 20 years of teaching experience within the University of Delhi. Her research specialization is in the field of mathematical programming and its application to allocation problems. She is currently working as a Professor in the Department of Mathematics, Keshav Mahavidyalaya, University of Delhi.

Prof. Shalini Arora is presently working as Professor in Mathematics at Applied Sciences and Humanities Department, IGDTUW. She has more than 20 years of Teaching experience. She did her Masters and Ph.D in Mathematics from IIT Delhi. She is a recipient of the 'Young Scientist Award' by the SERC division of DST. Her areas of research interest include Mathematical Programming, Allocation Problems viz., Transportation and Assignment Problems, Combinatorial optimization etc.

Dr. Anand J. Kulkarni holds a PhD in Distributed Optimization from Nanyang Technological University, Singapore. He worked as a Research Fellow at the Odette School of Business, University of Windsor, Canada. He is currently working as a Professor and Associate Director of the Institute of AI at the MITWPU, Pune, India.

Dr. Patrick Siarry received a PhD degree from the University Paris 6 in 1986 and a Doctorate of Sciences (Habilitation) from the University Paris 11 in 1994. Since 1995, he has been working as a professor in automatics and informatics. His main research interests are the design of new stochastic global optimization heuristics and their application to various engineering fields.

1 Estimation of Uncertainties for Multiserver Queuing Systems with Bernoulli Feedback

Divya Agarwal, Shweta Upadhyay and Ankita Bansal
Amity Institute of Applied Sciences, Amity University, Noida, UP, India

CONTENTS

1.1 INTRODUCTION

A queuing model is constructed in a way to predict the waiting time and the length of the queue, which are the two important measures that help in making business decisions. The first work on queuing models was done by A.K. Erlang in 1909. He is acknowledged to be the father of queueing theory. Many engineers and mathematicians have been inspired by his work to solve queueing problems using probabilistic methods [1, 2].

Telephonic communiques are viable when the customer is served by another person or the server. Also, it may happen while calling via telephone that when the first person calls the second person, the second person is busy on another call with a third person. In that case the first person receives a message that the second person is busy on the

DOI: 10.1201/9781003329039-1

other call, the first person enters the virtual retrial orbit and as soon as the server or the second person becomes free the first person gets served.

In this paper retrial queues have been taken into consideration. Retrial queues or queues with repeated orders are queuing systems in which a customer arrives for a service and discovers that all of the servers are busy, so they enter the virtual orbit and can try for the service again later. If there is any server free then the customer arriving takes the service instantly and departs the system. If the arriving customer discovers that all the servers are occupied then he/she joins the retrial group, known as the orbit, and can try again for the service after a certain duration of time. Call centres, telephone systems, etc. are some examples of retrial queues, which can be seen in [2–4].

A retrial queueing system with one server subject to starting failures and with working vacations was studied by Yang et al. [5]. They applied a matrix geometric method for computing the orbit size after obtaining the system's stability condition. The purpose was to calculate the optimal service rates during the vacation period and the normal period to minimize the expected cost per unit time. Tuan Phung Duc [6] provided a comprehensive study of applications on retrial queues. Kumar et al. [7] studied the modeling and optimization of a queue with one server with feedback and collision and analyzed the model using the technique of supplementary variables.

The examination of queues with multi servers is harder than that for queues with one server. However, retrial queues with multi servers are far more versatile and relevant in practise than single server retrial queues. Batch arrival is also an important feature of queuing modeling. Its real-life applications can be seen in elevators, banks, shopping malls, etc. Many authors have shown interest in this field, which can be seen in [8–16].

The mechanism known as the Bernoulli feedback also has been studied by many authors. After receiving the service, if one is displeased or the service is unsuccessful, the consumer may instantly join the back (tail) of the earlier queue as a feedback customer in order to receive again regular service. The Bernoulli distribution is followed by the customers joining the tail of the queue for another service. There are situations where re-service is required, for instance in information transmission: A bundle sent by the source can be sent back to the destination, and this process can continue until the bundle is finally sent, while visiting a doctor a patient may require re-service for further investigations. Takacs [17] investigated a single server queue with feedback. He calculated the overall time spent by the customer in the system, and for a stationary process the queue size distribution. Some interesting literature can be found in [7, 18–20].

Many authors have discussed retrial queues but very few have discussed retrial queues with feedback and geometric loss. However, in practical life if the customer arrives and sees that the queue is overlong, they might not want to stay in the track and will leave the system after finding each server busy (loss), or the customer may require the service again after taking the service onc e(feedback). M/M/c retrial queues when c = 1,2 with feedback and geometric loss were considered by Choi et al. [21]. Lin and Ke [22] discussed a model with geometric loss and feedback where rate matrix and stationary probabilities were computed using a matrix analytical approach. More related literature is given in Table 1.1.

TABLE 1.1
Summary of Literature Review

Author	Study	Reference
T. Hanschke	The underlying queueing process of a multi server queue with repeated order has been investigated and it can be represented as a level-dependent process. This paper shows the usage of criteria based on Lyapunov functions.	[23]
Srinivas R.Chakravarthy et al.	A queuing model containing c servers where customers are coming in accordance to the arrival process in groups has been discussed. By structuring the coefficient matrices further analysis of the model is done.	[24]
B.Krishna Kumar et al.	The paper discusses a queuing syestem with retrial, feedback and containing more than one server. They have considered a system in which the waiting position is limited. Matrix geometric methods have been used to derive the system performance measures.	[25]
Shweta Upadhyaya	She investigated performance analysis of a queue with a single server, retrial, Bernoulli feedback and modified vacation policy. Batch arrival has been considered and a supplementary variable technique has been used to derive the formulas for system size mean and distribution.	[26]
Jinting Wang et al.	A queue with retrial, batch arrival, starting failures and feedback has been considered. The impact of different parameters like μ, ν on the performance measures are also presented.	[27]
Tien Van Do et al.	A retrial queue with many servers and impatient customers was studied. Relations for the conditional mean no. of customers and the rate matrix have been obtained. Performance measures versus N (threshold) have been inspected and then they have derived N.	[28]

The novelty of this paper is that it deals with a multi server retrial system with both geometric loss and feedback where arrivals are in bulk. An application example of this model can be observed at stores of shopping malls. Customers arrive in batches of random sizes. We see numerous servers there, e.g. at the cash counter, customer care counter, counter for refunds and exchange etc. When customers enter and find the server free the service is taken immediately. When customers walk in and see that all the servers are occupied they may join the virtual orbit (retrial orbit) and make recurrent attempts from the orbit to take the service. Another scenario that can occur is that after seeing a long queue the customer decides not to join the queue and makes an exit from the system (geometric loss) without taking the service. Following the completion of the service, if the customer is displeased with the service or the service failed or may require a re service, he/she may either decide not to come again or to come again and join the queue for another service (feedback).

The rest of the paper is arranged in the following manner. In section 1.2, a detailed model description, state diagram and steady-state equations have been discussed. Section 1.3 is devoted to analysis of the model using the matrix geometric method (MGM), wherein we obtain the rate matrix of the infinitesimal generator of the Q and D process. Section 1.4 provides formulas for various useful performance measures obtained using MGM. In section 1.5 we explore the numerical results using tables and graphs which depict the impact of various parameters like μ, ν, λ on the length of the queue and waiting time. Section 1.6 provides the various applications of the model and Section 1.7 is the conclusion of the paper.

1.2 PORTRAYAL OF MODEL

Consider a queueing system with retrial, Bernoulli feedback and geometric loss. The arrival of the customers is in groups that follow the Poisson process with parameter λ. Batch size distribution is independent and identical to $(c_j; j = 1, 2, ...\infty)$; $\sum_{j=1}^{\infty} c_j = 1$. If a group of customers arrives there can be two options

- A customer from a group finds a free server and obtains the service while the remaining customers in the group join the retrial track.
- None of the servers is free so they may leave the system with uncertainty (probability) $\overline{\alpha}$ or join the orbit with the uncertainty α.

Let the customers service time following exponential distribution with mean $\frac{1}{\mu}$ and be independent. After the service is completed, the customer may not be satisfied with the service so may either exit the system with uncertainty (or chance) $\overline{\beta}$ or rejoin for another service at the end (tail) of the retrial orbit with uncertainty β. Every customer who remains within the orbit conducts recurrent tries at arbitrary intervals whose length is distributed exponentially with parameter ν. When the customer requests service from orbit and still finds all c servers occupied then the customer can either make an exit with $\overline{\alpha}$ or rejoin the orbit again with α. Every customer in orbit is served in an identical manner as the primary customer. Let the maximum number of customers staying in orbit who are permitted to perform retrials be F. Assume $\nu_j dt_1 + o(dt_1)$ where j is the no. of customers in the orbit at time t_1 as the probability of a recurring attempt during (t_1, t_1+dt_1), $\nu_j = min(j, F)\nu$. The newly arriving customers and those that are fed back, both receive service in the sequence in which they are joining the back (tail) of the the primary queue. All the processes, i.e. primary arrivals, inter retrial times and service times, are assumed to be independent of one another.

The system state can be expressed as $Q(t_1) = (J_1(t_1), X_1(t_1)); t_1 \geq 0$ where $J_1(t_1)$, $X_1(t_1)$ denotes the number of servers that are busy at time t_1 and the number of customers in the orbit at time t_1, respectively. The state space of $Q(t_1)$ can be given by $\Omega = [(i, j)$: i = 0,1,2,3,4,...c; j = 0,1,...∞] whose state transition diagram is given in

Figure 1.1, providing the wider view of how the model is working. It may be noted that the values λ and μ symbolize the arrival and service rates of customers, F has been replaced by N and in every node (c,d) c indicates busy servers whereas d indicates the number of customers in the orbit at time t_1.

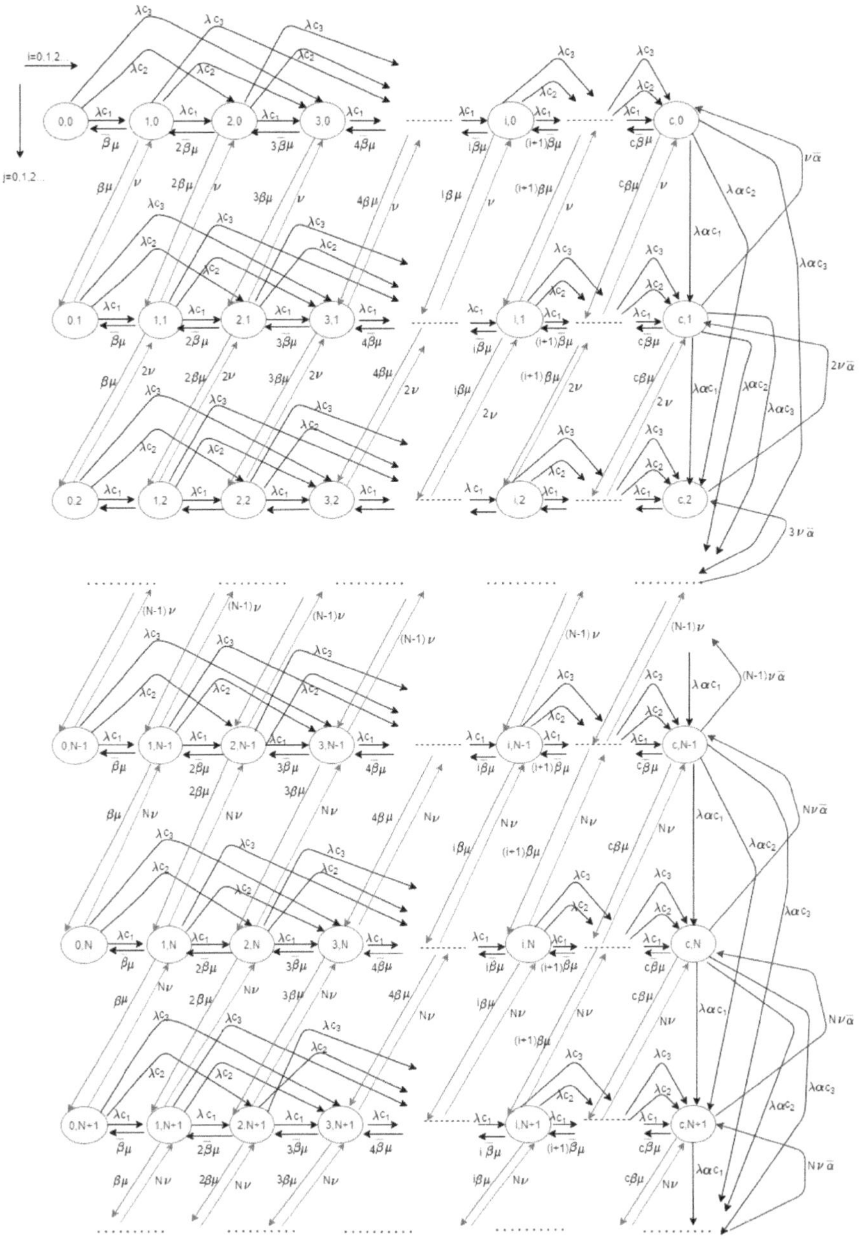

FIGURE 1.1 State diagram

By using appropriate transitions rates we have constructed the following transient state differential equations:

- *At* $j = 0$

$$P'_{(0,0)}(t_1) = \overline{\beta}\mu P_{(1,0)}(t_1) - (\lambda c_1 + \lambda c_2 + \lambda c_3)P_{(0,0)}(t_1) \tag{1.1}$$

$$\begin{aligned} P'_{(1,0)}(t_1) = &-(\lambda c_1 + \lambda c_2 + \lambda c_3)P_{(1,0)}(t_1) + \lambda c_1 P_{(0,0)}(t_1) - \overline{\beta}\mu P_{(1,0)}(t_1) \\ &- \beta\mu P_{(1,0)}(t_1) + 2\overline{\beta}\mu P_{(2,0)}(t_1) + \nu P_{(0,1)}(t_1) \end{aligned} \tag{1.2}$$

⋮

$$\begin{aligned} P'_{(3,0)}(t_1) = &-(\lambda c_1 + \lambda c_2 + \lambda c_3)P_{(3,0)}(t_1) - 3\overline{\beta}\mu P_{(3,0)}(t_1) - 3\beta\mu P_{(3,0)}(t_1) \\ &+ \lambda c_1 P_{(2,0)}(t_1) + \lambda c_2 P_{(1,0)}(t_1) + \lambda c_3 P_{(0,0)}(t_1) \\ &+ 4\overline{\beta}\mu P_{(4,0)}(t_1) + \nu P_{(2,1)}(t_1) \end{aligned} \tag{1.3}$$

$$\begin{aligned} P'_{(i,0)}(t_1) = &-(\lambda c_1 + \lambda c_2 + \lambda c_3)P_{(i,0)}(t_1) - i\overline{\beta}\mu P_{(i,0)}(t_1) \\ &- i\beta\mu P_{(i,0)}(t_1) + \lambda c_1 P_{(i-1,0)}(t_1) + \lambda c_2 P_{(i-2,0)}(t_1) \\ &+ \lambda c_3 P_{(i-3,0)}(t_1) + (i+1)\overline{\beta}\mu P_{(i+1,0)}(t_1) + \nu P_{(i-1,1)}(t_1) \end{aligned} \tag{1.4}$$

$$\begin{aligned} P'_{(c,0)}(t_1) = &-(\lambda\alpha c_1 + \lambda\alpha c_2 + \lambda\alpha c_3)P_{(c,0)}(t_1) - c\overline{\beta}\mu P_{(c,0)}(t_1) - c\beta\mu P_{(c,0)}(t_1) \\ &+ \lambda c_1 P_{(c-1,0)}(t_1) + \lambda c_2 P_{(c-2,0)}(t_1) + \lambda c_3 P_{(c-3,0)}(t_1) \\ &+ \nu P_{(c-1,1)}(t_1) + \nu\overline{\alpha} P_{(c,1)}(t_1) \end{aligned} \tag{1.5}$$

- *At* $j = 1$

$$P'_{(0,1)}(t_1) = -(\lambda c_1+\lambda c_2+\lambda c_3)P_{(0,1)}(t_1)+\overline{\beta}\mu P_{(1,1)}(t_1)-\nu P_{(0,1)}(t_1)+\beta\mu P_{(1,0)}(t_1) \tag{1.6}$$

$$\begin{aligned} P'_{(1,1)}(t_1) = &-(\lambda c_1 + \lambda c_2 + \lambda c_3)P_{(1,1)}(t_1) + \lambda c_1 P_{(0,1)}(t_1) - \overline{\beta}\mu P_{(1,1)}(t_1) - \beta\mu P_{(1,1)}(t_1) \\ &- \nu P_{(1,1)}(t_1) + 2\overline{\beta}\mu P_{(2,1)}(t_1) + 2\nu P_{(0,2)}(t_1) + 2\beta\mu P_{(2,0)}(t_1) \end{aligned} \tag{1.7}$$

⋮

$$\begin{aligned} P'_{(i,1)}(t_1) = &-(\lambda c_1 + \lambda c_2 + \lambda c_3)P_{(i,1)}(t_1) - i\overline{\beta}\mu P_{(i,1)}(t_1) - i\beta\mu P_{(i,1)}(t_1) \\ &+ \lambda c_1 P_{(i-1,1)}(t_1) + \lambda c_2 P_{(i-2,1)}(t_1) + \lambda c_3 P_{(i-3,1)}(t_1) + (i+1)\overline{\beta}\mu P_{(i+1,1)}(t_1) \\ &+ 2\nu P_{(i-1,2)}(t_1) - \nu P_{(i,1)}(t_1) + (i+1)\beta\mu P_{(i+1,0)}(t_1) \end{aligned} \tag{1.8}$$

$$P'_{(c,1)}(t_1) = -(\lambda\alpha c_1 + \lambda\alpha c_2 + \lambda\alpha c_3)P_{(c,1)}(t_1) - c\bar{\beta}\mu P_{(c,1)}(t_1) - c\beta\mu P_{(c,1)}(t_1) + \lambda c_1 P_{(c-1,1)}(t_1) + \lambda c_2 P_{(c-2,1)}(t_1) + \lambda c_3 P_{(c-3,1)}(t_1) + 2\nu P_{(c-1,2)}(t_1) + 2\nu\bar{\alpha} P_{(c,2)}(t_1) + \lambda\alpha c_1 P_{(c,0)}(t_1) - \nu\bar{\alpha} P_{(c,1)}(t_1) \tag{1.9}$$

- *At* $j = F - 1$

$$P'_{(0,F-1)}(t_1) = -(\lambda c_1 + \lambda c_2 + \lambda c_3)P_{(0,F-1)}(t_1) + \bar{\beta}\mu P_{(1,F-1)}(t_1) - (F-1)\nu P_{(0,F-1)}(t_1) + \beta\mu P_{(1,F-2)}(t_1) \tag{1.10}$$

$$P'_{(1,F-1)}(t_1) = -(\lambda c_1 + \lambda c_2 + \lambda c_3)P_{(1,F-1)}(t_1) + \lambda c_1 P_{(0,F-1)}(t_1) - \bar{\beta}\mu P_{(1,F-1)}(t_1) - \beta\mu P_{(1,F-1)}(t_1) - (F-1)\nu P_{(1,F-1)}(t_1) + 2\bar{\beta}\mu P_{(2,F-1)}(t_1) + F\nu P_{(0,F)}(t_1) + 2\beta\mu P_{(2,F-2)}(t_1) \tag{1.11}$$

$$P'_{(i,F-1)}(t_1) = -(\lambda c_1 + \lambda c_2 + \lambda c_3)P_{(i,F-1)}(t_1) - i\bar{\beta}\mu P_{(i,F-1)}(t_1) - i\beta\mu P_{(i,F-1)}(t_1) + \lambda c_1 P_{(i-1,F-1)}(t_1) + \lambda c_2 P_{(i-2,F-1)}(t_1) + \lambda c_3 P_{i-3,F-1}(t_1) + (i+1)\bar{\beta}\mu P_{(i+1,F-1)}(t_1) + F\nu P_{(i-1,F)}(t_1) - (F-1)\nu P_{i,F-1}(t_1) + (i+1)\beta\mu P_{i+1,F-2}(t_1) \tag{1.12}$$

$$P'_{(c,F-1)}(t_1) = -(\lambda\alpha c_1 + \lambda\alpha c_2 + \lambda\alpha c_3)P_{(c,F-1)}(t_1) - c\bar{\beta}\mu P_{(c,F-1)}(t_1) - c\beta\mu P_{(c,F-1)}(t_1) + \lambda c_1 P_{(c-1,F-1)}(t_1) + \lambda c_2 P_{(c-2,F-1)}(t_1) + \lambda c_3 P_{(c-3,F-1)}(t_1) + F\nu P_{(c-1,F)}(t_1) + F\nu\bar{\alpha} P_{(c,F)}(t_1) - (F-1)\nu\bar{\alpha} P_{(c,F-1)}(t_1) + \lambda\alpha c_1 P_{(c,F-2)}(t_1) + \lambda\alpha c_2 P_{(c,F-3)}(t_1) + \lambda\alpha c_3 P_{(c,F-4)}(t_1) \tag{1.13}$$

- *At* $j = F$

$$P'_{0,F}(t_1) = -(\lambda c_1 + \lambda c_2 + \lambda c_3)P_{(0,F)}(t_1) + \bar{\beta}\mu P_{(1,F)}(t_1) - (F)\nu P_{(0,F)}(t_1) + \beta\mu P_{(1,F-1)}(t_1) \tag{1.14}$$

$$P'_{(1,F)}(t_1) = -(\lambda c_1 + \lambda c_2 + \lambda c_3)P_{(1,F)}(t_1) + \lambda c_1 P_{(0,F)}(t_1) - \bar{\beta}\mu P_{(1,F)}(t_1) - \beta\mu P_{(1,F)}(t_1) - (F)\nu P_{(1,F)}(t_1) + 2\bar{\beta}\mu P_{(2,F)}(t_1) + F\nu P_{(0,F+1)}(t_1) + 2\beta\mu P_{(2,F-1)}(t_1) \tag{1.15}$$

$$P'_{(i,F)}(t_1) = -(\lambda c_1 + \lambda c_2 + \lambda c_3)P_{(i,F)}(t_1) - i\overline{\beta}\mu P_{i,F}(t_1) - i\beta\mu P_{i,F}(t_1) + \lambda c_1 P_{(i-1,F)}(t_1) + \lambda c_2 P_{i-2,F}(t_1) + \lambda c_3 P_{i-3,F}(t_1) + (i+1)\overline{\beta}\mu P_{i+1,F}(t_1) + F\nu P_{i-1,F+1}(t_1) - (F)\nu P_{i,F}(t_1) + (i+1)\beta\mu P_{i+1,F-1}(t_1) \quad (1.16)$$

$$P'_{(c,F)}(t_1) = -(\lambda\alpha c_1 + \lambda\alpha c_2 + \lambda\alpha c_3)P_{(c,F)}(t_1) - c\overline{\beta}\mu P_{(c,F)}(t_1) - c\beta\mu P_{(c,F)}(t_1) + \lambda c_1 P_{(c-1,F)}(t_1) + \lambda c_2 P_{(c-2,F)}(t_1) + \lambda c_3 P_{(c-3,F)}(t_1) + F\nu P_{(c-1,F+1)}(t_1) + F\nu\overline{\alpha} P_{(c,F+1)}(t_1) - (F)\nu\overline{\alpha} P_{(c,F)}(t_1) + \lambda\alpha c_1 P_{(c,F-1)}(t_1) + \lambda\alpha c_2 P_{(c,F-2)}(t_1) + \lambda\alpha c_3 P_{(c,F-3)}(t_1) \quad (1.17)$$

Since we are studying the steady state equations so we will develop the steady state difference equations for the model. These equations are time independent.

$$\overline{\beta}\mu P_{(1,0)} - (\lambda c_1 + \lambda c_2 + \lambda c_3)P_{(0,0)} = 0 \quad (1.18)$$

$$-(\lambda c_1 + \lambda c_2 + \lambda c_3)P_{(1,0)} + \lambda c_1 P_{(0,0)} - \overline{\beta}\mu P_{(1,0)} - \beta\mu P_{(1,0)} + 2\overline{\beta}\mu P_{(2,0)} + \nu P_{(0,1)} = 0 \quad (1.19)$$

$$-(\lambda c_1 + \lambda c_2 + \lambda c_3)P_{(3,0)} - 3\overline{\beta}\mu P_{(3,0)} - 3\beta\mu P_{(3,0)} + \lambda c_1 P_{(2,0)} + \lambda c_2 P_{(1,0)} + \lambda c_3 P_{(0,0)} + 4\overline{\beta}\mu P_{(4,0)} + \nu P_{(2,1)} = 0 \quad (1.20)$$

$$-(\lambda c_1 + \lambda c_2 + \lambda c_3)P_{(i,0)} - i\overline{\beta}\mu P_{(i,0)} - i\beta\mu P_{(i,0)} + \lambda c_1 P_{(i-1,0)} + \lambda c_2 P_{(i-2,0)} + \lambda c_3 P_{(i-3,0)} + (i+1)\overline{\beta}\mu P_{(i+1,0)} + \nu P_{(i-1,1)} = 0 \quad (1.21)$$

$$-(\lambda\alpha c_1 + \lambda\alpha c_2 + \lambda\alpha c_3)P_{(c,0)} - c\overline{\beta}\mu P_{(c,0)} - c\beta\mu P_{(c,0)} + \lambda c_1 P_{(c-1,0)} + \lambda c_2 P_{(c-2,0)} + \lambda c_3 P_{(c-3,0)} + \nu P_{(c-1,1)} + \nu\overline{\alpha} P_{(c,1)} = 0 \quad (1.22)$$

$$-(\lambda c_1 + \lambda c_2 + \lambda c_3)P_{(0,1)} + \overline{\beta}\mu P_{(1,1)} - \nu P_{(0,1)} + \beta\mu P_{(1,0)} = 0 \quad (1.23)$$

$$-(\lambda c_1 + \lambda c_2 + \lambda c_3)P_{(1,1)} + \lambda c_1 P_{(0,1)} - \overline{\beta}\mu P_{(1,1)} - \beta\mu P_{(1,1)} - \nu P_{(1,1)} + 2\overline{\beta}\mu P_{(2,1)} + 2\nu P_{(0,2)} + 2\beta\mu P_{(2,0)} = 0 \quad (1.24)$$

⋮

$$-(\lambda c_1 + \lambda c_2 + \lambda c_3)P_{(i,1)} - i\overline{\beta}\mu P_{(i,1)} - i\beta\mu P_{(i,1)} + \lambda c_1 P_{(i-1,1)} + \lambda c_2 P_{(i-2,1)} + \lambda c_3 P_{(i-3,1)} + (i+1)\overline{\beta}\mu P_{(i+1,1)} + 2\nu P_{(i-1,2)} - \nu P_{(i,1)} + (i+1)\beta\mu P_{(i+1,0)} = 0 \quad (1.25)$$

$$-(\lambda\alpha c_1+\lambda\alpha c_2+\lambda\alpha c_3)P_{(c,1)}-c\overline{\beta}\mu P_{(c,1)}-c\beta\mu P_{(c,1)}+\lambda c_1P_{(c-1,1)}+\lambda c_2P_{(c-2,1)}$$
$$+\lambda c_3P_{(c-3,1)}+2\nu P_{(c-1,2)}+2\nu\overline{\alpha}P_{(c,2)}+\lambda\alpha c_1P_{(c,0)}-\nu\overline{\alpha}P_{(c,1)}=0 \quad (1.26)$$

$$-(\lambda c_1+\lambda c_2+\lambda c_3)P_{(0,F-1)}+\overline{\beta}\mu P_{(1,F-1)}-(F-1)\nu P_{(0,F-1)}+\beta\mu P_{(1,F-2)}=0 \quad (1.27)$$

$$-(\lambda c_1+\lambda c_2+\lambda c_3)P_{(1,F-1)}+\lambda c_1P_{(0,F-1)}-\overline{\beta}\mu P_{(1,F-1)}-\beta\mu P_{(1,F-1)}$$
$$-(F-1)\nu P_{(1,F-1)}+2\overline{\beta}\mu P_{(2,F-1)}+F\nu P_{(0,F)}+2\beta\mu P_{(2,F-2)}=0 \quad (1.28)$$

$$-(\lambda c_1+\lambda c_2+\lambda c_3)P_{(i,F-1)}-i\overline{\beta}\mu P_{(i,F-1)}-i\beta\mu P_{(i,F-1)}+\lambda c_1P_{(i-1,F-1)}$$
$$+\lambda c_2P_{(i-2,F-1)}+\lambda c_3P_{i-3,F-1}+(i+1)\overline{\beta}\mu P_{(i+1,F-1)}+F\nu P_{(i-1,F)}$$
$$-(F-1)\nu P_{i,F-1}+(i+1)\beta\mu P_{i+1,F-2}=0 \quad (1.29)$$

$$-(\lambda\alpha c_1+\lambda\alpha c_2+\lambda\alpha c_3)P_{(c,F-1)}-c\overline{\beta}\mu P_{(c,F-1)}-c\beta\mu P_{(c,F-1)}+\lambda c_1P_{(c-1,F-1)}$$
$$+\lambda c_2P_{(c-2,F-1)}+\lambda c_3P_{(c-3,F-1)}+F\nu P_{(c-1,F)}+F\nu\overline{\alpha}P_{(c,F)}-(F-1)\nu\overline{\alpha}P_{(c,F-1)}$$
$$+\lambda\alpha c_1P_{(c,F-2)}+\lambda\alpha c_2P_{(c,F-3)}+\lambda\alpha c_3P_{(c,F-4)}=0 \quad (1.30)$$

$$-(\lambda c_1+\lambda c_2+\lambda c_3)P_{(0,F)}+\overline{\beta}\mu P_{(1,F)}-(F)\nu P_{(0,F)}+\beta\mu P_{(1,F-1)}=0 \quad (1.31)$$

$$-(\lambda c_1+\lambda c_2+\lambda c_3)P_{(1,F)}+\lambda c_1P_{(0,F)}-\overline{\beta}\mu P_{(1,F)}-\beta\mu P_{(1,F)}$$
$$-(F)\nu P_{(1,F)}+2\overline{\beta}\mu P_{(2,F)}+F\nu P_{(0,F+1)}+2\beta\mu P_{(2,F-1)}=0 \quad (1.32)$$

$$-(\lambda c_1+\lambda c_2+\lambda c_3)P_{(i,F)}-i\overline{\beta}\mu P_{i,F}-i\beta\mu P_{i,F}+\lambda c_1P_{(i-1,F)}+\lambda c_2P_{(i-2,F)}+\lambda c_3P_{(i-3,F)}$$
$$+(i+1)\overline{\beta}\mu P_{(i+1,F)}+F\nu P_{(i-1,F+1)}-(F)\nu P_{(i,F)}+(i+1)\beta\mu P_{(i+1,F-1)}=0 \quad (1.33)$$

$$-(\lambda\alpha c_1+\lambda\alpha c_2+\lambda\alpha c_3)P_{(c,F)}-c\overline{\beta}\mu P_{(c,F)}-c\beta\mu P_{(c,F)}+\lambda c_1P_{(c-1,F)}$$
$$+\lambda c_2P_{(c-2,F)}+\lambda c_3P_{(c-3,F)}+F\nu P_{(c-1,F+1)}+F\nu\overline{\alpha}P_{(c,F+1)}-(F)\nu\overline{\alpha}P_{(c,F)}$$
$$+\lambda\alpha c_1P_{(c,F-1)}+\lambda\alpha c_2P_{(c,F-2)}+\lambda\alpha c_3P_{(c,F-3)}=0 \quad (1.34)$$

where $P_{(a,b)}(t)$: uncertainty for a busy servers and b customers in the orbit at time t_1.

1.3 MATRIX GEOMETRIC METHOD

When a Markov process has a repetitive structure, we can sometimes readily derive the stationary probabilities by using MGM. This approach has been explored in detail by Neuts (1981) and can be applied when the generator matrix of the Markov process has identical repeating sub matrices. The infinitesimal generator P of the QBD process is of the form

$$
P=\begin{bmatrix}
A_0 & B_1 & B_2 & B_3 & 0 & 0\ldots & 0 & \cdots \\
C_1 & A_1 & B_1 & B_2 & B_3 & 0\ldots & 0 & \cdots \\
0 & C_2 & A_2 & B_1 & B_2 & B_3\ldots & 0 & \cdots \\
\vdots & \vdots & \vdots & \ddots & \vdots & \cdots & 0 & \cdots \\
0 & 0 & 0 & C_{F-1} & A_{F-1} & B_1 & B_2 & \cdots \\
0 & 0 & 0 & 0 & C & A & B_1 & \cdots \\
0 & 0 & 0 & 0 & 0 & C & A & \cdots \\
\vdots & \vdots & \vdots & \vdots & \vdots & \vdots & \vdots & \ddots
\end{bmatrix}
$$

where

$$
C_k=\begin{bmatrix}
0 & k\nu & 0 & 0 & \cdots & 0 & 0 & 0 \\
0 & 0 & k\nu & 0 & \cdots & 0 & 0 & 0 \\
0 & 0 & 0 & k\nu & \cdots & 0 & 0 & 0 \\
0 & 0 & 0 & 0 & \cdots & 0 & 0 & 0 \\
\vdots & \vdots & \vdots & \vdots & \ddots & \vdots & \vdots & \vdots \\
0 & 0 & 0 & 0 & \cdots & 0 & k\nu & 0 \\
0 & 0 & 0 & 0 & \cdots & 0 & 0 & k\nu \\
0 & 0 & 0 & 0 & \cdots & 0 & 0 & k\nu\overline{\alpha}
\end{bmatrix}
$$

$$
A_n=\left[\begin{matrix}
-\lambda-n\nu & \lambda c_1 & \lambda c_2 & \lambda c_3 & \cdots \\
\overline{\beta}\mu & -\lambda-\mu-n\nu & \lambda c_1 & \lambda c_2 & \cdots \\
0 & 2\overline{\beta}\mu & -\lambda-2\mu-n\nu & \lambda c_1 & \cdots \\
0 & 0 & 3\overline{\beta}\mu & -\lambda-3\mu-n\nu & \cdots \\
\vdots & \vdots & \vdots & \vdots & \ddots \\
0 & 0 & 0 & 0 & \cdots \\
0 & 0 & 0 & 0 & \cdots \\
0 & 0 & 0 & 0 & \cdots
\end{matrix}\right.
$$

$$
\left.\begin{matrix}
0 & 0 & 0 \\
0 & 0 & 0 \\
0 & 0 & 0 \\
0 & 0 & 0 \\
\vdots & \vdots & \vdots \\
-\lambda-(c-2)\mu-n\nu & \lambda c_1 & \lambda c_2 \\
(c-1)\overline{\beta}\mu & -\lambda-(c-1)\mu-n\nu & \lambda c_1 \\
0 & c\overline{\beta}\mu & -\lambda\alpha-c\mu-n\nu\overline{\alpha}
\end{matrix}\right]
$$

$$B_1=\begin{bmatrix} 0 & 0 & 0 & 0 & \cdots & 0 & 0 & 0 \\ \beta\mu & 0 & 0 & 0 & \cdots & 0 & 0 & 0 \\ 0 & 2\beta\mu & 0 & 0 & \cdots & 0 & 0 & 0 \\ 0 & 0 & 3\beta\mu & 0 & \cdots & 0 & 0 & 0 \\ \vdots & \vdots & \vdots & \vdots & \ddots & \vdots & \vdots & \vdots \\ 0 & 0 & 0 & 0 & \cdots & 0 & 0 & 0 \\ 0 & 0 & 0 & 0 & \cdots & (c-1)\beta\mu & 0 & 0 \\ 0 & 0 & 0 & 0 & \cdots & 0 & c\beta\mu & \lambda\alpha c_1 \end{bmatrix}$$

$$B_2=\begin{bmatrix} 0 & 0 & 0 & 0 & \cdots & 0 & 0 & 0 \\ 0 & 0 & 0 & 0 & \cdots & 0 & 0 & 0 \\ 0 & 0 & 0 & 0 & \cdots & 0 & 0 & 0 \\ 0 & 0 & 0 & 0 & \cdots & 0 & 0 & 0 \\ \vdots & \vdots & \vdots & \vdots & \ddots & \vdots & \vdots & \vdots \\ 0 & 0 & 0 & 0 & \cdots & 0 & 0 & 0 \\ 0 & 0 & 0 & 0 & \cdots & 0 & 0 & 0 \\ 0 & 0 & 0 & 0 & \cdots & 0 & 0 & \lambda\alpha c_2 \end{bmatrix}$$

$$B_3=\begin{bmatrix} 0 & 0 & 0 & 0 & \cdots & 0 & 0 & 0 \\ 0 & 0 & 0 & 0 & \cdots & 0 & 0 & 0 \\ 0 & 0 & 0 & 0 & \cdots & 0 & 0 & 0 \\ 0 & 0 & 0 & 0 & \cdots & 0 & 0 & 0 \\ \vdots & \vdots & \vdots & \vdots & \ddots & \vdots & \vdots & \vdots \\ 0 & 0 & 0 & 0 & \cdots & 0 & 0 & 0 \\ 0 & 0 & 0 & 0 & \cdots & 0 & 0 & 0 \\ 0 & 0 & 0 & 0 & \cdots & 0 & 0 & \lambda\alpha c_3 \end{bmatrix}$$

A_n; B_1, B_2 and B_3; C_k sub matrices are of the order $(c+1)*(c+1)$ where n = 0,1,2,..., F; k = 1,2,...,F; A = A_F; C =C_F. As P is a continuous time Markov chain and the system is modeled to accommodate infinite customers, the matrix P is $\infty * \infty$ matrix.

Q(t) is a Quasi Birth Death Process as in Neuts.

We can see that Markov chain P has a repetitive structure. Let π=[$\pi_0, \pi_1, \pi_2, ...$] be its steady-state uncertainty vector which satisfies

$$\pi\text{P} = 0 \text{ where each } \pi_j=[p_{0,j}\ p_{1,j}\ \cdots p_{c-1,j}\ p_{c,j}]\ j \geq 0.$$

π_j can be evaluated using a matrix geometric solution as $\pi_{j+1} = \pi_j R$ for$j \geq F$. The matrix R is the minimal non-negative solution of the equation

$$C + AR + B_1R^2 + B_2R^3 + B_3R^4 = 0$$

R is given by $\lim_{n\to\infty} R_n$ from Neuts and Latouche Ramaswami [29]. Sequence R_n is defined as $R_0 = 0$ and

$$R_{n+1} = -[A^{-1}C + R_n + A^{-1}B_1R_n^2 + A^{-1}B_2R_n^3 + A^{-1}B_3R_n^4] \text{ for } n = 0, 1, 2, ...$$

The sequence R_n converges to R monotonically. Therefore rate matrix R can be determined from the above equation by successive substitutions.

1.4 SYSTEM PERFORMANCE MEASURES

In this section some performance measures have been established. The performance measures are as follows:

- Expected System Size

$$L_s = \sum_{n=1}^{\infty} nP_n$$

- Expected Queue Length

$$L_q = \sum_{n=1}^{\infty} (n-1)P_n$$

- Expected Waiting Time in the system

$$W_s = \frac{L_s}{\lambda}$$

- Expected Waiting Time in the queue

$$W_q = \frac{L_q}{\lambda}$$

1.5 ILLUSTRATION

In this section, we have analyzed the model. We have evaluated various performance measures by taking fixed values of various parameters. Also, the steady-state probabilities have been obtained by coding a program in MATLAB® software based on MGM.

The impact of various parameters on the performance measures, e.g. waiting time and queue length, have been depicted in the tables below. Tables 1.2 to 1.4 show the effect on queue length and Table 1.5 shows the effect on waiting time.

Table 1.2 shows the impact of retrial rate and arrival rate on queue length. It can be seen that as retrial rate increases, queue length also increases significantly. Also at a specific retrial rate if the arrival rate increases, queue length increases notably. So in both the cases the model aligns perfectly with real-life situations.

Table 1.3 investigates the influence of arrival rate and service rate on queue length. As we increase the service rate, it can be seen that the queue length decreases, which

TABLE 1.2
Retrial Rate versus Queue Length

λ/ν	0.3	0.35	0.4	0.45	0.5
λ=0.1	24.755	27.9885	32.264	37.534	43.8001
λ=0.2	38.1792	43.6827	50.606	58.931	68.6826
λ=0.3	49.2241	56.7741	65.9624	76.8085	89.3614

TABLE 1.3
Service Rate versus Queue Length

λ/μ	0.4	0.45	0.5	0.55	0.6
λ=0.1	26.5648	22.3428	19.7325	18.0912	17.0739
λ=0.2	41.3076	33.9986	29.35	26.299	24.2682
λ=0.3	53.5571	43.7369	37.3535	33.0606	30.1113

TABLE 1.4
Retrial Rate versus Queue Length

μ/ν	0.3	0.35	0.4	0.45	0.5
μ=0.4	38.1792	43.6827	50.606	58.931	68.6826
μ=0.5	28.2382	30.2904	33.264	37.0546	41.6186
μ=0.6	24.1673	24.5167	25.6589	27.4263	29.732

TABLE 1.5
Retrial Rate versus Waiting Time

μ/ν	0.3	0.35	0.4	0.45	0.5
μ=0.4	164.0805	189.2469	219.8748	256.0282	297.8712
μ=0.5	118.5327	129.2	143.2144	160.3534	180.5416
μ=0.6	98.5379	102.1813	108.4917	117.0562	127.6574

is always observed in real life. Also, at a fixed value of service rate, on increasing the arrival rate, the queue length also increases. Table 1.4 shows that when the retrial rate rises, so does the length of the queue. Also, at a specific retrial rate, if the service rate increases, the length of queue decreases. Thus the effect on queue length of two parameters is the same as we observe in many service systems.

In Table 1.5 the effect of retrial rate and service rate on waiting time is presented. As we increase the retrial rate it is observed that waiting time also increases. Also, at a specific retrial rate, as service rate increases the waiting time decreases. So it can be concluded that the impact on waiting time of both the parameters is the same as what is seen in reall life.

For a better understanding, the results obtained are shown graphically in Figures 1.2–1.5. Figure 1.2 shows retrial rate versus queue length. It can be seen clearly that with the growth in retrial rate the queue length is also increasing. Figure 1.3 shows the service rate versus queue length; with an increase in service rate the queue length decreases. In Figure 1.4 as retrial rate increases there is an increase in queue length also. Figure 1.5 presents retrial rate versus waiting time and it can be seen that on increasing retrial rate, there is an increase in waiting time also.

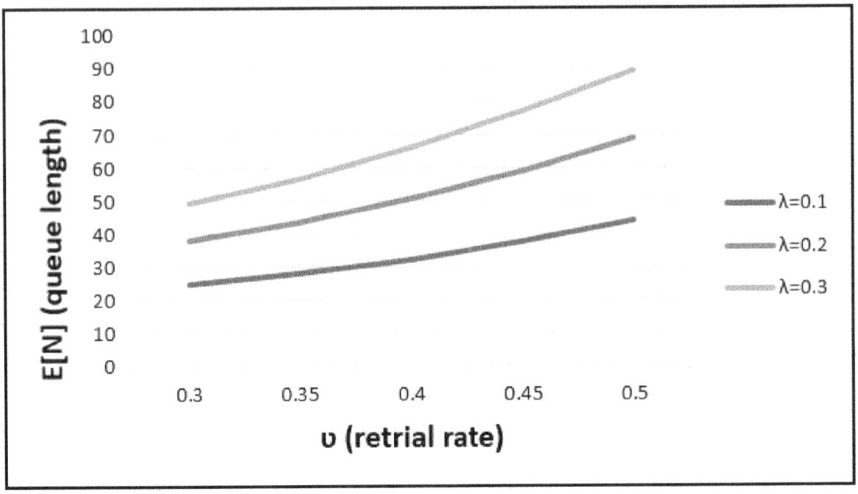

FIGURE 1.2 Retrial rate vs queue length

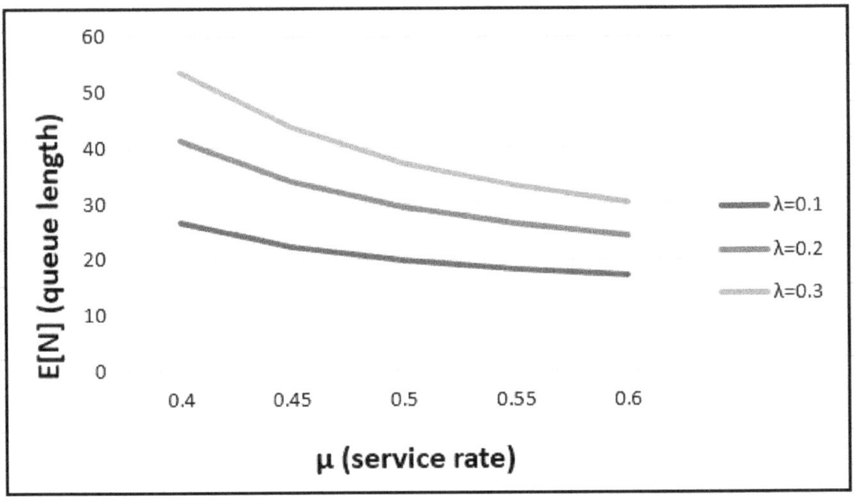

FIGURE 1.3 Service rate vs queue length

1.6 APPLICATION EXAMPLE

The model under consideration will be applicable for many computer and communication systems, manufacturing systems and many more. We can see various applications of the $M^X/M/c$ retrial queue with geometric loss and feedback model in real life.

One application may be a call centre that has numerous servers. In a call centre there are many agents who are available to receive the calls. The agents may receive the calls from a batch of customers at the same time. If few agents are available to

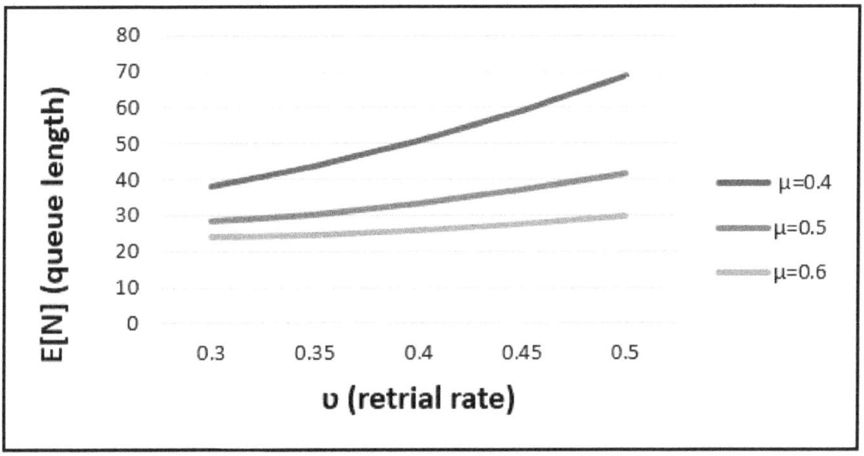

FIGURE 1.4 Retrial rate vs queue length

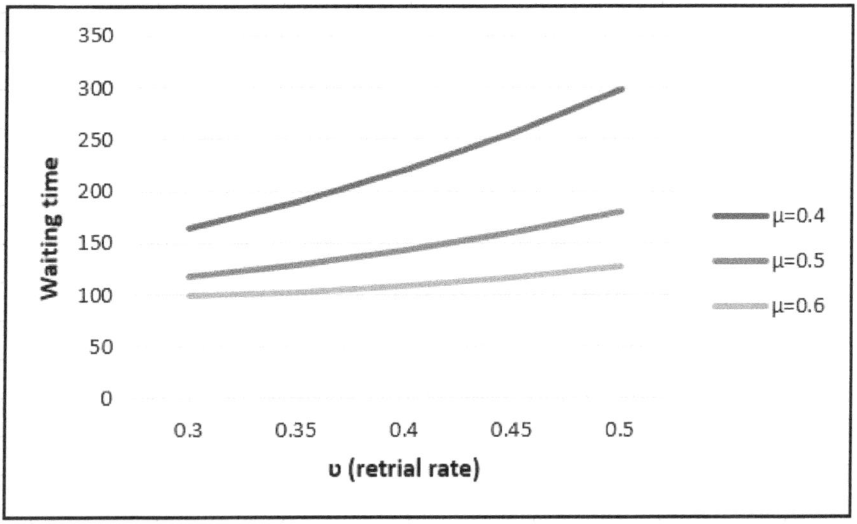

FIGURE 1.5 Retrial rate vs waiting time

receive the calls then some members of the batch receive the service right away and others may join the retrial orbit. If agents are not available to receive the calls and are busy at that time then the customers may either not wish to call again or may join the retrial orbit with uncertainty α and can retry for the service after some amount of time. When the service is completed, the customer may or may not call again for another service (i.e. they may either depart the system with uncertainty $\overline{\beta}$ or rejoin the orbit for re-service with uncertainty β).

A banking system can be another example of our model. There are several servers that provide different services in a bank. When a batch of clients enters the bank there are numerous possibilities: they may get the service, they may not take the service due to one or more reasons, or they may come for the service again (feedback). When a client enters and finds that all servers are free they will take the service immediately. On the other hand if they see that all servers are occupied they may not wish to take the service and will make an exit from the bank. Another scenario that can occur is that the client will enter the retrial orbit (and will attempt to obtain the service later) after finding all servers busy. After the completion of the bank work the client may decide either to leave the bank or to come again for another service.

Similarly another application can be observed in a ticket booking service over the telephone. Several staff members are available to handle ticket booking requests. At the same time different callers can make a call to book a ticket. If the signal is not busy the caller will take the service immediately. If a busy signal is received, the caller either balks or makes multiple attempts until the connection is established. After the booking of the ticket the caller will either end the conversation or if the ticket of the caller has been booked and they want to reschedule the booking, they may use the service again.

1.7 CONCLUSION

In this paper we have investigated a $M^X/M/c$ model with retrial queue, geometric loss and feedback. This model has been constructed as a quasi birth and death process. The matrix geometric method has been used to solve the steady-state governing equations numerically. Performance measures have been derived using MATLAB. The impact of various parameters like μ, ν, λ on queue length and waiting time have been shown through graphs. This model might be beneficial and can be considered for modeling banking services systems, control service channels, computer job processing and various other applications. The model could help in making decisions so that the availability and quality of the system can be improved.

REFERENCES

1. Sztrik J. Basic queueing theory. University of Debrecen, Faculty of Informatics. 193:60–7. 2012 Nov.
2. Yang T, Templeton JG. A survey on retrial queues. Queueing Systems. 2(3):201–33. 1987 Sep.
3. Yang DY, Ke JC, Wu CH. The multi-server retrial system with Bernoulli feedback and starting failures. International Journal of Computer Mathematics. 92(5):954–69. 2015 May.
4. Singla N, Kalra S. A Two-State Multiserver Queueing System with Retrials. International Journal of Open Problems in Computer Science and Mathematics. 12(3):62–74. 2019 Sep.
5. Yang DY, Wu CH. Performance analysis and optimization of a retrial queue with working vacations and starting failures. Mathematical and Computer Modelling of Dynamical Systems. 25(5):463–81. 2019 Sep.

6. Phung-Duc T. Retrial queueing models: A survey on theory and applications. arXiv preprint arXiv:1906.09560.1– 18.2019 Jun.
7. Kumar BK, Rukmani R, Thangaraj V, Krieger UR. A single server retrial queue with Bernoulli feedback and collisions. Journal of Statistical Theory and Practice. 4(2):243–60. 2010 Jun.
8. Chang FM, Ke JC. On a batch retrial model with J vacations. Journal of Computational and Applied Mathematics. 232(2):402–14. 2009 Oct.
9. Ke JC, Liu TH, Su S, Zhang ZG. On retrial queue with customer balking and feedback subject to server breakdowns. Communications in Statistics-Theory and Methods. 1–7. 2020 Nov.
10. Dudin AN, Dudina OS. Analysis of multiserver retrial queueing system with varying capacity and parameters. Mathematical Problems in Engineering. 1–12. 2015 Jan.
11. Abramov VM. Analysis of multiserver retrial queueing system: A martingale approach and an algorithm of solution. Annals of Operations Research. 141(1):19–50. 2006 Jan.
12. Neuts MF, Rao BM. Numerical investigation of a multiserver retrial model. Queueing Systems. 7(2):169–89. 1990 Jun.
13. Kulkarni VG. Expected waiting times in a multiclass batch arrival retrial queue. Journal of Applied uncertainty. 144–54. 1986 Mar.
14. Baruah M, Madan KC, Eldabi T. Balking and re-service in a vacation queue with batch arrival and two types of heterogeneous service. Journal of Mathematics Research. 4(4):114. 2012 Aug.
15. Bouchentouf AA, Guendouzi A. The $M^X/M/c$ Bernoulli feedback queue with variant multiple working vacations and impatient customers: performance and economic analysis. Arabian Journal of Mathematics. 9(2):309–27. 2020 Aug.
16. Nila M, Sumitha DD. $M^X/G/1$ Retrial Queue with Priority, Collisions and Feedback Customers, Proceedings of the First International Conference on Computing, Communication and Control System, I3CAC 2021. 1–12. 2021 June.
17. Takacs L. A single-server queue with feedback. Bell System Technical Journal. 42(2):505–19. 1963 Mar.
18. Disney RL, McNickle DC, Simon B. The M/G/1 queue with instantaneous Bernoulli feedback. Naval Research Logistics Quarterly. 27(4):635–44. 1980 Dec.
19. Choudhury G, Paul M. A two phase queueing system with Bernoulli feedback. International journal of information and management sciences. 16(1):35. 2005 Jan.
20. Mokaddis GS, Metwally SA, Zaki BM. A feedback retrial queuing system with starting failures and single vacation. Journal of Applied Science and Engineering. 10(3):183–92. 2007 Sep.
21. Choi BD, Kim YC, Lee YW. The M/M/c retrial queue with geometric loss and feedback. Computers & Mathematics with Applications. 36(6):41–52. 1998 Sep.
22. Lin CH, Ke JC. On the multi-server retrial queue with geometric loss and feedback: computational algorithm and parameter optimization. International Journal of Computer Mathematics. 88(5):1083–101. 2011 Mar.
23. Hanschke T. A matrix continued fraction algorithm for the multiserver repeated order queue. Mathematical and Computer Modelling. 30(3–4):159–70. 1999 Aug.
24. Chakravarthy SR, Dudin AN. A multi-server retrial queue with BMAP arrivals and group services. Queueing Systems. 42(1):5–31. 2002 Sep.
25. Kumar BK, Rukmani R, Thangaraj V. On multiserver feedback retrial queue with finite buffer. Applied Mathematical Modelling. 33(4):2062–83. 2009 Apr.
26. Upadhyaya S. Performance analysis of a batch arrival retrial queue with Bernoulli feedback. International Journal of Mathematics in Operational Research. 6(6):680–703. 2014 Jan.

27. Wang J, Zhou PF. A batch arrival retrial queue with starting failures, feedback and admission control. Journal of Systems Science and Systems Engineering. 19(3):306–20. 2010 Sep.
28. Van Do T, Do NH, Zhang J. An enhanced algorithm to solve multiserver retrial queueing systems with impatient customers. Computer & Industrial Engineering. 65(4):719–28. 2013 Aug.
29. Latouche G,Ramaswami V. Introduction to matrix analytic methods in stochastic modeling. Society for Industrial and Applied Mathematics; 1999 Jan.

2 Optimality for Fuzzy Transportation Problem under Ranking Method

P. Anukokila[1] *and B. Radhakrishnan*[2]
[1]Department of Mathematics, PSG College of Arts & Science College, Coimbatore, TN, India
[2]Department of Mathematics, PSG College of Technology, Coimbatore, Tamil Nadu, India

CONTENTS

2.1 INTRODUCTION

Mathematical optimization can be used to solve a variety of modeling, design, control, and decision-making challenges. The minimizing (or maximizing) of the objectives, given the constraints for the issue to be solved, is the traditional framework for optimization. Many design issues, on the other hand, are characterized by many objectives, which need a trade-off between various purposes, resulting in under- or over-achievement of various objectives. Bector et al. [1] proposed the principle of optimization theory. Although optimization approaches such as linear programming

DOI: 10.1201/9781003329039-2

have been shown to be beneficial in solving a narrow set of operational problems, they are not generally applied to address higher-level management issues. Because of the availability of time-sharing services, non-optimizing techniques (such as simulation) may now be used to handle a broader range of issues than analytic models could. Optimization is frequently confused with operations research and decision theory. The quest for optimal solutions is also a primary goal in other fields such as engineering design, regional politics, logistics, and many more.

Furthermore, considerable flexibility may be present for identifying the problem's restrictions, and some of the decision-making objectives may be understood only roughly. Several writers researched fuzzy optimization for multi-objective programming issues, citing [2–5]. Many of the objectives in management choices, for example, can be defined roughly in linguistic terms, but there is no precise mathematical formula accessible. Furthermore, in some cases, decision limitations may be loosened if the decision objectives can be improved.

It is not always possible or required to precisely quantify numerous system performance criteria, parameters, and decision variables. Variables are said to be uncertain or fuzzy when their values cannot be properly determined. Probability distributions can be used to quantify values that are uncertain. Alternatively, fuzzy membership functions can be used to quantify them if they are better represented by qualitative adjectives such as dry or wet, hot or cold, clean or unclean, and high or low. Quantitative optimization models can include both probability distributions and fuzzy membership functions for these uncertain or qualitative variables. Fuzzy set theory is a useful tool for resolving real-world decision-making problems. It has been used in a variety of Operations Research approaches, including linear programming, non-linear programming, and queuing theory. The most essential technique among these is linear programming, which will be addressed in this research.

Zadeh, on the other hand, coined the term "fuzzy" in 1962, and formally published the classic article "Fuzzy Set Theory" in 1965. The goal of the fuzzy set theory was to improve the oversimplified model, resulting in a more resilient and flexible model that could be used to address real-world complex systems with human characteristics. When a complex physical system does not provide a set of differential or difference equations as a precise or reasonably accurate mathematical model, especially when the system description necessitates certain human experience in linguistic terms, it is now widely accepted that fuzzy systems and fuzzy control theories have some distinguishing features and merits over many other approaches. Mathematicians only defined the ideas of sets and functions to represent issues in the 20th century. This method of problem representation is more strict. In many cases, the solutions based on this principle are useless. The fuzzy notion helped to solve this problem. Fuzzy sets have been used to redefine almost all mathematical, engineering, medical, and other concepts. As a result, popularizing these ideals for our next generation is critical.

2.1.1 Classic Set

A classic set is a grouping of items in a certain location, as shown in Figure 2.1. An object can either be a part of the set or it can't be. As a result, the contrast between individuals who are a part of the set and those who are not is obvious. A set in classical

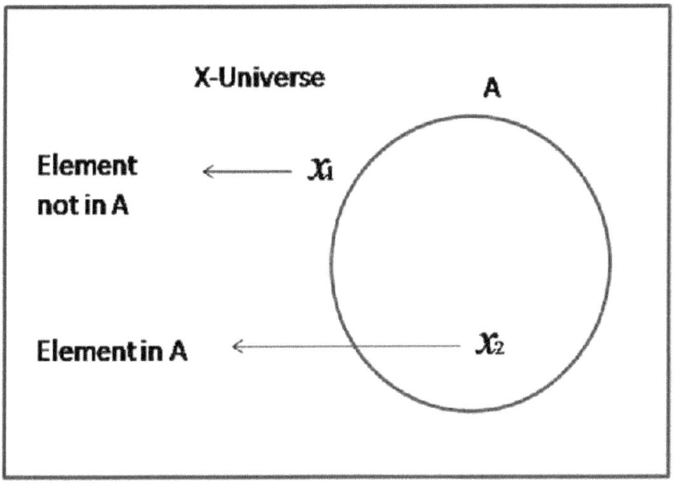

FIGURE 2.1 Classic set

set theory always has a clear boundary since belonging to a set is a binary concept. In a real scenario, there are two different kinds of sets. While some sets (such a collection of married people) have apparent bounds, others do not (e.g. the set of happily married couples, the set of best graduate schools, etc.). The conventional set theory cannot be used to investigate sets with non-sharp boundaries.

2.1.2 Fuzzy Set

A set with "un-sharp" and uncertain limits is referred to as fuzzy. By allowing for partial membership, it changes the notion of membership from a binary categorization inherent in conventional set theory. By allowing membership in a set to be a question of degree, fuzzy set theory gets beyond the limitations of classical set theory. A value in the range of 0 and 1 denotes the degree of membership in a fuzzy set; 0 denotes wholly non-membership and 1 denotes complete and utter membership. It is exhibited in Figure 2.2.

The idea of a fuzzy set continues to serve as an incredibly simple starting point for the conception of a conceptual framework that is comparable to the framework used for ordinary sets in many ways, but is more general and, potentially, may prove to have a much broader scope of applicability, particularly in the fields of pattern classification and information processing. In one of his earliest works on fuzzy set theory, Zadeh penned the following: a framework like this essentially "provides a logical way to dealing with instances where the reason of imprecision is the absence of finely characterized class membership requirements rather than the presence of random variables." From the standpoint of application in science and engineering, Zimmermann's book [6] decided to make a major contribution to the development of the dominant model as fuzzy sets decision-making and expert systems.

Applications of this theory encompass artificial intelligence, computer science, medicine, control engineering, decision theory, expert systems, logic, management

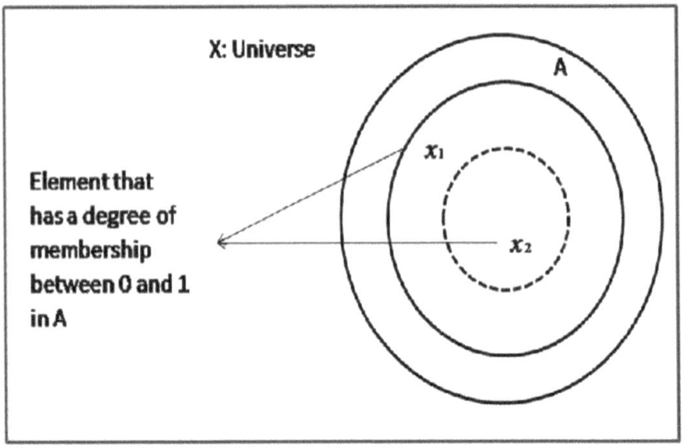

FIGURE 2.2 Fuzzy set

science, operations research, pattern recognition, and robotics. Mathematical progress has acquired a very significant level and is still progressing today. This topic centers on optimization theory, with a focus on fuzzy linear programming in particular. Despite the availability of other alternatives, including queuing theory, heuristic methods, nonlinear programming, and others, an examination of the operations research literature suggests that linear programming is the strategy most frequently used in practical settings. The methods and applications of fuzzy mathematical programming have been established by Zimmermann [7]. Moreover he studied [8] the optimization of fuzzy systems.

2.1.3 Membership Function

Membership function is the most important element of the fuzzy set theory and it allows the fuzzy approach to evaluate uncertain and ambiguous matters. The role of the membership function is to represent an individual and subjective human perception as a member of a fuzzy set. A fuzzy set has several membership functions $\mu_{\mathcal{A}}$, defined as a function from a well-defined universe, X into a unit interval, 0 through 1, as shown in the following equation. The function $\mu_{\mathcal{A}} : X \to [0,1]$ is defined by

$$\mu_{\mathcal{A}}(x) = \begin{cases} 1 & if \ x \in \mathcal{A} \\ 0 & if \ x \notin \mathcal{A}. \end{cases}$$

The graphical representation of membership function is shown in Figure 2.3.

Determining the membership function is the essential component in a fuzzy analytical method. When employing fuzzy sets, assessing the membership function accurately is one of the most challenging tasks. Depending on the topic at hand, multiple metrics are applied to the membership function. Fuzzy membership functions include triangles, trapezoids, bell-shaped curves, S-shaped curves, π-shaped curves, Gaussian functions, and sigmoid functions.

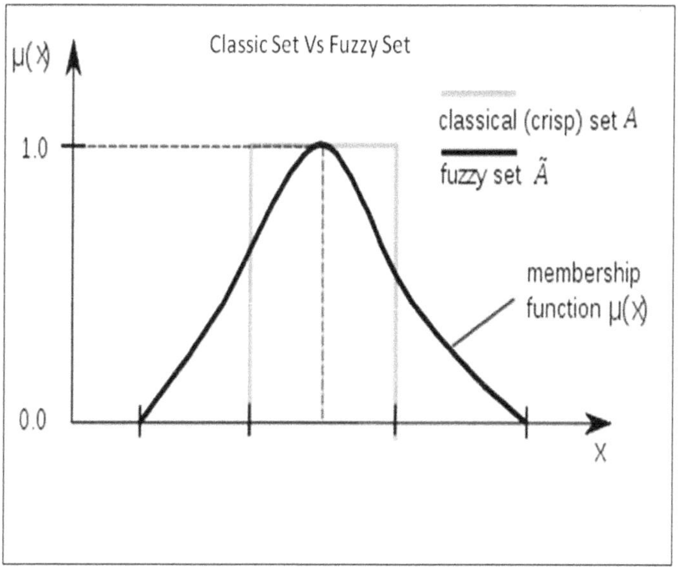

FIGURE 2.3 Membership function

The functions that are used the most frequently in practise are the triangle, trapezoid, bell-curve, Gaussian, and sigmoid functions. Due to the wide range of membership functions, picking the appropriate one is essential to accomplishing accurate fuzzy set application. The analyst bases their choice of membership function type on a thorough review of the data. Here are some core rules to assist users get to choose the membership functions to utilize:

(i) Triangular and trapezoidal shapes are the basic types. Both membership functions have been extensively utilized in fuzzy set applications due to their uncomplicated formulations and significant computational efficiency.
(ii) If there is more fluctuation in the data sets, the trapezoidal function is preferable to the triangle function.
(iii) If there is a large number of observations, it is preferable to use curved membership functions such bell curves, Gaussian, and sigmoid. They can bring results that are much more precise.

2.1.4 Motivation

The creation of a mathematical model constitutes a need for the majority of operations empirical investigations. Models that depict fundamental relationships between variables usually comprise an objective function for evaluating various solutions as well as constraints that restrict solutions to workable values. Although the analyst would prefer to utilize a systems approach to examine into the general implications of the issue, from the perspective of the initial issue, the model's solution is valuable.

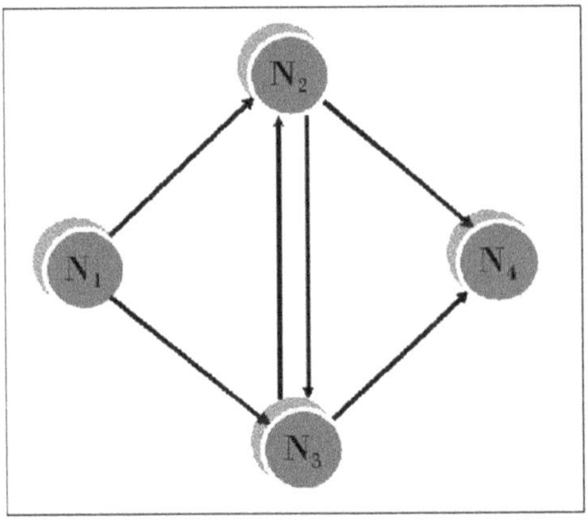

FIGURE 2.4 Network flow

Due to the prevalence of complex systems in several fields of study, this chapter's major goal is to investigate how to handle a fuzzy transportation problem using a ranking algorithm.

2.1.4.1 Network Flow Programming

Network flow programs are a subset of the more general linear programs and are referred to as "network flow programs." It is shown in Figure 2.4. Examples of network flow problems include the transportation problem, the assignment problem, the shortest path problem, the maximum flow problem, the pure minimum cost flow problem, and the diversified minimum cost flow problem. It is an important class because many aspects of real-world situations may be modeled as networks, and the representation of the model is much more condensed than that of a general linear program. When a scenario can be completely characterized as a network, there exist incredibly effective methods for addressing the optimization problem that, in terms of computer time and space consumption, are many times more effective than linear programming.

Figure 2.5 shows the relationships between the various network flow programming models and linear programming.

2.1.4.2 Shortest Path Problem

In this circumstance, which tends to make use of a modular network structure, just the arc cost is essential. As shown in Figure 2.6, the length of a path is equal to the sum of the arc costs along the path. Exploring the shortest route between a network element and another node, or perhaps all other nodes, is the objective. The fact that the collection of all shortest paths from a particular node forms a graph structure known as a tree gives rise to the name of the shortest path tree problem. Since finding paths to every node is not noticeably more complicated than finding a path to a single node, the shortest path tree problem is systematically solved.

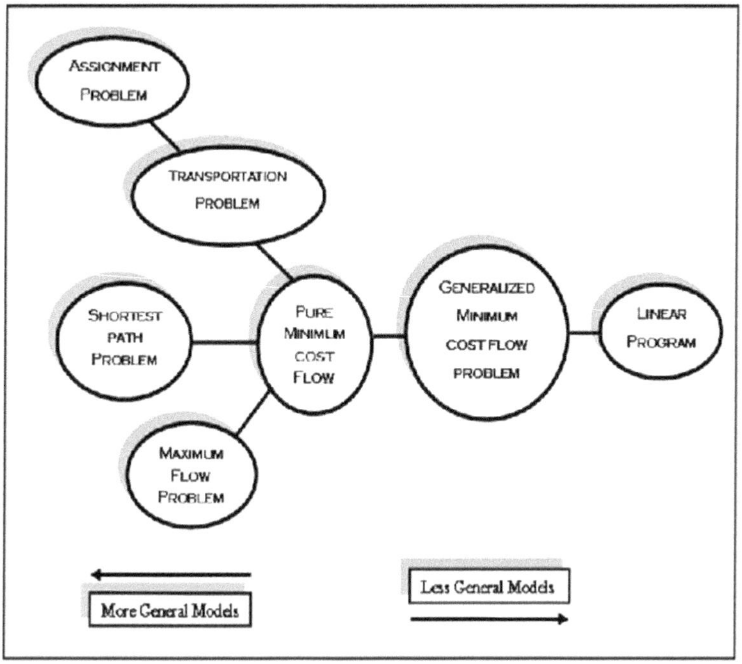

FIGURE 2.5 The relationships between network problems

2.1.4.3 Mines and Factories

Suppose we have a collection of n mines mining iron ore, and a collection of n factories that consume the iron ore that the mines produce. For the sake of argument, these mines and factories form two disjoint subsets M and F of the Euclidean plane $\mathbb{R}^2$. Also we have a cost function $c : \mathbb{R}^2 \times \mathbb{R}^2 \to [0, \infty)$, so that $c(x, y)$ is the cost of transporting one shipment of iron from x to y. We disregard the time frame of the transporting for the sake of simplicity. We also assume that no cargoes can be split across the factories that each mine can feed, and that each facility can only operational with a full container (factories cannot work at half- or double-capacity). Having made the above assumptions, a transportation plan is a bi-jection $T : M \to F$ i.e., an arrangement whereby each mine $m \in M$ supplies precisely one factory $T(m) \in F$. We wish to find the optimal transport plan, the plan T whose total cost $c(T) := \sum_{m \in M} c(m, T(m))$ is the least of all possible transportation plans from M to F. In practice, the assignment problem is a motivating specific case of the transportation problem.

2.1.4.4 Moving Books (The Importance of the Cost Function)

Suppose we have n books of equal width on a shelf (the real line), arranged in a single contiguous block. We wish to rearrange them into another contiguous block, but shifted one book-width to the right. Two obvious candidates for the optimize transport plan present themselves:

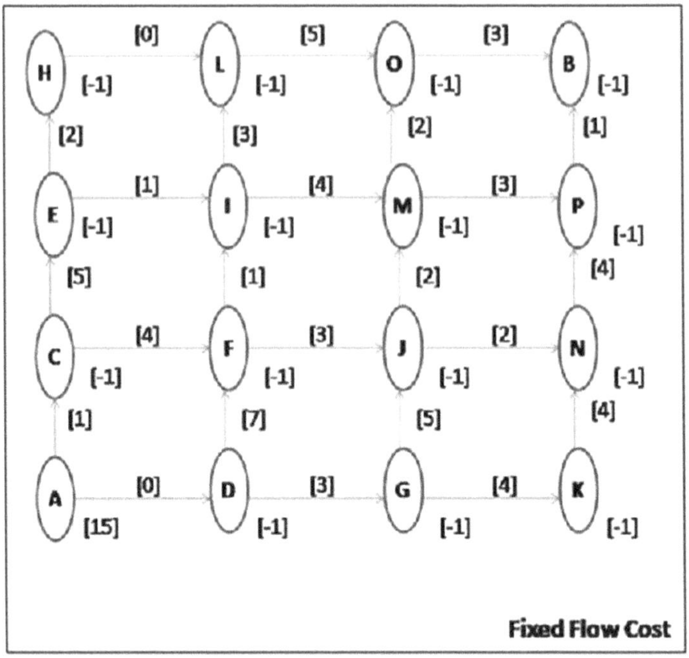

FIGURE 2.6 Network model for the shortest path problem

- move all n books one book-width to the right; ("many small moves")
- moves the leftmost book n book-widths to the right and leaves all other books fixed. ("one big move").

If the cost function is proportional to Euclidean distance $c(x, y) = \alpha|x-y|$ then these two candidates are both optimal. If, on the other hand, we choose the strictly convex cost function proportional to the square of Euclidean distance $c(x, y) = \alpha|x - y|^2$, then the "many small moves" option becomes the unique minimizer.

Mathematicians appreciate convex cost functions, whereas economists support concave cost functions. The logical explanation for this is that mechanical waves of 200 kilometers costs a fraction of what transferring them 100 kilometers would cost after they have been loaded onto, for example, a goods train. Concave cost functions are used to reflect the economies of scale.

2.1.4.5 Abstract Formulation (Monge and Kantorovich Formulations)

The transportation problem appears to be slightly different in modern or more technical literature because of the progression of Riemannian geometry and measure theory. Even if it is uncomplicated, the mines-factories example is a useful point of comparison when taking the abstract scenario into account. We set up mines to supply several manufacturers and factories to accept iron from various miners, considering the potential possibility that we may not want all mines and factories open for business.

Let X and Y be two separable metric spaces such that any probability measure on X (or Y) is a Radon measure (i.e., they are Radon spaces). Let $c : X \times Y \to [0, \infty]$ be a Borel-measurable function. Given probability measures μ on X and ν on Y, Monge's formulation of the optimal transportation problem is to find a transport map $T : X \to Y$ that realizes the infimum

$$\inf \int_X (c(x, T(x)))d\mu(x)|T_*(\mu) = \nu$$

where $T_*(\mu)$ denotes the push forward of μ by T. A map T that attains this infimum (i.e., makes it a minimum instead of an infimum) is called an "optimal transport map." Monge's formulation of the optimal transportation problem can be ill-posed, because sometimes there is no T satisfying $T_*(\mu) = \nu$: this happens, for example, when μ's a Dirac measure but ν is not). We can improve on this by adopting Kantorovich's formulation of the optimal transportation problem, which is to find a probability measure γ on $X \times Y$ that attains the infimum

$$\inf \int_X (c(x, y))d_\gamma(x, y)|\ \gamma \in \Gamma(\mu, \nu).$$

2.1.5 Methods

2.1.5.1 North-West Corner Rule

The fuzzy initial basic feasible solution to a transportation problem can be found in a wide range of approaches. The north-west corner rule (NWCR) method has the benefit of being easy to understand and apply. Its desensitization to cost, which leads to inadequate early solutions, is one of its main downsides. The required steps to arrive at an initial solution utilizing NWCR are as follows:

(a) Choose the cell in the fuzzy transportation table's north-west (upper left) corner, and then allocate as many units as designed to obtain the minimum balance between the available fuzzy supply and fuzzy demand.
(b) Adjust the associated rows' and columns' fuzzy supply and fuzzy demand numbers judiciously.
(c) Move horizontally to the subsequent cell in the second column if the blurred requirements for the first cell are achieved.
(d) Move down to the first cell in the second row if the former row's fuzzy supply is exhausted.
(e) The next allocation can be made in a cell in the adjacent row or column if fuzzy supply and fuzzy demand seem to be equal for any particular frame.
(f) Continue until no more fuzzy supply or fuzzy demand quantities remain.

2.1.5.2 Vogel's Approximation Method

In most circumstances, but not always, Vogel's approximation technique (VAM), an improved version of the least-cost strategy, enhances the overall initial solutions. The primary idea of VAM is to reduce opportunity (or penalty) costs. The opportunity cost is the distinction between the lowest and next-lowest cost options for a specific supply row or demand column. Given that it regularly yields ideal or almost ideal starting solutions, this strategy is preferred over the preceding one. Therefore, the time required to reach the optimal solution is significantly reduced if we start with VAM's initial solution and work our way there. The steps needed to choose an initial solution using VAM are as follows:

(a) The fuzzy difference between the smallest and next-smallest fuzzy costs in each row and column seems to be the fuzzy penalty cost.
(b) Among the fuzzy penalties are found in step (**a**), choose the fuzzy maximum penalty, by ranking methods. If this maximum penalty is more than one, choose any one arbitrary.
(c) In the selected row or column as by step (**b**) find out the cell having the least fuzzy cost. Allocate to this cell as much as possible depending on the fuzzy capacity and fuzzy demand.
(d) Delete the row or column which is fully exhausted. Again compute column and row fuzzy penalties for the reduced fuzzy transportation table and then go to step (**b**), repeat the procedure until all the rim demands are satisfied refer to Tables 2.1 to 2.3.

Once the fuzzy initial basic feasible solution has been computed, the following step in the debate is to assess if the resulting solution is fuzzy optimal or not. The first basic fuzzy transportation solution that seems to have exactly $m+n-1$ non-negative allocations, where m is the number of fuzzy origins and n is the number of fuzzy destinations, can be tested for fuzzy optimality.

2.1.5.3 *uv*−Method

uv−method has been introduced to find the optimal solution of a fuzzy transportation problem with the help of an initial basic feasible solution. The steps for finding the optimal solution are as follows:

(a) Find the initial basic feasible solution of the fuzzy transportation problem using fuzzy NWCR or fuzzy VAM.
(b) Introduce the fuzzy dual variables $\tilde{u}_i$ and $\tilde{v}_j$ corresponding to each ith row and j^{th} column, respectively. Write $\tilde{u}_i$ in front of each ith row and $\tilde{v}_j$ at the bottom of each jth column. Take any one of the $\tilde{u}_i$ or $\tilde{v}_j$ to the zero ranked fuzzy number.
(c) For basic cells, $\tilde{u}_i + \tilde{v}_j = \tilde{c}_{ij}$. This relation assigns values to all $\tilde{u}_i$ and v_j.
(d) For non-basic cells find the rank of $\tilde{d}_{ij} = \tilde{u}_i + \tilde{v}_j - \tilde{c}_{ij}$, for all $i,\ j$, and write them in corner of the concerned cell.
    (i) If $\tilde{d}_{ij} \leq_R 0$, for all i, j, then the obtained IBFS is the fuzzy optimal solution.
    (ii) If there exist at least one $\tilde{d}_{ij}$ such that $\tilde{d}_{ij} >_R 0$, then the IBFS is not optimal.
(e) In the FTP choose that $\tilde{d}_{ij}$ whose rank is most positive.

2.1.5.4 Ranking Method

Ranking fuzzy numbers is a powerful technique for decision-making. In fuzzy decision analysis, fuzzy quantities are used to illustrate how well different options perform when mimicking a real-world issue. The bulk of ranking algorithms that have been suggested in the literature so far have been not able to differentiate between fuzzy values, and some of them are illogical. It is impossible to organize fuzzy numbers because they may overlap since fuzzy numbers are represented by possibility distributions. It is true that fuzzy numbers cannot be compared in the same manner that real numbers can be ranked linearly even though they are frequently partial orders. Each fuzzy quantity is switched to a real number, and each fuzzy number with a natural order is given a real number in order to rank them. Since Zadeh [9] originally suggested the theory of fuzzy sets in 1976, other ranking techniques have been created. According to Wang and Kerre (citing [10, 11]), a wide range of existing fuzzy ranking systems were rigorously examined, and various axioms were presented as suitable properties to determine the rationality of a fuzzy ranking approach. Liou [12] and Su [13] derived the fuzzy programming based on interval-valued fuzzy numbers and ranking, which in turn was obtained from ranking fuzzy numbers with integral values. Numerous fuzzy number ranking techniques currently in use are based on integral area measurements of the fuzzy number membership function.

As encouraged by the results of [14–20] we apply the resilient ranking approach to solve the fuzzy transportation problem as well as to obtain an optimal solution by making continuous improvements in the original basic feasible solution.

2.2 PROBLEM FORMULATION

Consider transportation problem with m fuzzy origins (rows) and n fuzzy destination (columns). Let $\mathcal{C}_{ij} = \left[c_{ij}^{(1)}, c_{ij}^{(2)}, c_{ij}^{(3)}, c_{ij}^{(4)}\right]$ be the cost of transporting one unit of the product from i^{th} fuzzy origin to j^{th} fuzzy destination.
$a_i = \left[a_i^{(1)}, a_i^{(2)}, a_i^{(3)}, a_i^{(4)}\right]$ be the quantity of commodity available at fuzzy origin i,
$b_j = \left[b_j^{(1)}, b_j^{(2)}, b_j^{(3)}, b_j^{(4)}\right]$ be the quantity of commodity needed at fuzzy destination j.
$X_{ij} = \left[x_{ij}^{(1)}, x_{ij}^{(2)}, x_{ij}^{(3)}, x_{ij}^{(4)}\right]$ is quantity transported problem can be stated in Table 2.1.

The linear programming model representing the fuzzy transportation is given by

$$\left\{\begin{aligned} &\text{minimize } \mathcal{Z} = \sum_{i=1}^{m}\sum_{j=1}^{n}\left[c_{ij}^{(1)}, c_{ij}^{(2)}, c_{ij}^{(3)}, c_{ij}^{(4)}\right]\left[x_{ij}^{(1)}, x_{ij}^{(2)}, x_{ij}^{(3)}, x_{ij}^{(4)}\right] \\ &\text{subject to} \\ &\sum_{j=1}^{n}\left[x_{ij}^{(1)}, x_{ij}^{(2)}, x_{ij}^{(3)}, x_{ij}^{(4)}\right] = \left[a_i^{(1)}, a_i^{(2)}, a_i^{(3)}, a_i^{(4)}\right], \text{ for } i = 1, 2, \dots, m \\ &\sum_{i=1}^{m}\left[x_{ij}^{(1)}, x_{ij}^{(2)}, x_{ij}^{(3)}, x_{ij}^{(4)}\right] = \left[b_j^{(1)}, b_j^{(2)}, b_j^{(3)}, b_j^{(4)}\right], \text{ for } j = 1, 2, \dots, n \\ &\left[x_{ij}^{(1)}, x_{ij}^{(2)}, x_{ij}^{(3)}, x_{ij}^{(4)}\right] \geq 0. \end{aligned}\right. \tag{2.1}$$

TABLE 2.1
Transportation Problem

	1		2...		n		Fuzzy Capacity
1	c_{11}	x_{11}	$c_{12}\ldots$	$x_{12}\ldots$	c_{1n}	x_{1n}	a_1
2	c_{21}	x_{21}	$c_{22}\ldots$	$x_{22}\ldots$	c_{2n}	x_{2n}	a_2
⋮							
m	c_{m1}	x_{m1}	$c_{m2}\ldots$	$x_{m2}\ldots$	c_{mn}	x_{mn}	a_m
Demand	b_1		$b_2\ldots$		b_n		$\sum_{i=1}^{m} a_i = \sum_{j=1}^{n} b_j$

The given fuzzy transportation problem is said to be balanced

$$\sum_{i=1}^{m}\left[a_i^{(1)}, a_i^{(2)}, a_i^{(3)}, a_i^{(4)}\right] = \sum_{j=1}^{n}\left[b_j^{(1)}, b_j^{(2)}, b_j^{(3)}, b_j^{(4)}\right]$$

that is, if the total fuzzy capacity is equal to the total fuzzy demand. When the costs or time $\widetilde{\mathcal{C}}_{ij}$ are fuzzy numbers then the total cost becomes a fuzzy number

$$\widetilde{\mathcal{Z}} = \sum_{i=1}^{m}\sum_{j=1}^{n}\left[\tilde{c}_{ij}^{(1)}, \tilde{c}_{ij}^{(2)}, \tilde{c}_{ij}^{(3)}, \tilde{c}_{ij}^{(4)}\right]\left[x_{ij}^{(1)}, x_{ij}^{(2)}, x_{ij}^{(3)}, x_{ij}^{(4)}\right].$$

Hence it cannot be minimized directly. For solving the problem we defuzzify the fuzzy cost coefficients into crisp ones by a fuzzy number ranking method.

2.2.1 Preliminaries

Definition 2.2.1. *A fuzzy set is characterized by a membership function mapping elements of a domain, space or universe of discourse X to the unit interval* $[0, 1]$, *that is,*

$$\mathcal{A} = \{x,\ \mu_{\mathcal{A}}(x) : x \in X\}.$$

Here $\mu_{\mathcal{A}} : x \to [0, 1]$ *is a mapping called the membership value of* $x \in X$ *in a fuzzy set* $\mathcal{A}$. *These membership grades are often represented by real numbers ranging from* $[0, 1]$.

Definition 2.2.2. *A fuzzy set* $\mathcal{A}$ *of the universe of discourse X is called a normal fuzzy set, implying that there exist at least one* $x \in X$ *such that* $\mu_{\mathcal{A}}(x) = 1$.

Definition 2.2.3. *A fuzzy set* $\mathcal{A}$ *is convex if and only if, for any* $x_1, x_2 \in X$, *the membership function of* $\mathcal{A}$ *satisfies the inequality*

$$\mu_{\mathcal{A}}\{\lambda x_1 + (1-\lambda)x_2\} \geq \min\{\mu_{\mathcal{A}}(x_1),\ \mu_{\mathcal{A}}(x_2)\},\ 0 \leq \lambda \leq 1.$$

Definition 2.2.4. *(Fuzzy Number) Fuzzy number $\mathcal{A}$ is a fuzzy set defined on the set of real numbers $\mathbb{R}$ characterized by means of a membership function $\mu_{\mathcal{A}}(x)$, $\mu_{\mathcal{A}} : \mathbb{R} \to [0,1]$, which is upper semi-continuous and which fulfills the following conditions:*

(i) *Normality, that is,* $\sup_{x \in \mathbb{R}} \mu_{\mathcal{A}}(x) = 1.$

(ii) *Convexity, that is, for any $x, y \in \mathbb{R}$, $\lambda \in [0,1]$ it holds*
$\mu_{\mathcal{A}}(\lambda x + (1-\lambda)y) \geq \min(\mu_{\mathcal{A}}(x),\ \mu_{\mathcal{A}}(y)).$

Definition 2.2.5. *(Triangular Fuzzy Number) For a triangular fuzzy number $\mathcal{A}(x)$, it can be represented by $\mathcal{A}(a,b,c;1)$ with the membership function given by*

$$\mu_{\mathcal{A}}(x) = \begin{cases} \dfrac{(x-a)}{(b-a)}, & a \leq x \leq b \\ 1, & x = b \\ \dfrac{(c-x)}{(c-b)}, & c \leq x \leq b \\ 0, & otherwise. \end{cases}$$

Definition 2.2.6. *(Trapezoidal Fuzzy Number) For a trapezoidal number $\mathcal{A}(x)$, it can be represented by $\mathcal{A}(a,b,c,d;1)$ with the membership function given by*

$$\mu_{\mathcal{A}}(x) = \begin{cases} \dfrac{(x-a)}{(b-a)}, & a \leq x \leq b \\ 1, & b \leq x \leq c \\ \dfrac{(d-x)}{(d-c)}, & c \leq x \leq d \\ 0, & otherwise. \end{cases}$$

Definition 2.2.7. *($\alpha-$cut of a Trapezoidal Fuzzy Number) The $\alpha-$cut of a fuzzy number $\mathcal{A}(x)$ is defined as $\mathcal{A}(\alpha) = \{x : \mu(x) \geq \alpha,\ \alpha \in [0,1]\}$.*

Definition 2.2.8. *(Ranking Function) An efficient approach for comparing the fuzzy numbers is the use of a ranking function [18, 19], $\mathfrak{R} : F(\mathbb{R}) \to (\mathbb{R})$, where $F(\mathbb{R})$ is a set of fuzzy numbers defined on set of real numbers, which maps each fuzzy number into a real line, where a natural order exists, that is,*

$$\begin{aligned} &(i)\ \tilde{A} >_{\mathfrak{R}} \tilde{B}\ \ iff\ \ \mathfrak{R}(\tilde{A}) > \mathfrak{R}(\tilde{B}); \\ &(ii)\ \tilde{A} <_{\mathfrak{R}} \tilde{B}\ \ iff\ \ \mathfrak{R}(\tilde{A}) < \mathfrak{R}(\tilde{B}); \\ &(iii)\ \tilde{A} =_{\mathfrak{R}} \tilde{B}\ \ iff\ \ \mathfrak{R}(\tilde{A}) = \mathfrak{R}(\tilde{B}). \end{aligned}$$

Let $\tilde{A} = (a_1, b_1, c_1, d_1; w_1)$, $\tilde{B} = (a_2, b_2, c_2, d_2; w_2)$ be two generalized trapezoidal fuzzy numbers and w = minimum (w_1, w_2), then $\mathfrak{R}(\tilde{A}) = w(a_1 + b_1 + c_1 + d_1)/4$ and $\mathfrak{R}(\tilde{B}) = w(a_2 + b_2 + c_2 + d_2)/4$.

Remark: 2.2.1. *Let $(\tilde{A}_i)$, $i = 1, 2, \ldots, n$ be a set of generalized trapezoidal fuzzy numbers. If $\mathfrak{R}(\tilde{A}_k) \leq \mathfrak{R}(\tilde{A}_i)$, for all i, then the generalized trapezoidal fuzzy number $\tilde{A}_k$ is the minimum of $\tilde{A}_i$; $i = 1, 2, \ldots, n$ and if $\mathfrak{R}(\tilde{A}_k) \geq \mathfrak{R}(\tilde{A}_i)$, for all i, then the generalized trapezoidal fuzzy number $\tilde{A}_k$ is the maximum of $\tilde{A}_i$; $i = 1, 2, \ldots, n$.*

2.3 FUZZY TRANSPORTATION PROBLEM

Consider transportation problem with m fuzzy origins (rows) and n fuzzy destination (columns). Let $\mathcal{C}_{ij} = \left[c_{ij}^{(1)}, c_{ij}^{(2)}, c_{ij}^{(3)}, c_{ij}^{(4)}\right]$ be the cost of transporting one unit of the product from i^{th} fuzzy origin to j^{th} fuzzy destination.
$a_i = \left[a_i^{(1)}, a_i^{(2)}, a_i^{(3)}, a_i^{(4)}\right]$ is the quantity of commodity available at fuzzy origin i,
$b_j=\left[b_j^{(1)}, b_j^{(2)}, b_j^{(3)}, b_j^{(4)}\right]$ is the quantity of commodity needed at fuzzy destination j.
$X_{ij} = \left[x_{ij}^{(1)}, x_{ij}^{(2)}, x_{ij}^{(3)}, x_{ij}^{(4)}\right]$ is the quantity transported problem, which can be stated in Table 2.2.

The linear programming model representing the fuzzy transportation is given by

$$\begin{cases} \text{minimize } \mathcal{Z} = \sum_{i=1}^{m}\sum_{j=1}^{n}\left[c_{ij}^{(1)}, c_{ij}^{(2)}, c_{ij}^{(3)}, c_{ij}^{(4)}\right]\left[x_{ij}^{(1)}, x_{ij}^{(2)}, x_{ij}^{(3)}, x_{ij}^{(4)}\right] \\ \text{subject to} \\ \sum_{j=1}^{n}\left[x_{ij}^{(1)}, x_{ij}^{(2)}, x_{ij}^{(3)}, x_{ij}^{(4)}\right] = \left[a_i^{(1)}, a_i^{(2)}, a_i^{(3)}, a_i^{(4)}\right], \text{ for } i = 1, 2, \dots, m \\ \sum_{i=1}^{m}\left[x_{ij}^{(1)}, x_{ij}^{(2)}, x_{ij}^{(3)}, x_{ij}^{(4)}\right] = \left[b_j^{(1)}, b_j^{(2)}, b_j^{(3)}, b_j^{(4)}\right], \text{ for } j = 1, 2, \dots, n \\ \left[x_{ij}^{(1)}, x_{ij}^{(2)}, x_{ij}^{(3)}, x_{ij}^{(4)}\right] \geq 0. \end{cases} \tag{2.2}$$

The given fuzzy transportation problem is said to be balanced

$$\sum_{i=1}^{m}\left[a_i^{(1)}, a_i^{(2)}, a_i^{(3)}, a_i^{(4)}\right] = \sum_{j=1}^{n}\left[b_j^{(1)}, b_j^{(2)}, b_j^{(3)}, b_j^{(4)}\right]$$

that is, if the total fuzzy capacity is equal to the total fuzzy demand. When the costs or time $\widetilde{\mathcal{C}}_{ij}$ are fuzzy numbers then the total cost becomes a fuzzy number

$$\widetilde{\mathcal{Z}} = \sum_{i=1}^{m}\sum_{j=1}^{n}\left[\tilde{c}_{ij}^{(1)}, \tilde{c}_{ij}^{(2)}, \tilde{c}_{ij}^{(3)}, \tilde{c}_{ij}^{(4)}\right]\left[x_{ij}^{(1)}, x_{ij}^{(2)}, x_{ij}^{(3)}, x_{ij}^{(4)}\right].$$

TABLE 2.2
Transportation Problem

	1	2...	n	Fuzzy Capacity
1	c_{11} x_{11}	$c_{12}\cdots$ $x_{12}\cdots$	c_{1n} x_{1n}	a_1
2 ⋮	c_{21} x_{21}	$c_{22}\cdots$ $x_{22}\cdots$	c_{2n} x_{2n}	a_2
m	c_{m1} x_{m1}	$c_{m2}\ldots$ $x_{m2}\ldots$	c_{mn} x_{mn}	a_m
Demand	b_1	$b_2\ldots$	b_n	$\sum_{i=1}^{m} a_i = \sum_{j=1}^{n} b_j$

Hence it cannot be minimized directly. For solving the problem we defuzzify the fuzzy cost coefficients into crisp ones by a fuzzy number ranking method.

2.4 ROBUST RANKING TRANSPORTATION PROBLEM

Results employing a robust ranking technique, which satisfies the compensation, linearity, and additivity qualities, are what most of us intuitively would expect. Srivastava et al. [22] and Kalifa [23] discussed the fuzzy LPP with $\alpha-$ cut robust ranking methods and heptagonal fuzzy transportation problem. Given a convex fuzzy number $\widetilde{\mathcal{C}}$, the robust ranking index is defined by

$$\Re(\widetilde{\mathcal{C}}) = \int_0^1 0.5(c_\alpha^L + c_\alpha^U)d\alpha.$$

In this chapter, this approach is used for ranking the objective values. The robust ranking index $\Re(\widetilde{\mathcal{C}})$ gives the representative value of the fuzzy number $\widetilde{\mathcal{C}}$. It satisfies the linearity and additive property. If $\widetilde{\mathcal{A}} = a\widetilde{\mathcal{B}} + b\widetilde{\mathcal{C}}$ and $\widetilde{\mathcal{D}} = k\widetilde{\mathcal{E}} - t\widetilde{\mathcal{F}}$ where a, b, k, t are constants, then we have

$$\Re(\widetilde{\mathcal{A}}) = a\Re(\widetilde{\mathcal{B}}) + b\Re(\widetilde{\mathcal{C}}),$$
$$\text{and} \quad \Re(\widetilde{\mathcal{D}}) = k\Re(\widetilde{\mathcal{E}}) - t\Re(\widetilde{\mathcal{F}}).$$

The fuzzy transportation problem can be transformed into a crisp transportation problem in the LPP form on the basis of this property.

The Robust Ranking Technique is:
If $\Re(\widetilde{U}) \le \Re(\widetilde{V})$ then $\widetilde{U} \le \widetilde{V}$ that is, $\min(\widetilde{U}, \widetilde{V}) = \widetilde{U}$. For the transportation problem (2.3.1), with fuzzy objective function

$$\text{minimize } \widetilde{\mathcal{Z}} = \sum_{i=1}^{m}\sum_{j=1}^{n}\left[\tilde{c}_{ij}^{(1)}, \tilde{c}_{ij}^{(2)}, \tilde{c}_{ij}^{(3)}, \tilde{c}_{ij}^{(4)}\right]\left[x_{ij}^{(1)}, x_{ij}^{(2)}, x_{ij}^{(3)}, x_{ij}^{(4)}\right].$$

We apply the robust ranking method to get the minimum objective value $\widetilde{\mathcal{Z}}^*$ from the formulation:

$$\begin{cases} \Re(\widetilde{\mathcal{Z}}^*) = \text{minimize} \quad \mathcal{Z} = \sum_{i=1}^{m}\sum_{j=1}^{n} \Re(\widetilde{\mathcal{C}}_{ij})x_{ij} \\ \text{subject to} \\ \sum_{j=1}^{n}\left[x_{ij}^{(1)}, x_{ij}^{(2)}, x_{ij}^{(3)}, x_{ij}^{(4)}\right] = \left[a_{ij}^{(1)}, a_{ij}^{(2)}, a_{ij}^{(3)}, a_{ij}^{(4)}\right], \text{ for } i = 1, 2, \ldots, m \\ \sum_{i=1}^{m}\left[x_{ij}^{(1)}, x_{ij}^{(2)}, x_{ij}^{(3)}, x_{ij}^{(4)}\right] = \left[b_{ij}^{(1)}, b_{ij}^{(2)}, b_{ij}^{(3)}, b_{ij}^{(4)}\right], \text{ for } j = 1, 2, \ldots, n \\ \left[x_{ij}^{(1)}, x_{ij}^{(2)}, x_{ij}^{(3)}, x_{ij}^{(4)}\right] \ge 0 \end{cases} \quad (2.3)$$

The decision variable denotes the transportation of the person i to job j. $\widetilde{\mathcal{C}}_{ij}$ is the cost of assigning the jth job to the ith person. Since $\Re(\widetilde{\mathcal{C}}_{ij})$ are crisp values, this problem

(2.3) is obviously the crisp transportation problem of the form (2.2) which can be solved by the conventional methods, namely the Vogel's approximation method to solve the LPP form of the problem. Once the optimal solution x^* of (2.3) is found, the optimal fuzzy objective value of the original problem can be calculated as

$$\tilde{z}^* = \sum_{i=1}^{m}\sum_{j=1}^{n}\left[\tilde{c}_{ij}^{(1)}, \tilde{c}_{ij}^{(2)}, \tilde{c}_{ij}^{(3)}, \tilde{c}_{ij}^{(4)}\right]\left[x_{ij}^{(1)}, x_{ij}^{(2)}, x_{ij}^{(3)}, x_{ij}^{(4)}\right].$$

2.5 APPLICATION

2.5.1 Least Time Fuzzy Transportation Model

There seem to be transportation challenges where conserving computation time rather than moving as soon as it arrives is the goal. These fundamental problems are typically seen in military services, hospital administration, etc., where the fuzzy cost is less extremely important than the fuzzy delivery speed or fuzzy time of supply. Now, while solving problems where the objective is to fuzzy minimize time for each route, the cost per unit is replaced by the fuzzy time required to ship the quantity x_{ij} from the origin i to the destination j, where $i = 1, 2, 3, \dots, m$ and $j = 1, 2, 3, \dots, n$. The corresponding fuzzy transportation matrix is given in Table 2.3.

The linear programming model representing the fuzzy transportation problem is given by

$$\left\{\begin{array}{l} \text{minimize } \tilde{z} = \sum_{i=1}^{m}\sum_{j=1}^{n}\left[f\tilde{t}_{ij}^{(1)}, f\tilde{t}_{ij}^{(2)}, f\tilde{t}_{ij}^{(3)}, f\tilde{t}_{ij}^{(4)}\right]\left[x_{ij}^{(1)}, x_{ij}^{(2)}, x_{ij}^{(3)}, x_{ij}^{(4)}\right] \\ \text{subject to} \\ \sum_{j=1}^{n}\left[x_{ij}^{(1)}, x_{ij}^{(2)}, x_{ij}^{(3)}, x_{ij}^{(4)}\right] = \left[\tilde{a}_{ij}^{(1)}, \tilde{a}_{ij}^{(2)}, \tilde{a}_{ij}^{(3)}, \tilde{a}_{ij}^{(4)}\right], \text{ for } i = 1, 2, \dots, m \\ \sum_{i=1}^{m}\left[x_{ij}^{(1)}, x_{ij}^{(2)}, x_{ij}^{(3)}, x_{ij}^{(4)}\right] = \left[\tilde{b}_{ij}^{(1)}, \tilde{b}_{ij}^{(2)}, \tilde{b}_{ij}^{(3)}, \tilde{b}_{ij}^{(4)}\right], \text{ for } j = 1, 2, \dots, n \\ \left[x_{ij}^{(1)}, x_{ij}^{(2)}, x_{ij}^{(3)}, x_{ij}^{(4)}\right] \geq 0. \end{array}\right. \quad (2.4)$$

TABLE 2.3
Least Time Fuzzy Transportation Problem

	1	2 ⋯	n	Supply
1	ft_{11}	$ft_{12} \cdots$	ft_{1n}	a_1
2	ft_{21}	$ft_{22} \cdots$	ft_{2n}	a_2
⋮				
m	ft_{m1}	$ft_{m2} \cdots$	ft_{mn}	a_m
Demand	b_1	$b_2 \cdots$	b_n	$\sum_{i=1}^{m} a_i = \sum_{j=1}^{n} b_j$

The fuzzy transportation problem is balanced, if

$$\sum_{i=1}^{m}\left[\tilde{a}_i^{(1)},\tilde{a}_i^{(2)},\tilde{a}_i^{(3)},\tilde{a}_i^{(4)}\right]=\sum_{j=1}^{n}\left[\tilde{b}_j^{(1)},\tilde{b}_j^{(2)},\tilde{b}_j^{(3)},\tilde{b}_j^{(4)}\right].$$

That is, the total fuzzy capacity is equal to the total fuzzy demand. When the costs or time units $F\widetilde{T}_k$ $(= [f\tilde{t}_{ij}^{(1)},f\tilde{t}_{ij}^{(2)},f\tilde{t}_{ij}^{(3)},f\tilde{t}_{ij}^{(4)}])$ are fuzzy numbers, then the total unit time cost becomes a fuzzy number

$$\widetilde{z}=\sum_{i=1}^{m}\sum_{j=1}^{n}\left[f\tilde{t}_{ij}^{(1)},f\tilde{t}_{ij}^{(2)},f\tilde{t}_{ij}^{(3)},f\tilde{t}_{ij}^{(4)}\right]\left[x_{ij}^{(1)},x_{ij}^{(2)},x_{ij}^{(3)},x_{ij}^{(4)}\right].$$

Consequently, it cannot be directly lowered significantly. The trapezoidal fuzzy number robust ranking algorithm transforms the fuzzy coefficient into crisp ones. No matter how many units are whisked away, the imprecise time of shipment remains constant. Additionally, while shipments from origins to destinations can occur at the same hazy time on various routes, the total plan's shipment time is not equal to the sum of the times for all of the routes. Let $F\widetilde{T}_k$ be the largest fuzzy time associated k^{th} feasible plan. Our objective is therefore, to find out a plan for which $F\widetilde{T}_k$ is the minimum of all values of k.

Procedure for Minimum $F\widetilde{T}_k$:

(I) Find a fuzzy IBFS. This is obtained by using the same method as for the normal fuzzy transportation technique.

(II) Find $F\widetilde{T}_k$ corresponding to the current feasible solution and cross all the fuzzy non-basic cells for which $ft_{ij} \geq F\widetilde{T}_k$.

(III) Draw a closed path (as in the normal fuzzy transportation technique) for the fuzzy basic variable associated with $F\widetilde{T}_k$ such that when the values of the corner elements are shifted around, the fuzzy basic variable reduces to zero and no variable becomes negative. This procedure ends if no such closed path can be traced out, otherwise go to II.

2.6 NUMERICAL EXAMPLES

Example 2.6.1. *The following fuzzy transportation problem can be addressed by starting with the fuzzy initial basic workable resolution acquired using the fuzzy north-west corner and fuzzy Vogel's approximation methods:*

	1	2	3	4	Supply
A	[−2,0,2,8]	[−2,0,2,8]	[−2,0,2,8]	[−1,0,1,4]	[0,2,4,6]
B	[4,8,12,16]	[4,7,9,12]	[2,4,6,8]	[1,3,5,7]	[2,4,9,13]
C	[2,4,9,13]	[0,6,8,10]	[0,6,8,10]	[4,7,9,12]	[2,4,6,8]
Demand	[1,3,5,7]	[0,2,4,6]	[1,3,5,7]	[1,3,5,7]	[4,10,19,27]

Since $\sum_{i=1}^{m} a_i = \sum_{j=1}^{n} b_j$, *the problem is a balanced fuzzy transportation problem. There exists a fuzzy initial basic feasible solution.*

	1	2	3	4	Supply
A	[-2,0,2,8] [0,2,4,6]	[-2,0,2,8]	[-2,0,2,8]	[-1,0,1,4]	[0,2,4,6]
B	[4,8,12,16]	[4,7,9,12]	[2,4,6,8] [-5,-1,6,12]	[1,3,5,7] [1,3,5,7]	[2,4,9,13]
C	[2,4,9,13] [-5,-1,3,7]	[0,6,8,10] [0,2,4,6]	[0,6,8,10] [-11,-3,6,12]	[4,7,9,12]	[2,4,6,8]
Demand	[1,3,5,7]	[0,2,4,6]	[1,3,5,7]	[1,3,5,7]	[4,10,19,27]

Since the number of occupied cells having $m + n - 1 = 6$ *and are also independent, there exist non-degenerate fuzzy basic feasible solutions. Therefore, the initial fuzzy transportation minimum cost is*

$$\left[z^{(1)}, z^{(2)}, z^{(3)}, z^{(4)}\right] = [0,2,4,6][-2,0,2,8] + [-5,-1,6,12][2,4,6,8]$$
$$+ [1,3,5,7][1,3,5,7] + [2,4,9,13][-5,-1,3,7]$$
$$+ [0,6,8,10][0,2,4,6] + [0,6,8,10][-11,-3,6,12]$$
$$\left[z^{(1)}, z^{(2)}, z^{(3)}, z^{(4)}\right] = [-122,-2,139,257].$$

To Find the Optimal Solution:

Applying the fuzzy uv–method, we determine a set of numbers $\left[u_i^{(1)}, u_i^{(2)}, u_i^{(3)}, u_i^{(4)}\right]$ *and* $\left[v_j^{(1)}, v_j^{(2)}, v_j^{(3)}, v_j^{(4)}\right]$ *each row and column such that*

$$\left[c_{ij}^{(1)}, c_{ij}^{(2)}, c_{ij}^{(3)}, c_{ij}^{(4)}\right] = \left[u_i^{(1)}, u_i^{(2)}, u_i^{(3)}, u_i^{(4)}\right] + \left[v_j^{(1)}, v_j^{(2)}, v_j^{(3)}, v_j^{(4)}\right]$$

for each occupied cell. Since 3rd *row has maximum number of allocations, we give fuzzy number* $\left[u_3^{(1)}, u_3^{(2)}, u_3^{(3)}, u_3^{(4)}\right] = [-2,-1,1,2]$. *The remaining numbers can be obtained as given below:*

$$\left[c_{31}^{(1)}, c_{31}^{(2)}, c_{31}^{(3)}, c_{31}^{(4)}\right] = \left[u_3^{(1)}, u_3^{(2)}, u_3^{(3)}, u_3^{(4)}\right] + \left[v_1^{(1)}, v_1^{(2)}, v_1^{(3)}, v_1^{(4)}\right]$$
$$\left[v_1^{(1)}, v_1^{(2)}, v_1^{(3)}, v_1^{(4)}\right] = [0,3,10,15]$$

$$\left[c_{33}^{(1)}, c_{33}^{(2)}, c_{33}^{(3)}, c_{33}^{(4)}\right] = \left[u_3^{(1)}, u_3^{(2)}, u_3^{(3)}, u_3^{(4)}\right] + \left[v_3^{(1)}, v_3^{(2)}, v_3^{(3)}, v_3^{(4)}\right]$$
$$\left[v_3^{(1)}, v_3^{(2)}, v_3^{(3)}, v_3^{(4)}\right] = [-2,5,9,12]$$

$$\left[c_{11}^{(1)}, c_{11}^{(2)}, c_{11}^{(3)}, c_{11}^{(4)}\right] = \left[u_1^{(1)}, u_1^{(2)}, u_1^{(3)}, u_1^{(4)}\right] + \left[v_1^{(1)}, v_1^{(2)}, v_1^{(3)}, v_1^{(4)}\right]$$
$$\left[u_1^{(1)}, u_1^{(2)}, u_1^{(3)}, u_1^{(4)}\right] = [-17,-10,-1,8]$$

$$\left[c_{23}^{(1)}, c_{23}^{(2)}, c_{23}^{(3)}, c_{23}^{(4)}\right] = \left[u_2^{(1)}, u_2^{(2)}, u_2^{(3)}, u_2^{(4)}\right] + \left[v_3^{(1)}, v_3^{(2)}, v_3^{(3)}, v_3^{(4)}\right]$$
$$\left[u_2^{(1)}, u_2^{(2)}, u_2^{(3)}, u_2^{(4)}\right] = [-10, -5, 1, 10]$$

$$\left[c_{24}^{(1)}, c_{24}^{(2)}, c_{24}^{(3)}, c_{24}^{(4)}\right] = \left[u_2^{(1)}, u_2^{(2)}, u_2^{(3)}, u_2^{(4)}\right] + \left[v_4^{(1)}, v_4^{(2)}, v_4^{(3)}, v_4^{(4)}\right]$$
$$\left[v_4^{(1)}, v_4^{(2)}, v_4^{(3)}, v_4^{(4)}\right] = [-9, 2, 10, 17].$$

We find for each empty cell of the sum $\left[u_i^{(1)}, u_i^{(2)}, u_i^{(3)}, u_i^{(4)}\right]$ *and* $\left[v_j^{(1)}, v_j^{(2)}, v_j^{(3)}, v_j^{(4)}\right]$.

Next we find the net evaluation $\left[z_{ij}^{(1)}, z_{ij}^{(2)}, z_{ij}^{(3)}, z_{ij}^{(4)}\right]$ *is given by*

$$\left[z_{ij}^{(1)}, z_{ij}^{(2)}, z_{ij}^{(3)}, z_{ij}^{(4)}\right] = \left[c_{ij}^{(1)}, c_{ij}^{(2)}, c_{ij}^{(3)}, c_{ij}^{(4)}\right] - \left[u_i^{(1)}, u_i^{(2)}, u_i^{(3)}, u_i^{(4)}\right] + \left[v_j^{(1)}, v_j^{(2)}, v_j^{(3)}, v_j^{(4)}\right].$$

Therefore the fuzzy optimal solution in terms of trapezoidal fuzzy numbers are:

$$\left[x_{11}^{(1)}, x_{11}^{(2)}, x_{11}^{(3)}, x_{11}^{(4)}\right] = [0, 2, 4, 6]$$
$$\left[x_{23}^{(1)}, x_{23}^{(2)}, x_{23}^{(3)}, x_{23}^{(4)}\right] = [-5, -1, 6, 12]$$
$$\left[x_{24}^{(1)}, x_{24}^{(2)}, x_{24}^{(3)}, x_{24}^{(4)}\right] = [1, 3, 5, 7]$$
$$\left[x_{31}^{(1)}, x_{31}^{(2)}, x_{31}^{(3)}, x_{31}^{(4)}\right] = [-5, -1, 3, 7]$$
$$\left[x_{32}^{(1)}, x_{32}^{(2)}, x_{32}^{(3)}, x_{32}^{(4)}\right] = [0, 2, 4, 6]$$
$$\left[x_{33}^{(1)}, x_{33}^{(2)}, x_{33}^{(3)}, x_{33}^{(4)}\right] = [-11, -3, 6, 12].$$

Hence the total fuzzy transportation minimum cost is

$$\begin{aligned}\left[z^{(1)}, z^{(2)}, z^{(3)}, z^{(4)}\right] &= [0, 2, 4, 6][-2, 0, 2, 8] + [-5, 1, 6, 12][2, 4, 6, 8] \\ &\quad + [1, 3, 5, 7][1, 3, 5, 7] + [2, 4, 9, 13][-5, -1, 3, 7] \\ &\quad + [0, 6, 8, 10][0, 2, 4, 6] + [0, 6, 8, 10][-11, -3, 6, 12].\end{aligned}$$
$$\left[z^{(1)}, z^{(2)}, z^{(3)}, z^{(4)}\right] = [-122, -2, 139, 257].$$

Applying the fuzzy robust ranking method. Now we calculate $\Re(-2, 0, 2, 8)$. *The membership function of the trapezoidal number* $(-2, 0, 2, 8)$ *is*

$$\mu(x) = \begin{cases} \dfrac{(x+2)}{2}, & -2 \leq x \leq 0 \\ 1, & x = 0 \\ \dfrac{(8-x)}{6}, & 2 \leq x \leq 8 \\ 0, & otherwise. \end{cases}$$

The $\alpha-$cut of the fuzzy number $(-2,0,2,8)$ is

$$
\begin{aligned}
(c_\alpha^L,\ c_\alpha^U) = (2\alpha - 2,\ 2 - 2\alpha) \textit{ for which } \Re(\widetilde{\mathfrak{C}}_{ij}) &= \Re(-2,0,2,8) \\
&= \int_0^1 0.5(c_\alpha^L, c_\alpha^U)d\alpha \\
&= \int_0^1 0.5(2\alpha - 2 + 2 - 2\alpha)d\alpha \\
&= 0.
\end{aligned}
$$

Proceeding similarly, the robust ranking indices for the fuzzy costs c_{ij} *are calculated as:*

$$
\Re(\tilde{c}_{11}) = 0,\ \Re(\tilde{c}_{12}) = 0,\ \Re(\tilde{c}_{13}) = 0,\ \Re(\tilde{c}_{14}) = 0,\ \Re(\tilde{c}_{21}) = 10,\ \Re(\tilde{c}_{22}) = 6.75,
$$
$$
\Re(\tilde{c}_{23}) = 4,\ \Re(\tilde{c}_{24}) = 3,\ \Re(\tilde{c}_{31}) = 4.75,\ \Re(\tilde{c}_{32}) = 5,\ \Re(\tilde{c}_{33}) = 5,\ \Re(\tilde{c}_{34}) = 6.75.
$$

We replace these values for their corresponding $\widetilde{\mathfrak{C}}_{ij}$ *in* (2.4.1)*, which results in a convenient transportation problem in the linear programming problem. We solve it by VAM to get the following optimal solution*

$$
x_{11}^* = x_{23}^* = x_{24}^* = x_{31}^* = x_{32}^* = x_{33}^* = 1
$$

$$
x_{12}^* = x_{13}^* = x_{14}^* = x_{21}^* = x_{22}^* = x_{34}^* = 0
$$

with the optimal objective value $\Re(\tilde{z}^*) = 21.75$*, which represents the optimal total cost. In other words the optimal transportation problem is*

$$
A \to 1,\ A \to 3,\ B \to 4,\ C \to 1,\ C \to 2,\ C \to 3.
$$

The fuzzy optimal total cost is calculated as

$$
\begin{aligned}
&(\tilde{c}_{11}) + (\tilde{c}_{23}) + (\tilde{c}_{24}) + (\tilde{c}_{31}) + (\tilde{c}_{32}) + (\tilde{c}_{33}) \\
&\quad = [0,2,4,6][-2,0,2,8] + [-5,-1,6,12][2,4,6,8] + [1,3,5,7][1,3,5,7] \\
&\qquad + [2,4,9,13][-5,-1,3,7] + [0,6,8,10][0,2,4,6] + [0,6,8,10][-11,-3,6,12].
\end{aligned}
$$

Also we find that

$$
\Re(\widetilde{\mathcal{Z}}^*) = \Re(-122,-2,139,257) = 21.75.
$$

In the above examples it has been shown that the total optimal cost obtained by our method remains the same as that obtained by defuzzifying the total fuzzy optimal cost by applying the robust ranking method [21].

Example 2.6.2. *Consider Example 6.1. Solve this problem by using least time transportation model:*

	1	2	3	4	Supply
A	[-2,0,2,8] [0,2,4,6]	[-2,0,2,8]	[-2,0,2,8]	[-1,0,1,4]	[0,2,4,6]
B	[4,8,12,16]	[4,7,9,12]	[2,4,6,8] [-5,-1,6,12]	[1,3,5,7] [1,3,5,7]	[2,4,9,13]
C	[2,4,9,13] [-5,-1,3,7]	[0,6,8,10] [0,2,4,6]	[0,6,8,10] [-11,-3,6,12]	[4,7,9,12]	[2,4,6,8]
Demand	[1,3,5,7]	[0,2,4,6]	[1,3,5,7]	[1,3,5,7]	[4,10,19,27]

In this table

$$x_{11} = [0, 2, 4, 6],\ x_{23} = [-5, 1, 6, 2],\ x_{24} = [1, 3, 5, 7],$$
$$x_{31} = [-5, 1, 3, 7],\ x_{32} = [0, 2, 4, 6],\ x_{33} = [-1, -3, 6, 12]$$

and all other variables are zero. The shipping fuzzy times are

$$ft_{11} = [-2, 0, 2, 8],\ ft_{23} = [2, 4, 6, 8],\ ft_{24} = [1, 3, 5, 7],$$
$$ft_{31} = [2, 4, 9, 13],\ ft_{32} = [0, 6, 8, 10],\ ft_{33} = [0, 6, 8, 10].$$

Therefore all the fuzzy shipments of this plan will be complete after

$$\begin{aligned} FT_1 &= \max(ft_{11},\ ft_{23},\ ft_{24},\ ft_{31},\ ft_{32},\ ft_{33}) \\ &= ft_{31} \\ &= [2, 4, 9, 13] \textit{ time units.} \end{aligned}$$

Therefore, the cell $(3, 1)$ *is crossed out according to VAM, since it has* $ft_{31} = [2, 4, 9, 13]$. *It is clear from the closed path that* X_{31} *can be decreased by only* $[-5, -1, 3, 7]$ *units. The next closed path solution is*

$$x_{11} = [0, 2, 4, 6],\ x_{21} = [-5, -1, 3, 7], x_{23} = [-7, -7, 4, 12],$$
$$x_{24} = [1, 3, 5, 7],\ x_{32} = [0, 2, 4, 6],\ x_{33} = [7, 5, 0, 6].$$

The shipping fuzzy times are

$$ft_{11} = [-2, 0, 2, 8],\ ft_{21} = [4, 8, 12, 16],\ ft_{23} = [2, 4, 6, 8],$$
$$ft_{24} = [1, 3, 5, 7],\ ft_{32} = [0, 6, 8, 10],\ ft_{33} = [0, 6, 8, 10].$$

$$\begin{aligned} FT_2 &= \max(ft_{11},\ ft_{21},\ ft_{23},\ ft_{24},\ ft_{32},\ ft_{33}) \\ &= ft_{21} \\ &= [4, 8, 12, 16] \textit{ time units.} \end{aligned}$$

Therefore, the cell $(2, 1)$ *is crossed out according to Vogel's approximation method, since it has* $ft_{21} = [4, 8, 12, 16]$. *Similarly we find the closed path is*

$$x_{11} = [1, 3, 5, 7],\ x_{14} = [-5, -1, 3, 7],\ x_{23} = [-7, -7, 4, 2],$$
$$x_{24} = [2, 4, 6, 8],\ x_{32} = [0, 2, 4, 6],\ x_{33} = [7, 5, 0, 6].$$

The shipping fuzzy times are

$$ft_{11} = [-2, 0, 2, 8],\ ft_{14} = [-1, 0, 1, 4],\ ft_{23} = [2, 4, 6, 8],$$
$$ft_{24} = [1, 3, 5, 7],\ ft_{32} = [0, 6, 8, 10],\ ft_{33} = [0, 6, 8, 10].$$

$$\begin{aligned} FT_3 &= \max(ft_{11},\ ft_{14},\ ft_{23},\ ft_{24},\ ft_{32},\ ft_{33}) \\ &= ft_{23} \\ &= [2, 4, 6, 8] \text{ time units.} \end{aligned}$$

Therefore the cell $(2, 3)$ *is crossed out.*

$$x_{11} = [1, 3, 5, 7],\ x_{14} = [-5, -1, 3, 7],\ x_{24} = [1, -1, 8, 4],$$
$$x_{32} = [0, 2, 4, 6],\ x_{33} = [-1, -7, 9, 9],\ x_{34} = [7, 7, -4, -2].$$

The shipping fuzzy times are

$$ft_{11} = [-2, 0, 2, 8],\ ft_{14} = [-1, 0, 1, 4],\ ft_{24} = [1, -1, 8, 4],$$
$$ft_{32} = [0, 2, 4, 6],\ ft_{33} = [-1, -7, 9, 9],\ ft_{34} = [7, -7, 4, 2].$$

$$\begin{aligned} FT_4 &= \max(ft_{11},\ ft_{14},\ ft_{24},\ ft_{32},\ ft_{33},\ ft_{34}) \\ &= ft_{34} \\ &= [4, 7, 9, 12] \text{ time units.} \end{aligned}$$

Therefore the cell $(3, 4)$ *is crossed out.*

$$x_{11} = [1, 3, 5, 7],\ x_{14} = [-5, -1, 3, 7],\ x_{22} = [-7, -7, 4, 2],$$
$$x_{24} = [11, 15, -5, -1],\ x_{32} = [13, 11, -2, -2],\ x_{33} = [-1, -7, 9, 9].$$

The shipping fuzzy times are

$$ft_{11} = [-2, 0, 2, 8],\ ft_{14} = [-1, 0, 1, 4],\ ft_{22} = [4, 7, 9, 12],$$
$$ft_{24} = [1, 3, 5, 7],\ ft_{32} = [0, 6, 8, 10],\ ft_{33} = [-1, -7, 9, 9].$$

$$\begin{aligned} FT_5 &= \max(ft_{11},\ ft_{14},\ ft_{22},\ ft_{24},\ ft_{32},\ ft_{33}) \\ &= ft_{22} \\ &= [4, 7, 9, 12] \text{ time units.} \end{aligned}$$

Therefore the cell $(2, 2)$ *is crossed out.*

$$x_{11} = [1, 3, 5, 7],\ x_{12} = [-7, 7, 4, 8],\ x_{14} = [-14, -10, 5, 13],$$
$$x_{24} = [-8, -12, 19, 9],\ x_{32} = [13, 11, -2, -2],\ x_{33} = [-1, -7, 9, 9].$$

The shipping fuzzy times are

$$ft_{11} = [-2, 0, 2, 8],\ \ ft_{12} = [-2, 0, 2, 8],\ \ ft_{14} = [-1, 0, 1, 4],$$
$$ft_{24} = [1, 3, 5, 7],\ \ ft_{32} = [0, 6, 8, 10],\ \ ft_{33} = [0, 6, 8, 10].$$

$$\begin{aligned} FT_6 &= \max(ft_{11},\ ft_{12},\ ft_{14},\ ft_{24},\ ft_{32},\ ft_{33}) \\ &= ft_{33} \\ &= [0, 6, 8, 10]\ \textit{time units}. \end{aligned}$$

Therefore the cell $(3, 3)$ *is crossed out. The solution of the final table will be of the form*

	1	2	3	4	Supply
A	[-2,0,2,8] [1,3,5,7]	[-2,0,2,8] [-9,-11,2,16]	[-2,0,2,8] [-1,-7,9,9]	[-1,0,1,4] [-14,-10,5,13]	[0,2,4,6]
B	[4,8,12,16]	[4,7,9,12]	[2,4,6,8] [-5,-1,6,12]	[1,3,5,7] [1,3,5,7]	[2,4,9,13]
C	[2,4,9,13] [-5,-1,3,7]	[0,6,8,10] [0,2,4,6]	[0,6,8,10] [-11,-3,6,12]	[4,7,9,12]	[2,4,6,8]
Demand	[1,3,5,7]	[0,2,4,6]	[1,3,5,7]	[1,3,5,7]	[4,10,19,27]

Since no closed path can be traced for x_{24}, *the iterative procedure ends here. Thus the above plan is optimal and the total shipment time is* $[1, 3, 5, 7]$ *units. That is, the optimal solution for the least time costs is*

$x_{11} = [1, 3, 5, 7]$ *with* $ft_{11} = [-2, 0, 2, 8]$, $x_{12} = [-9, -11, 2, 16]$ *with* $ft_{12} = [-2, 0, 2, 8]$,
$x_{13} = [-1, -7, 9, 9]$ *with* $ft_{13} = [-2, 0, 2, 8]$, $x_{14} = [-14, -10, 5, 13]$ *with* $ft_{14} = [-1, 0, 1, 4]$,
$x_{24} = [-8, -12, 19, 19]$ *with* $ft_{24} = [1, 3, 5, 7]$, $x_{32} = [-3, -9, 20, 22]$ *with* $ft_{32} = [0, 6, 8, 10]$.

To Find Optimal Solution by uv−Method:

Applying the fuzzy uv−method, we determine a set of numbers $\left[u_i^{(1)}, u_i^{(2)}, u_i^{(3)}, u_i^{(4)}\right]$ *and* $\left[v_j^{(1)}, v_j^{(2)}, v_j^{(3)}, v_j^{(4)}\right]$, *with each row and column such that*

$$\left[c_{ij}^{(1)}, c_{ij}^{(2)}, c_{ij}^{(3)}, c_{ij}^{(4)}\right] = \left[u_i^{(1)}, u_i^{(2)}, u_i^{(3)}, u_i^{(4)}\right] + \left[v_j^{(1)}, v_j^{(2)}, v_j^{(3)}, v_j^{(4)}\right]$$

for each occupied cell. Since the first row has the maximum number of allocations, we give the fuzzy number $\left[u_1^{(1)}, u_1^{(2)}, u_1^{(3)}, u_1^{(4)}\right] = [-2, -1, 1, 2]$. *The remaining numbers can be obtained as given below.*

$$\left[c_{11}^{(1)}, c_{11}^{(2)}, c_{11}^{(3)}, c_{11}^{(4)}\right] = \left[u_1^{(1)}, u_1^{(2)}, u_1^{(3)}, u_1^{(4)}\right] + \left[v_1^{(1)}, v_1^{(2)}, v_1^{(3)}, v_1^{(4)}\right]$$
$$\left[v_1^{(1)}, v_1^{(2)}, v_1^{(3)}, v_1^{(4)}\right] = [-4, -1, 3, 10]$$

$$\left[c_{12}^{(1)}, c_{12}^{(2)}, c_{12}^{(3)}, c_{12}^{(4)}\right] = \left[u_1^{(1)}, u_1^{(2)}, u_1^{(3)}, u_1^{(4)}\right] + \left[v_2^{(1)}, v_2^{(2)}, v_2^{(3)}, v_2^{(4)}\right]$$
$$\left[v_2^{(1)}, v_2^{(2)}, v_2^{(3)}, v_2^{(4)}\right] = [-4, -1, 3, 10]$$

$$\left[c_{13}^{(1)}, c_{13}^{(2)}, c_{13}^{(3)}, c_{13}^{(4)}\right] = \left[u_1^{(1)}, u_1^{(2)}, u_1^{(3)}, u_1^{(4)}\right] + \left[v_3^{(1)}, v_3^{(2)}, v_3^{(3)}, v_3^{(4)}\right]$$
$$\left[v_3^{(1)}, v_3^{(2)}, v_3^{(3)}, v_3^{(4)}\right] = [-4, 1, 3, 0]$$

$$\left[c_{14}^{(1)}, c_{14}^{(2)}, c_{14}^{(3)}, c_{14}^{(4)}\right] = \left[u_1^{(1)}, u_1^{(2)}, u_1^{(3)}, u_1^{(4)}\right] + \left[v_4^{(1)}, v_4^{(2)}, v_4^{(3)}, v_4^{(4)}\right]$$
$$\left[v_4^{(1)}, v_4^{(2)}, v_4^{(3)}, v_4^{(4)}\right] = [6, -2, 1, 3]$$

$$\left[c_{24}^{(1)}, c_{24}^{(2)}, c_{24}^{(3)}, c_{24}^{(4)}\right] = \left[u_2^{(1)}, u_2^{(2)}, u_2^{(3)}, u_2^{(4)}\right] + \left[v_4^{(1)}, v_4^{(2)}, v_4^{(3)}, v_4^{(4)}\right]$$
$$\left[u_2^{(1)}, u_2^{(2)}, u_2^{(3)}, u_2^{(4)}\right] = [-2, 2, 7, 6]$$

$$\left[c_{32}^{(1)}, c_{32}^{(2)}, c_{32}^{(3)}, c_{32}^{(4)}\right] = \left[u_3^{(1)}, u_3^{(2)}, u_3^{(3)}, u_3^{(4)}\right] + \left[v_2^{(1)}, v_2^{(2)}, v_2^{(3)}, v_2^{(4)}\right]$$
$$\left[u_3^{(1)}, u_3^{(2)}, u_3^{(3)}, u_3^{(4)}\right] = [-10, 3, 9, 14].$$

To Find Ranking Optimal Solution:

Now we calculate $\Re(-2,-1,1,2)$ *by applying the robust ranking method. The membership function of the trapezoidal fuzzy number* $[-2,-1,1,2]$ *is*

$$\mu(x) = \begin{cases} (x+2), & -2 \le x \le -1 \\ 1, & -1 \le x \le 1 \\ (2-x), & 1 \le x \le 2 \\ 0, & otherwise. \end{cases}$$

The α*−cut of a fuzzy number* $[-2,-1,1,2]$ *is* $(c_\alpha^L,\ c_\alpha^U) = (\alpha - 2,\ 2 - \alpha)$ *for which*

$$\begin{aligned} \Re(u_1) &= \int_0^1 0.5(c_\alpha^L, c_\alpha^U)d\alpha \\ &= \int_0^1 0.5(\alpha - 2, 2 - \alpha)d\alpha \\ &= 0. \end{aligned}$$

Similarly for

$$\Re(u_2) = 3.25,\ \Re(u_3) = 10.5,\ \Re(v_1) = 2,\ \Re(v_2) = 2,\ \Re(v_3) = 2,\ \Re(v_4) = 2.$$

Hence the total transportation fuzzy optimal minimum cost is

$$\Re(\tilde{z}) = \Re(u_i + v_j) = 21.75.$$

2.7 CONCLUSION

In supply chains and logistics, transportation models are frequently employed to cut costs. In order to meet transportation needs in a way that is both cost-effective and efficient, the solid transportation problem takes into consideration both supply and demand as well as transportation capacity. The membership function of the fuzzy total transportation costs is generated in this chapter by deriving the robust fuzzy transportation problem under specific circumstances. The idea is premised on the idea of extension. When doing robust optimizations, certain considerations must be taken before learning the actual values of the unknown parameters, while others—known as recourse choices—must be made after acquiring the information. There are two variations of the robust fuzzy transportation problem: one with inequality constraints and the other with equality requirements. One of the future objectives could be to use epidemic models to produce the best results for complex domains.

ACKNOWLEDGEMENTS

The authors would like to specially thank the editor and the anonymous referees for their valuable suggestions that led to the improvement of the chapter.

CONFLICT OF INTEREST

The authors declare that they have no conflicts of interest.

REFERENCES

1. C.R. Bector, S. Chandra, J. Dutta, *Principles of Optimization Theory*, Narosa Publishing House, 2005.
2. G.S. Mahapatra, T.K. Roy, Fuzzy multi objective mathematical programming on reliability optimization model, *Applied Mathematics and Computation*, 174 (2006), 643–659.
3. S.A. Orlovski, Multi objective programming problems with fuzzy parameters, *Control and Cybernetics*, 4 (1984), 175–184.
4. D. Peidro, J. Mula, R. Poler, J.L. Verdegay, Fuzzy optimization for supply chain planning under supply, demand and process uncertainties, *Fuzzy Sets and Systems*, 160 (2009), 2640–2657.
5. G. Wang, G. Ziyou and W. Zhongping, A global optimization algorithm for solving the bi-level linear functional programming problem, *Computers & Industrial Engineering*, 63 (2012), 428–432.
6. H.J. Zimmermann, *Fuzzy Sets Decision – Making and Expert Systems*, Kluwer Academic Publisher, Boston, 1987.
7. H.J. Zimmermann, Methods and applications of fuzzy mathematical programming in: R. R. Yager, and L.A. Zadeh eds., An Introduction to Fuzzy Logic Applications in Intelligent Systems, Kluwer Academic Publisher, Boston, 1992, 97–120.
8. H.J. Zimmermann, Description and optimization to fuzzy systems, *International Journal of General Systems*, 2 (1976), 209–215.
9. L.A. Zadeh, Fuzzy sets, *Information and Control*, 8 (1965), 338–353.

10. E. Lee, R.J. Li, Comparision of fuzzy numbers based on the probability measure of fuzzy events, *Computers and Mathematical Application*, 15 (1988), 887–896.
11. X. Wang, E.E. Kerre, Reasonable properties for the ordering of fuzzy quantities(I), *Fuzzy Sets and Systems*, 118 (2001), 375–385.
12. S.T. Liou, M.J. Wang, Ranking fuzzy number with integral values, *Fuzzy Sets and Systems*, 50 (1992), 247–255.
13. J.S. Su, Fuzzy programming based on interval-valued fuzzy numbers and ranking, *International Journal of Contemporary Mathematical Sciences*, 2 (2007), 27–39.
14. H. Abdollahnejad Barough, A linear programming priority method for a fuzzy transportation problem with non-linear constraints, *Fuzzy Information and Engineering*, 3 (2011), 193–208.
15. J.J. Buckly, Possibilistic linear programming with triangular fuzzy numbers, *Fuzzy Sets and Systems*, 26 (1988), 135–138.
16. S. Chanas, D. Kuchta, A concept of solution of the transportation problem with fuzzy cost co-efficients, *Fuzzy Sets and Systems*, 82 (1996), 299–305.
17. A. Kumar, P. Singh, A. Kaur, P. Kaur, A new approach for ranking non normal p-norm trapezoidal fuzzy numbers, *Computers and Mathematics with Applications*, 61 (2011), 881–887.
18. S.T. Liu, C. Kao, Network flow problems with fuzzy arc lengths, *IEEE Transactions on System, Man, and Cybernetics – Part B*, 34 (2004), 765–769.
19. H.R. Maleki, M. Mashinchi, Fuzzy number linear programming: A probabilistic approach(3), *Journal of Applied Mathematics and Computation*, 15 (2004), 333–341.
20. V. Pandian, G. Natarajan, A new algorithm for finding a fuzzy optimal solution for fuzzy transportation problem, *Applied Mathematical Science*, 4 (2010), 79-90.
21. R.R. Yager, A procedure for ordering fuzzy subsets of the unit interval, *Information Science*, 24 (1981), 143–161.
22. H. ABD EL-Wahed Kalifa, Goal programming approach for solving heptagonal fuzzy transportation problem under budgetary constraint, *Operations Research and Decisions*, 1 (2020), 85–96.
23. B. Srivastava, B. Agarwal, S. Kumar, Fuzzy linear programming problem with $\alpha-$cut and robust ranking methods, *International Journal of Statistics and Applied Mathematics*, 2 (2022), 57–62.

3 Solution of Bilevel Linear Fractional Transportation Problem with Pythagorean Fuzzy Numbers

Ritu Arora[1] *and Shalini Arora*[2]

[1]Professor, Keshav Mahavidyalaya, University of Delhi, India

[2]Professor, Indira Gandhi Delhi Technical University for Women, Kashmere Gate, India

CONTENTS

3.1 INTRODUCTION

Transportation has always been an alluring field for researchers. Distinct models and techniques to solve transportation problems have been presented by several authors. Hitchcock [1] instigated the notion of transportation. Brigden [2] examined the transportation problem with mixed constraints. Chanas et al. [3] proposed the models on fuzzy transportation problem and also analyzed its solution methodology. Behbahani

DOI: 10.1201/9781003329039-3

et al. [4] explored the impact of transportation decisions on varied groups of society. Transportation has its own vital role to play in an economic system. It helps the producers to reach the end consumers, thus satisfying their requirements. It has been observed during the lockdown period that there was a need to transport the maximum number of units of goods like grocery, medications, medicinal articles like PPE kits, oxygen cylinders, etc. to the affected areas within sufficient time. Thus, the concept of Linear Fractional Transportation Problem (LFTP) comes into picture so as to minimize the cost of transportation and delivering the maximum number of units of goods simultaneously. LFTP has been studied by various authors and distinct solution modes have been proffered by them. Miansian [5] defined a fractional transportation problem and proposed solution methodology for it. Joshi and Gupta [6] solved LFTP with varying demand and supply. Cetin and Tiryaki [7] proposed generalized Dinkelbach's algorithm for multiobjective LFTP. Mahmoodirad et al. [8] considered uncertain parameters of belief degree in a LFTP and solved it. Liu [9] contemplated the fractional transportation problem (FTP) with fuzzy parameters and solved it. Radhakrishnan and Anukokila [10] solved the fuzzy FTP by Werner's fuzzy operator. Javaid [11] considered fractional objective functions for an uncertain multi-objective transportation problem, which was solved by fuzzy goal programming by defining an equivalent deterministic model. Kaushal et al. [12] applied a bilevel fractional transportation problem to food chain industry and Indore city waste, thus exhibiting its application. Agarwal and Ganesan [13] solved FTP with stochastic parameters that follow the exponential distribution.

3.1.1 Literature Review

Bilevel Programming Problem (BLPP) is a hierarchical non-convex programming problem. It manages the decision-maker at two levels, namely, the upper level and lower level. The upper level is called the leader and the other is called the follower. The leader puts forth the course of action and the follower executes the working accordingly. Mathematically, BLPP is presented as [14]:

$$(BLPP): \quad \min_{Z_1} F_{11}(Z_1, Z_2)$$

where for a given Z_1, Z_2 solves

$$\min_{Z_2} F_{12}(Z_1, Z_2)$$
$$\text{subject to } Z_1, Z_2 \in S.$$

Here, $S = [(Z_1, Z_2) : AZ_1 + BZ_2 \leq b; (Z_1, Z_2) \geq 0]$, where Z_1 and Z_2 can be linear or non-linear.

The diverse implementation of BLPP can be seen in innumerable realistic scenarios like the education sector, agriculture, finance, medicine, economics, and hydro and power energy sectors. BLPP has its relevance in the domain of transportation also. Sakawa et al. [15] formulated the transportation problem as a two-level integer programming problem and obtained its satisfactory solution by the method of interactive fuzzy programming. Bagloee et al. [16] applied hybrid method based machine learning to a discrete network design problem. Lv et al. [17] established the road pricing problem and its pertinence to bilevel programming method under stochastic and fuzzy uncertainties. Guo et al. [18] developed a bilevel mixed integer nonlinear programming problem by analyzing an integrated production and transportation scheduling problem.

Fuzzy programming enables the decision-makers to solve and to perceive a satisfactory solution for an optimization problem. The methodology of fuzzy programming was incited by Tanaka [19]. Decision-making problems can also be solved by a goal programming method. The correspondence between fuzzy programming and goal programming was given by Mohamed [20]. Lu et al. [21] proposed two-phase fuzzy programming for management of municipal solid waste. Turgay and Taskin [22] employed an exponential membership function in fuzzy goal programming approach for health care planning. Bal and Satoglu [23] applied goal programming for reverse logistics operations planning. A goal programming model was formulated by Chen et al. [24] for bicriteria solid transportation problem with uncertain parameters. Shih [25] used fuzzy linear programming to determine the transportation planning for cement. Zheng and Ling [26] proposed fuzzy optimization for transportation in an emergency in disaster relief supply chains.

In a classical transportation problem, parameters examined by the researchers are specific in nature. Due to fluctuating market conditions and cost, supply and demand parameters are not constant. To cater this irregular market behavior, the concept of Pythagorean fuzzy numbers came into existence. Yager [27] conceptualized Pythagorean fuzzy membership grades and Pythagorean fuzzy subsets. The operational laws of Pythagorean fuzzy sets and their properties were discussed by Zhang and Xu [28]. Adhami and Ahmad [29] solved a multi-objective transportation problem by developing a Pythagorean hesitant fuzzy computational algorithm. Garg [30] presented the improved score function for interval valued Pythagorean fuzzy sets. Wan et al. [31] presented a three-phase technique with Pythagorean fuzzy numbers for applying to haze management. Jianping et al. [32] develops a Pythagorean fuzzy environment for qualitative evaluation of green supplier selection. Yucesan and Kahraman [33] applied a Pythagorean fuzzy analytical hierarchy process for analyzing hydroelectric power plants. Ejewa [34] applied max-min-max composition of Pythagorean fuzzy sets in career placements. Liu et al. [35] evolved a Pythagorean fuzzy-based method for managing medical waste.

In practical scenarios, the transportation problem not only deals with the minimization of transportation cost but it also deals with the minimization of time. It also takes account that a considerable amount of goods should be disbursed and commodities should reach the maximum number of people. To accomplish all these objectives, Bilevel Linear Fractional Transportation Problem (BLFTP) is expounded in this paper.

As we are all aware, during the Covid-19 pandemic, the market became unpredictable due to various lockdowns which led to uncertainty and that itself is the motivation of the current problem. Pythagorean fuzzy numbers (PFNs) exhibit the uncertainty in supply and demand for BLFTP. The Pythagorean parameters are analyzed using Score Function. The satisfactory solution for BLFTP is compared after solving the problem by two different modes, fuzzy programming and goal programming. The division of the paper is as follows: BLFTP is explained in section 3.2. Section 3.3 introduces

PFNs which form the basis of BLFTP under uncertainty. Section 3.4 presents BLFTP with Pythagorean fuzzy parameters and the approach to solve the problem. Section 3.5 exhibits the application of a fuzzy bilevel problem with Pythagorean parameters. The conclusions and future prospects are given in Section 3.6.

3.2 BILEVEL LINEAR FRACTIONAL TRANSPORTATION PROBLEM

This section determines a bilevel linear fractional transportation problem (BLFTP). The problem consists of two levels, upper level and lower level. It has been observed that during the supply of goods, some amount gets damaged or destroyed due to adverse weather conditions or large storage timings. Therefore, it is required that during transportation, cost should be minimized. Also, a considerable and adequate quantity should reach a large number of people. Mathematically, BLFTP is defined as:

$$(BLFTP): \quad \min_{Z_1} F_{11}(Z_1, Z_2) = \frac{p_1^T Z_1 + p_2^T Z_2}{q_1^T Z_1 + q_2^T Z_2}$$

where Z_2 solves

$$\min_{Z_2} F_{12}(Z_1, Z_2) = \frac{s_1^T Z_1 + s_2^T Z_2}{w_1^T Z_1 + w_2^T Z_2} \quad \text{(for a given } Z_1\text{)}$$

subject to

$$\begin{aligned}
&\sum_{l \in L_{11}} z_{rl} \leq u_r^{'} \ \forall\, r \in R_{11},\\
&\sum_{r \in R_{11}} z_{rl} \geq v_l^{'} \ \forall\, l \in L_{11},\\
&\sum_{l \in L_{12}} z_{rl} \leq u_r^{''} \ \forall\, r \in R_{12},\\
&\sum_{r \in R_{12}} z_{rl} \geq v_l^{''} \ \forall\, l \in L_{12},\\
&z_{rl} \geq 0 \ \forall\, (r, l) \in R_T \times L_T.
\end{aligned} \tag{3.1}$$

The objective function at both the levels are linear fractional, therefore, they are pseudoconcave. Hence, the minimum of the bilevel problem will be procured at an extreme point of the feasible region. Here, $(p_1^T Z_1 + p_2^T Z_2)$ depicts the cost of transportation of products from r^{th} source to l^{th} destination and $(q_1^T Z_1 + q_2^T Z_2)$ depicts the maximum amount of goods to be transported at upper level. Likewise, $(s_1^T Z_1 + s_2^T Z_2)$ represents the cost of the transportation of products and $(w_1^T Z_1 + w_2^T Z_2)$ are the maximal goods to be transported at lower level.

Notations

R_T = Total number of sources.
L_T = Total number of destinations.
R_{11} = Number of sources at upper level.

L_{11} = Number of destinations at upper level.
R_{12} = Number of sources at lower level.
L_{12} = Number of destinations at lower level.
$p_1 = [p'_{rl}]$; $q_1 = [q'_{rl}]$; $s_1 = [s'_{rl}]$; $w_1 = [w'_{rl}]$ $\forall r \in R_{11}$, $l \in L_{11}$.
$p_2 = [p''_{rl}]$; $q_2 = [q''_{rl}]$; $s_2 = [s''_{rl}]$; $w_2 = [w''_{rl}]$ $\forall r \in R_{12}$, $l \in L_{12}$.
$R_T = R_{11} \cup R_{12}$; $R_{11} = \{1, 2, ..., r_1\}$; $R_{12} = \{r_1 + 1, ..., r\}$.
$L_T = L_{11} \cup L_{12}$; $L_{11} = \{1, 2, ..., l_1\}$; $L_{12} = \{l_1 + 1, ..., l\}$.
$p'_{rl} > 0$, $s'_{rl} > 0$; $p''_{rl} > 0$; $s''_{rl} > 0$: cost parameters for the upper level and lower level problem respectively.
$q'_{rl} > 0$, $w'_{rl} > 0$; $q''_{rl} > 0$; $w''_{rl} > 0$: is the benefit earned by the upper and lower level by transporting maximum number of units of goods.
$Z_1 = [z_{rl}], r \in R_{11}, l \in L_{11}$ is the quantity transported from the rth origin to l^{th} destination at the upper level.
$Z_2 = [z_{rl}], r \in R_{12}, l \in L_{12}$ is the quantity transported from the rth origin to l^{th} destination at the lower level.
$u'_r > 0$, $v'_l > 0$; $r \in R_{11}, l \in L_{11}$: upper level supply and demand parameters.
$u''_r > 0$, $v''_l > 0$; $r \in R_{12}, l \in L_{12}$: lower level supply and demand parameters.

Assumption: For solving (BLFTP), $(q_1^T Z_1 + q_2^T Z_2) > 0$ and $(w_1^T Z_1 + w_2^T Z_2) > 0$ for all the values of the feasible region.
Feasibility condition:

$$\sum_{r \in R_T} u_r \geq \sum_{l \in L_T} v_l \; ; \; \sum_{r \in R_{11}} u'_r \geq \sum_{l \in L_{11}} v'_l \; \& \; \sum_{r \in R_{12}} u''_r \geq \sum_{l \in L_{12}} v''_l .$$

3.3 BASIC DEFINITIONS

This section describes the definitions that play a pivotal role in the formulation of a bilevel fractional transporation problem with Pythagorean fuzzy numbers.

Definition 3.3.1. *[28]: Let Q be a universal set. A Pythagorean fuzzy set in Q is defined by the set of three elements and it is represented as follows:*

$$\tilde{H} = \{< q, \mu_{\tilde{H}}(q), \gamma_{\tilde{H}}(q) >: q \in Q\},$$

where $\mu_{\tilde{H}}(q) : Q \to [0,1]$ *and* $\gamma_{\tilde{H}}(q) : Q \to [0,1]$ *are functions such that* $(\mu_{\tilde{H}}(q))^2 + (\gamma_{\tilde{H}}(q))^2 \leq 1$, $\forall\, q \in Q$. *For each* $q \in Q$, $\mu_{\tilde{H}}(q)$ *and* $\gamma_{\tilde{H}}(q)$ *specify the degree of membership and degree of non-membership respectively.*
Also, $\pi_{\tilde{H}}(q)$ *denotes the degree of uncertainty of the element q in the set* $\tilde{H}$ *and is presented by* $\pi_{\tilde{H}}(q) = \sqrt{(1 - (\mu_{\tilde{H}}(q)^2) - (\gamma_{\tilde{H}}(q))^2}$.

Definition 3.3.2. *[28]:Score Function: The Pythagorean fuzzy number* $\delta = (\mu_{\tilde{H}}, \gamma_{\tilde{H}})$, *such that* $\mu_{\tilde{H}} \geq 0$; $\gamma_{\tilde{H}} \geq 0$; *and* $\mu_{\tilde{H}}^2 + \gamma_{\tilde{H}}^2 \leq 1$; *its score function is defined as* $S(\delta) = (\mu_{\tilde{H}}^2 - \gamma_{\tilde{H}}^2), S(\delta) \in [-1, 1]$.

Definition 3.3.3. *[28]:Accuracy Function: The accuracy function for a Pythagorean fuzzy number* $\delta = (\mu_{\tilde{H}}, \gamma_{\tilde{H}})$, *is defined as* $S(\delta) = (\mu_{\tilde{H}}^2 + \gamma_{\tilde{H}}^2), S(\delta) \in [-1, 1]$.

Definition 3.3.4. *[30]:Improved Score Function: The improved score function for a PFN* $\delta = (\mu_{\tilde{H}}, \gamma_{\tilde{H}})$ *is denoted as* $P(\delta)$ *and is defined as*
$P(\delta) = (\mu_{\tilde{H}}^2 - \gamma_{\tilde{H}}^2)(1 + \sqrt{(1-\mu_{\tilde{H}}^2) - \gamma_{\tilde{H}}^2}), P(\delta) \in [-1, 1].$

3.4 BILEVEL FRACTIONAL TRANSPORTATION PROBLEM WITH PYTHAGOREAN FUZZY PARAMETERS

Unexpected lockdowns due to the pandemic led to unstable market conditions. In these difficult times, aid to pandemic-hit zones was not only provided by government but also through other organizations as well. The help was given to these areas by supplying the needful items such as food, vaccines, oxygen cylinders, etc. In this situation, supply and demand both varied. Therefore, to deal with these unpredictable conditions, cost, demand and supply parameters in BLFTP are rendered as PFNs. The resulting problem is defined as Bilevel Fractional Transportation Problem with Pythagorean Fuzzy Numbers and it is denoted by (BFTPPFN).The mathematical representation of BFTPPFN is given as follows:

$$(BFTPPFN): \quad \min_{Z_1} F_{11}(Z_1, Z_2)^{\tilde{H}} = \frac{(p_1^T Z_1 + p_2^T Z_2)^{\tilde{H}}}{(q_1^T Z_1 + q_2^T Z_2)^{\tilde{H}}}$$

where Z_2 solves

$$\min_{Z_2} F_{12}(Z_1, Z_2)^{\tilde{H}} = \frac{(s_1^T Z_1 + s_2^T Z_2)^{\tilde{H}}}{(w_1^T Z_1 + w_2^T Z_2)^{\tilde{H}}} \quad \text{(for a given } Z_1)$$

subject to

$$\begin{aligned}
&\sum_{l \in L_{11}} z_{rl} \leq u_r^{'\tilde{H}} \quad \forall\, r \in R_{11},\\
&\sum_{r \in R_{11}} z_{rl} \geq v_l^{'\tilde{H}} \quad \forall\, l \in L_{11},\\
&\sum_{l \in L_{12}} z_{rl} \leq u_r^{''\tilde{H}} \quad \forall\, r \in R_{12},\\
&\sum_{r \in R_{12}} z_{rl} \geq v_l^{''\tilde{H}} \quad \forall\, l \in L_{12},\\
&z_{rl} \geq 0 \;\forall\, (r, l) \in R_T \times L_T.
\end{aligned} \tag{3.2}$$

From equation (3.2), it is observed that

$$\sum_{l \in L_{11}} v_l^{'\tilde{H}} \leq \sum_{l \in L_{11}} \sum_{r \in R_{11}} z_{rl} \leq \sum_{r \in R_{11}} u_r^{'\tilde{H}}$$

Thus, we get $\sum_{r\in R_{11}} u_r^{'\tilde{H}} \geq \sum_{l\in L_{11}} v_l^{'\tilde{H}}$. Similarly, $\sum_{r\in R_{12}} u_r^{''\tilde{H}} \geq \sum_{l\in L_{12}} v_l^{''\tilde{H}}$. Since the feasibility condition at the upper level and lower level problem is described, BFTPPFN is a *well-defined problem*.

The objective functions, supply and the demand parameters at both levels in BFTPPFN are not certain, hence, they are represented as Pythagorean fuzzy numbers (PFN). In terms of PFNs, the (BFTPPFN) can be represented as follows:

$$\min_{Z_1} F_{11}(Z_1, Z_2)^{\tilde{H}} = \frac{\sum_r \sum_l (\xi'_{rl}, \phi'_{rl}) z_{rl} + \sum_r \sum_l (\xi''_{rl}, \phi''_{rl}) z_{rl}}{\sum_r \sum_l (\lambda'_{rl}, \psi'_{rl}) z_{rl} + \sum_r \sum_l (\lambda''_{rl}, \psi''_{rl}) z_{rl}}$$

where Z_2 solves

$$\min_{Z_2} F_{12}(Z_1, Z_2)^{\tilde{H}} = \frac{\sum_r \sum_l (\rho'_{rl}, \sigma'_{rl}) z_{rl} + \sum_r \sum_l (\rho''_{rl}, \sigma''_{rl}) z_{rl}}{\sum_r \sum_l (\kappa'_{rl}, \tau'_{rl}) z_{rl} + \sum_r \sum_l (\kappa''_{rl}, \tau''_{rl}) z_{rl}} \quad \text{(for a given } Z_1\text{)}$$

subject to

$$\begin{aligned}
&\sum_{l\in L_{11}} z_{rl} \leq (\alpha'_{rl}, \beta'_{rl}) \ \forall\, r \in R_{11},\\
&\sum_{r\in R_{11}} z_{rl} \geq (\theta'_{rl}, \chi'_{rl}) \ \forall\, l \in L_{11},\\
&\sum_{l\in L_{12}} z_{rl} \leq (\alpha''_{rl}, \beta''_{rl}) \ \forall\, r \in R_{12},\\
&\sum_{r\in R_{12}} z_{rl} \geq (\theta''_{rl}, \chi''_{rl}) \ \forall\, l \in L_{12},\\
&z_{rl} \geq 0 \ \forall\, (r, l) \in R_T \times L_T.
\end{aligned} \tag{3.3}$$

Here, $u_r^{'\tilde{H}} = (\alpha'_{rl}, \beta'_{rl})$ is a PFN where α'_{rl} and β'_{rl} represents the degree of acceptance and non-acceptance for supply parameters at upper level.

$u_r^{''\tilde{H}} = (\alpha''_{rl}, \beta''_{rl})$ is a PFN where α''_{rl} and β''_{rl} represents the degree of acceptance and non-acceptance for supply parameters at lower level. Also, $v_l^{'\tilde{H}} = (\theta'_{rl}, \chi'_{rl})$ is a PFN where θ'_{rl} and χ'_{rl} represents the degree of acceptance and non-acceptance for demand parameters at upper level. $v_l^{''\tilde{H}} = (\theta''_{rl}, \chi''_{rl})$ is a PFN where θ''_{rl} and χ''_{rl} represents the degree of acceptance and non-acceptance for demand parameters at lower level. $(\xi'_{rl}, \phi'_{rl}), (\lambda'_{rl}, \psi'_{rl}), (\rho'_{rl}, \sigma'_{rl}), (\kappa'_{rl}, \tau'_{rl})$ are the acceptance and non-acceptance for the parameters in the objective functions at the upper level. $(\xi''_{rl}, \phi''_{rl}), (\lambda''_{rl}, \psi''_{rl}), (\rho''_{rl}, \sigma''_{rl}), (\kappa''_{rl}, \tau''_{rl})$ are the acceptance and non-acceptance for the parameters in the objective functions at the lower level.

3.4.1 Solution Methodology for BFTPPFN

To solve BFTPPFN, Pythagorean fuzzy cost, supply and demand parameters are replaced by their respective score values. The problem so obtained is denoted by

BFTPSV and it is represented as follows:

$$\min_{Z_1} F_{11}(Z_1,Z_2)^{\tilde{H}} = \frac{\begin{matrix}\sum_r \sum_l ({\xi'_{rl}}^2 - {\phi'_{rl}}^2)[1+\sqrt{(1-{\xi'_{rl}}^2)-{\phi'_{rl}}^2}]z_{rl} \\ +\sum_r \sum_l ({\xi''_{rl}}^2 - {\phi''_{rl}}^2)[1+\sqrt{(1-{\xi''_{rl}}^2)-{\phi''_{rl}}^2}]z_{rl}\end{matrix}}{\begin{matrix}\sum_r \sum_l ({\lambda'_{rl}}^2 - {\psi'_{rl}}^2)[1+\sqrt{(1-{\lambda'_{rl}}^2)-{\psi'_{rl}}^2}]z_{rl} \\ +\sum_r \sum_l ({\lambda''_{rl}}^2 - {\psi''_{rl}}^2)[1+\sqrt{(1-{\lambda''_{rl}}^2)-{\psi''_{rl}}^2}]z_{rl}\end{matrix}}$$

where Z_2 solves

$$\min_{Z_2} F_{12}(Z_1,Z_2)^{\tilde{H}} = \frac{\begin{matrix}\sum_r \sum_l ({\rho'_{rl}}^2 - {\sigma'_{rl}}^2)[1+\sqrt{(1-{\rho'_{rl}}^2)-{\sigma'_{rl}}^2}]z_{rl} \\ +\sum_r \sum_l ({\rho''_{rl}}^2 - {\sigma''_{rl}}^2)[1+\sqrt{(1-{\rho''_{rl}}^2)-{\sigma''_{rl}}^2}]z_{rl}\end{matrix}}{\begin{matrix}\sum_r \sum_l ({\kappa'_{rl}}^2 - {\tau'_{rl}}^2)[1+\sqrt{(1-{\kappa'_{rl}}^2)-{\tau'_{rl}}^2}]z_{rl} \\ +\sum_r \sum_l ({\kappa''_{rl}}^2 - {\tau''_{rl}}^2)[1+\sqrt{(1-{\kappa''_{rl}}^2)-{\tau''_{rl}}^2}]z_{rl}\end{matrix}}$$

(for a given Z_1)
subject to

$$\begin{aligned}
&\sum_{l\in L_{11}} z_{rl} \le ({\alpha'_{rl}}^2 - {\beta'_{rl}}^2)[1+\sqrt{(1-{\alpha'_{rl}}^2)-{\beta'_{rl}}^2}] \ \forall\, r \in R_{11}, \\
&\sum_{r\in R_{11}} z_{rl} \ge ({\theta'_{rl}}^2 - {\chi'_{rl}}^2)[1+\sqrt{(1-{\theta'_{rl}}^2)-{\chi'_{rl}}^2}] \ \forall\, l \in L_{11}, \\
&\sum_{l\in L_{12}} z_{rl} \le ({\alpha''_{rl}}^2 - {\beta''_{rl}}^2)[1+\sqrt{(1-{\alpha''_{rl}}^2)-{\beta''_{rl}}^2}] \ \forall\, r \in R_{12}, \\
&\sum_{r\in R_{12}} z_{rl} \ge ({\theta''_{rl}}^2 - {\chi''_{rl}}^2)[1+\sqrt{(1-{\theta''_{rl}}^2)-{\chi''_{rl}}^2}] \ \forall\, l \in L_{12}, \\
&\qquad z_{rl} \ge 0 \ \forall\, (r,l) \in R_T \times L_T.
\end{aligned} \tag{3.4}$$

The satisfactory solution to the problem (BFTPPFN) is obtained by solving the problem (BFTPSV) by two methods, Fuzzy programming and Goal programming.

3.4.2 Fuzzy Programming Approach

To solve BFTPSV by a fuzzy programming technique, the objective functions at two levels are solved individually with respect to the constraint set (4) and their maximum and minimum values are deliberated. The membership and non-membership functions for the upper level are denoted as $\mu(F_{11}(Z_1,Z_2))$ and $\gamma(F_{11}(Z_1,Z_2))$, respectively.

They are defined as follows:

$$\mu(F_{11}(Z_1,Z_2)) = \begin{cases} 1 & F_{11} \le F_{11}{}^{min} \\ \frac{F_{11}{}^{max} - F_{11}(Z_1,Z_2)}{F_{11}{}^{max} - F_{11}{}^{min}} & F_{11}{}^{min} \le F_{11} \le F_{11}{}^{max} \\ 0 & F_{11} \ge F_{11}{}^{max} \end{cases}$$

$$\gamma(F_{11}(Z_1,Z_2)) = \begin{cases} 0 & F_{11} \le F_{11}{}^{min} \\ \frac{F_{11}(Z_1,Z_2) - F_{11}{}^{min}}{F_{11}{}^{max} - F_{11}{}^{min}} & F_{11}{}^{min} \le F_{11} \le F_{11}{}^{max} \\ 1 & F_{11} \ge F_{11}{}^{max} \end{cases}$$

Similarly, we can define membership and non-membership functions for the objective functions at the lower level. Let the minimum degree of satisfaction for the membership and non-membership functions at the upper level be denoted by ζ_{11} and $\zeta_{11}^{'}$, respectively. Likewise, let the minimum degree of satisfaction for the membership and non-membership functions at lower level be denoted by ζ_{12} and $\zeta_{12}^{'}$ respectively.

Let, $\varepsilon_1 = \text{Max}(\zeta_{11}, \zeta_{12})$ and $\varepsilon_2 = \text{Min}(\zeta_{11}^{'}, \zeta_{12}^{'})$.

The fuzzy programming model for (BFTPSV) is denoted as (PFPP) and is proposed as follows:

$$\text{Method-I} : (PFPP) : \quad \text{Max}(\varepsilon_1 - \varepsilon_2)$$

subject to

$$\begin{aligned} \mu(F_{11}) &\ge \varepsilon_1, \\ \mu(F_{12}) &\ge \varepsilon_1, \\ \gamma(F_{11}) &\le \varepsilon_2, \\ \gamma(F_{12}) &\le \varepsilon_2, \end{aligned} \tag{3.5}$$

$$\begin{aligned} &\sum_{l \in L_{11}} z_{rl} \le (\alpha_{rl}^{'\,2} - \beta_{rl}^{'\,2})[1 + \sqrt{(1 - \alpha_{rl}^{'\,2}) - \beta_{rl}^{'\,2}}] \ \forall\, r \in R_{11}, \\ &\sum_{r \in R_{11}} z_{rl} \ge (\theta_{rl}^{'\,2} - \chi_{rl}^{'\,2})[1 + \sqrt{(1 - \theta_{rl}^{'\,2}) - \chi_{rl}^{'\,2}}] \ \forall\, l \in L_{11}, \\ &\sum_{l \in L_{12}} z_{rl} \le (\alpha_{rl}^{''\,2} - \beta_{rl}^{''\,2})[1 + \sqrt{(1 - \alpha_{rl}^{''\,2}) - \beta_{rl}^{''\,2}}] \ \forall\, r \in R_{12}, \\ &\sum_{r \in R_{12}} z_{rl} \ge (\theta_{rl}^{''\,2} - \chi_{rl}^{''\,2})[1 + \sqrt{(1 - \theta_{rl}^{''\,2}) - \chi_{rl}^{''\,2}}] \ \forall\, l \in L_{12}, \end{aligned} \tag{3.6}$$

$$z_{rl} \ge 0 \ \forall\, (r,l) \in R_T \times L_T; \varepsilon_1^2 + \varepsilon_2^2 \le 1, \ \varepsilon_1, \varepsilon_2 \in [0,1].$$

PFPP is solved by the computing software LINGO 17.0. The obtained solution is the satisfactory solution for BFTPPFN. BFTPSV is also solved by a goal programming mode to procure the satisfactory solution to BFTPPFN.

3.4.3 Goal Programming Approach

To determine the strategy of goal programming, positive and negative deviational variables are assigned to the objective functions at both levels. The aspiration levels for the objective functions at two levels is designated as $F_{1i}^t = \frac{F_{1i}^{Max}+F_{1i}^{Min}}{2}; i = 1, 2.$

$$\text{Method-II} : (GPPFN) : \qquad Min\ \varrho$$

subject to

$$F_{11} - d_{P1}^+ + d_{N1}^- = F_{11}^t,$$
$$F_{12} - d_{P2}^+ + d_{N2}^- = F_{12}^t, \tag{3.7}$$

$$\sum_{l \in L_{11}} z_{rl} \leq ({\alpha'_{rl}}^2 - {\beta'_{rl}}^2)[1 + \sqrt{(1 - {\alpha'_{rl}}^2) - {\beta'_{rl}}^2}]\ \forall\ r \in R_{11},$$

$$\sum_{r \in R_{11}} z_{rl} \geq ({\theta'_{rl}}^2 - {\chi'_{rl}}^2)[1 + \sqrt{(1 - {\theta'_{rl}}^2) - {\chi'_{rl}}^2}]\ \forall\ l \in L_{11},$$

$$\sum_{l \in L_{12}} z_{rl} \leq ({\alpha''_{rl}}^2 - {\beta''_{rl}}^2)[1 + \sqrt{(1 - {\alpha''_{rl}}^2) - {\beta''_{rl}}^2}]\ \forall\ r \in R_{12}, \tag{3.8}$$

$$\sum_{r \in R_{12}} z_{rl} \geq ({\theta''_{rl}}^2 - {\chi''_{rl}}^2)[1 + \sqrt{(1 - {\theta''_{rl}}^2) - {\chi''_{rl}}^2}]\ \forall\ l \in L_{12},$$

$$z_{rl} \geq 0\ \forall\ (r, l) \in R_T \times L_T,$$
$$\varrho - d_{P1}^+ \geq 0;\ \varrho - d_{P2}^+ \geq 0;\ \ d_{P1}^+ d_{N1}^- = 0;\ d_{P2}^+ d_{N2}^- = 0;\ 0 \leq \varrho \leq 1. \tag{3.9}$$

3.5 APPLICATION OF BLFTP WITH PYTHAGOREAN FUZZY PARAMETERS

Example 3.5.1. *Consider the four states 1,2,3,4. These states have grocery, vaccines, oxygen cylinders in ample quantity that can not only fulfill their own requirements but can help the neighboring states also. Suppose that another four states A,B,C,D require these commodities for their population. These states have further created four nodal zones I,II,III,IV that are responsible for the distribution of goods to the needy.*

Let F_{11} be the objective function at the upper level (from states 1, 2, 3, 4 to states A,B,C,D) and F_{12} be the objective function at the lower level (states A,B,C,D to nodal zones). Table 3.1 and Table 3.2 represents the costs and quantity transported at the upper level and at lower level, respectively.

The supply and demand constraints for the upper level problem is represented as follows:

$$z_{11} + z_{12} + z_{13} + z_{14} \leq (0.7, 0.1) \tag{3.10}$$
$$z_{21} + z_{22} + z_{23} + z_{24} \leq (0.8, 0.3) \tag{3.11}$$
$$z_{31} + z_{32} + z_{33} + z_{34} \leq (0.7, 0.2) \tag{3.12}$$
$$z_{41} + z_{42} + z_{43} + z_{44} \leq (0.8, 0.2) \tag{3.13}$$

TABLE 3.1
Data for the Upper Level

States \ States	A	B	C	D	
1	(0.5,0.4)	(0.3,0.1)	(0.6,0.4)	(0.4,0.2) $\rightarrow p_{rl}$	
	(0.6,0.3)	(0.2,0.3)	(0.7,0.4)	(0.3,0.1)	$\rightarrow q_{rl}$
2	(0.4,0.5)	(0.3,0.1)	(0.7,0.5)	(0.5,0.1)	
	(0.6,0.4)	(0.4,0.3)	(0.6,0.5)	(0.7,0.4)	
3	(0.6,0.5)	(0.7,0.4)	(0.2,0.1)	(0.4,0.1)	
	(0.7,0.2)	(0.5,0.2)	(0.8,0.4)	(0.7,0.1)	
4	(0.6,0.4)	(0.4,0.2)	(0.6,0.3)	(0.4,0.3)	
	(0.5,0.3)	(0.5,0.4)	(0.5,0.2)	(0.7,0.4)	

TABLE 3.2
Data for the Lower Level

States \ Organizations	I	II	III	IV	
A	(0.6,0.3)	(0.6,0.4)	(0.7,0.5)	(0.5,0.4) $\rightarrow s_{rl}$	
	(0.5,0.2)	(0.3,0.2)	(0.5,0.4)	(0.3,0.2)	$\rightarrow w_{rl}$
B	(0.6,0.4)	(0.4,0.3)	(0.7,0.6)	(0.3,0.2)	
	(0.5,0.3)	(0.3,0.1)	(0.5,0.4)	(0.6,0.3)	
C	(0.7,0.5)	(0.5,0.1)	(0.7,0.1)	(0.7,0.2)	
	(0.6,0.1)	(0.4,0.1)	(0.6,0.3)	(0.2,0.1)	
D	(0.2,0.1)	(0.4,0.2)	(0.4,0.1)	(0.5,0.4)	
	(0.8,0.4)	(0.6,0.3)	(0.7,0.1)	(0.6,0.3)	

$$z_{11} + z_{21} + z_{31} + z_{41} \geq (0.7, 0.4) \tag{3.14}$$
$$z_{12} + z_{22} + z_{32} + z_{42} \geq (0.6, 0.1) \tag{3.15}$$
$$z_{13} + z_{23} + z_{33} + z_{43} \geq (0.6, 0.4) \tag{3.16}$$
$$z_{14} + z_{24} + z_{34} + z_{44} \geq (0.7, 0.2) \tag{3.17}$$

The supply and demand constraints for the lower level problem are represented as follows:

$$z_{11} + z_{12} + z_{13} + z_{14} \leq (0.7, 0.4) \tag{3.18}$$
$$z_{21} + z_{22} + z_{23} + z_{24} \leq (0.6, 0.1) \tag{3.19}$$
$$z_{31} + z_{32} + z_{33} + z_{34} \leq (0.6, 0.4) \tag{3.20}$$
$$z_{41} + z_{42} + z_{43} + z_{44} \leq (0.7, 0.2) \tag{3.21}$$
$$z_{11} + z_{21} + z_{31} + z_{41} \geq (0.7, 0.4) \tag{3.22}$$
$$z_{12} + z_{22} + z_{32} + z_{42} \geq (0.6, 0.4) \tag{3.23}$$
$$z_{13} + z_{23} + z_{33} + z_{43} \geq (0.6, 0.5) \tag{3.24}$$
$$z_{14} + z_{24} + z_{34} + z_{44} \geq (0.8, 0.5) \tag{3.25}$$

TABLE 3.3
Comparative Analysis

Methodology	Z_{11}	Z_{12}
Fuzzy programming	0.352725	1.11494
Programming	0.316715	1.80671

The Pythagorean cost, supply and demand parameters are replaced by their respective score values. Accordingly, the fuzzy problem is described, which is solved by computing software LINGO 17.0. The problem is also solved by a goal programming technique. The objective function values from two approaches are represented in Table 3.3.

It has been observed from Table 3.3 that, although the minimum value is obtained through fuzzy programming at the upper level, the difference is marginal. Moreover, the solution at the lower level is the minimum by fuzzy programming as compared to the goal programming approach.

3.6 CONCLUSIONS AND PROSPECTIVE WORK

A Bilevel Linear Fractional Transportation Problem (BLFTP) with uncertain parameters is defined in this paper. During the Covid-19 pandemic and especially at the time of lockdowns, the cost, demand and supply in markets showed erratic behavior. The paper attempts to explain the situation where the essential commodities are delivered from provider states to the needy ones and then to the people at large through the nodal zones. The uncertainty in the cost coefficients of the objective functions, supply and demand parameters at both the levels are PFNs. These fuzzy numbers are transformed into their respective crisp parameters using score function. The problem is designed so as to find a satisfactory solution for the objective functions at both the levels. The problem has been solved by two distinct procedures: fuzzy programming and goal programming using LINGO 17.0. A comparative analysis of the solutions obtained from the two methods is also dispensed. As a limitation of this paper, it will be difficult to manage multi-level fractional transportation problem with PFNs using exponential and hyperbolic membership functions.

BLFTP has been examined in this paper with PFNs. Indeterminacy of the paper can also be contemplated by varying the number of sources and destinations. The problem can also be explored by other algorithmic approaches, such as bat algorithm, monkey algorithm, etc. Although cost is minimized in the formulated problem, the minimization of time can also be considered in the objective function at two levels.

ACKNOWLEDGEMENTS

The authors are readily grateful to the reviewers for their valuable suggestions, incorporation of which have helped us in improving the quality of the paper to a great extent.

CONFLICT OF INTEREST

The authors declare that they have no conflict of interest. The authors declare that they have no financial or non-financial interests.

REFERENCES

1. Hitchcock, F.L.: The Distribution of a Product from Several Sources to Numerous Localities. Journal of Mathematics and Physics. **20**, 224–230 (1941).
2. Brigden, M.E.B.: A Variant of the Transportation Problem in which the Constraints are of Mixed Type. Journal of the Operational Research Society. **25**, 437–445 (1974).
3. Chanas, S., Delgado, M., Verdegay, J.L., Vila, M.A.: Interval and Fuzzy Extensions of Classical Transportation Problems. Transportation Planning and Technology. **17**, 203–218 (1993).
4. Behbahani,H., Nazari,S.,Jafari Kang, M., Litman, T.: A Conceptual Framework to Formulate Transportation Network Design Problem considering Social Equity Criteria. Transportation Research Part A: Policy and Practice. **125**, 171–183 (2019).
5. Minasian, I.M.S.: Fractional Programming. 1st edn.Springer, Dordrecht (1997).
6. Joshi, V.D., Gupta, N.: Linear Fractional Transportation Problem with Varying Demand and Supply. LeMatematiche. **66**, 3–12 (2011).
7. Cetin, N., Tiryaki, F.: A Fuzzy Approach using Generalized Dinkelbach's Algorithm for Multiobjective Linear Fractional Transportation Problem. Mathematical Problems in Engineering. **2014**, Article Id 702319 (2014).
8. Mahmoodirad, A., Dehghan, R., Niroomand, S.: Modelling Linear Fractional Transportation Problem in Belief Degree-based Uncertain Environment. Journal of Experimental & Theoretical Artificial Intelligence. **31**, 393–408 (2019).
9. Liu, S.T.: Fractional Transportation Problem with Fuzzy Parameters. Soft Computing. **20**, 3629–3636 (2016).
10. Radhakrishnan, B., Anukokila, P.: A Compensatory Approach to Fuzzy Fractional Transportation Problem. International Journal of Mathematics in Operations Research. **6**, 176–192 (2014).
11. Javaid, S., Jalil, S.A., Asim, Z.: A Model for Uncertain Multi-objective Transportation Problem with Fractional Objectives. International Journal of Operations Research. **14**, 11–25 (2017).
12. Kaushal, B., Arora, R., Arora, S.: An Aspect of Bilevel Fixed Charge Fractional Transportation Problem. International Journal of Applied and Computational Mathematics. **6**, 1–19 (2020).
13. Agrawal, P., Ganesh, T.: Fuzzy Fractional Stochastic Transportation Problem involving Exponential Distribution. Opsearch. **57**,1093–1114 (2020).
14. Arora, S.R., Arora, R.: A weighting method for 0–1 indefinite quadratic bilevel programming. Operational Research-An International Journal. **11**, 311–324 (2011).
15. Sakawa, M., Nishizaki, I., Uemura, Y.: A Decentralized Two-Level Transportation Problem in a Housing Material Manufacturer: Interactive Fuzzy Programming Approach. European Journal of Operational Research. **141**, 167–185 (2002).
16. Bagloee, S.A., Asadi, M., Sarvi, M., Patriksson, M.: A Hybrid Machine-Learning and Optimization Method to Solve Bi-level Problems. Expert Systems with Applications. **95**, 142–152 (2018).
17. Lv, Y., Wang, S., Gao, Z., Cheng, G., Huang, G., He, Z.: A Sustainable Road Pricing Oriented Bilevel Optimization Approach under Multiple Environmental Uncertainties. International Journal of Sustainable Transportation. online, 2021.

18. Guo, Z., Zhang,D., Leung,S.Y.S., Shi, L.: A Bi-level Evolutionary Optimization Approach for Integrated Production and Transportation Sheduling. Applied Soft computing. **42**, 215–228 (2016).
19. Tanaka, H., Asai, K.: Fuzzy Linear Programming Problems with Fuzzy Numbers. Fuzzy Sets and Systems.**13**, 1–10 (1984).
20. Mohamed, R.H.: The Relationship between Goal Programming and Fuzzy Programming. Fuzzy Sets and Systems. **89**, 215–222 (1997).
21. Lu, H.W., Huang, G.H., Xu, Y., He, L.: Inexact Two-Phase Fuzzy Programming and its Application to Municipal Solid Waste Management. Engineering Applications of Artificial Intelligence. **25**, 1529–1536 (2012).
22. Turgay, S., Taskin, H.: Fuzzy Goal Programming for Health-Care Organization. Computers & Industrial Engineering. **86**, 14–21 (2015).
23. Bal, A., Satoglu, S.I.: A Goal Programming Model for Sustainable Reverse Logistics Operations Planning and an Application. Journal of Cleaner Production. **201**, 1081–1091 (2018).
24. Chen, L., Peng, J., Zhang, B.: Uncertain Goal Programming Models for Bicriteria Solid Transportation Problem. Applied Soft Computing. **51**, 49–59 (2017).
25. Shih, L.H.: Cement Transportation Planning via Fuzzy Linear Programming. International Journal of Production Economics. **58**, 277–287 (1999).
26. Zheng, Y.J., Ling, H.F.: Emergency Transportation Planning in Disaster Relief Supply Chain Management: A Cooperative Fuzzy Optimization Approach. Soft Computing. **17**, 1301–1314 (2013).
27. Yager, R.R., Abbasov, A.M.: Pythagorean Membership Grades, Complex Numbers, and Decision Making. International Journal of Intelligent Systems. **28**, 436–452 (2013).
28. Zhang, X., Xu, Z.: Extension of TOPSIS to Multiple Criteria Decision-making with Pythagorean Fuzzy Sets. International Journal of Intelligent Systems. **29**, 1061–1078 (2014).
29. Adhami, A.Y., Ahamd, F.: Interactive Pythagorean-Hesitant Fuzzy Computational Algorithm for Multi-objective Transportation Problem under Uncertainty. International Journal of Management Science and Engineering Management. **15**, 1–10 (2020).
30. Garg, H.: A New Improved Score Function of an Interval-Valued Pythagorean Fuzzy Set Based TOPSIS Method. International Journal for Uncertainty Quantification. **7**, 463–474 (2017).
31. Wan, S.P., Li, S.Q., Dong, J.Y.: A Three-Phase Method for Pythagorean Fuzzy Multi-attribute Group Decision-making and Application to Haze Management. Computers & Industrial Engineering. **123**, 348–363 (2018).
32. Jianping, F., Xiaona, L. Meiqin, W., Zhan, W.: Green Supplier Selection with Undesirable Outputs DEA under Pythagorean Fuzzy Environment. Journal of Intelligent & Fuzzy Systems. **37**, 2443–2452 (2019).
33. Yucesan, M., Kahraman, G.: Risk Evaluation and Prevention in Hydropower Plant Operations: A Model based on Pythagorean Fuzzy AHP. Energy Policy. **126**, 343–351 (2019).
34. Ejegwa, P.A.: Pythagorean Fuzzy set and its Application in Career Placements based on Academic Performance using Max-Min-Max Composition. Complex & Intelligent Systems. **5**, 165–175 (2019).
35. Liu, P., Rani, P., Mishra, A.R.: A Novel Pythagorean Fuzzy Combined Compromise Solution Framework for the Assessment of Medical Waste Treatment Technology. Journal of Cleaner production. **292**, 126047 (2021).

4 Optimal Production Evaluation of Cotton in Different Soil and Water Conditions in Sundarban of West Bengal under Hesitant Interval Fuzzy Environment Using Projection Measures

Ankan Bhaumik[1] *and Sankar Kumar Roy*[2]
[1]Dhamcha Chhagulia Siddheswari High School, Dhamcha, Paschim Medinipur, West Bengal, India
[2]Department of Applied Mathematics with Oceanology and Computer Programming, Vidyasagar University, Paschim Medinipur, West Bengal, India

CONTENTS

DOI: 10.1201/9781003329039-4

4.1 INTRODUCTION

Agriculture is the development of organisms like plants, fungi, and other life-forms for food, fiber, bio-fuel, medicines and other products used for the sustenance of human life. Farming is the key development issue in the rise of sedentary human civilization, whereby farming of domesticated species allowed food excesses that supported the events of human progress and nurtured the development of civilization. Agriculture can be characterized as the precise and controlled use of living life forms and the environment to work on the human condition. In spite of the fact that the land is essentially needed for the production of food for human and other's consumption, agricultural exercises additionally incorporate the developiment of plants for fiber, fuels, and other natural and organic products like drugs, etc.

Physical, chemical, and organic inputs are crucial for agricultural systems, and these are mainly provided by the soil, water, sunlight, plants, animal and organic agents. Nonetheless, soil plays an important role in agriculture. Not all agricultural lands are proficient or reasonable for cultivation all agricultural products. The primary restricting variables are the environment and topography of the considered lands. Soils with all their fluctuations are a key restricting element. All these factors must be taken into account for sustainable agricultural production. The suitability of land alludes to the capacity of a portion of land to allow the production of yields in a sustainable way. The land suitability is classified through the cultivation of several crops considering scientific requirements. Effective decision-making with regard to suitable agricultural land and the corresponding yield gives optimum results. The term *land suitability* assessment could be deciphered as the evaluation of land execution when the land is utilized for an explicit yield. Land suitability evaluation is quite possibly one of the main issues in agriculture. Thus, land suitability for cultivation is one of the most interesting research topics nowadays.

4.1.1 Cotton Production Problem

Cotton, an important agricultural product and a fundamental cash-crop, is used as the main component in the textile and garment industry, medical sectors, oil industry, etc. It is a globally used fiber and makes significant contributions to the world's economy (USDA, 2018). The USA, China and Indian sub-continents produce more than half of the World's cotton. The total global cotton production was 124.77 million bales in 2018 (USDA, 2018). But the production of cotton varies due to different causes. It may be due to various environmental issues, like water supply to land's salinity; different manual factors, like the technique of seeding in fields, man-made irrigation system, etc. During pre- and post-budding situations, sufficient water is needed. 150 to 195 days are considered to be the seasonal growth period of cotton, and in this period two or three subsequent irrigations are required. These requirements vary region-wise and country-wise. In some regions, the tillage system fully depends upon rain, whereas in other regions it depends upon groundwater. Irrigation methods may be surface or subsurface or sprinkler or drip. Thus water is important in cotton production. Also, the form of the land indicates the production variation of cotton throughout the World. Mainly, land-salinity is vital for cotton yiled. Excessive salinity decreases soil's others nutritional ingredients. Sodium chloride, sodium sulfate, etc. are the minerals that cause soil salinity, and a concentration of these salts near the roots of cotton plants results in dehydration at the roots and ultimately the death of the plants. Salinity of land can decrease the quality of the cotton fiber. Thus, the evaluation of the optimal production of cotton is important.

4.1.2 MCDM and Uncertainty

Multi-criteria decision-making (MCDM) problems can be considered as a sub-discipline of operations research. MCDM problems consist of the following components: the alternatives (which decision-makers want to choose), the criteria by which the alternatives are judged/compared, the weights to related criteria, and the decision-makers whose preferences are counted. Basically, the implementation of MCDM depends upon (i) the specific choice of criteria and related weights and (ii) the accurate determination of weights. Firstly, in MCDM problems, objectives, alternatives, and decision-makers are identified. Secondly, criteria are specified and all information related to criteria are gathered. Based on the weighting of the criteria, the criteria values are re-evaluated. Thirdly, a ranking approach (depending on the considered problem) is applied to get a better alternative.

MCDM problems can be treated within a fuzzy environment. Fuzzy set (FS) (Zadeh, 1965) and the concept of the fuzziness of different aspects play important roles when we consider real-life problems. Obviously, the consideration of optimal choices is based on the fuzziness of the considered data. In real life, variables are expressed through languages, words, and sentences rather than numerical values and are more acceptable to the decision-makers. These types of variables, called linguistic variables, strengthened the relevancy of themselves in real-life uncertainties in various fields after the proposal of the linguistic information and linguistic approach in fuzzy-set theory made by Zadeh (1975). The two-dimensional representations (i.e., (member, membership degree)) and logic of FS sometimes become insufficient to consider all the uncertainties in the system. A single membership value corresponding to an element is inadequate to present complete information in reality. Torra (2010) coined the term hesitant fuzzy set (HFS) as another accretion of FS where every member can be defined with a set of membership degrees. HFS is considered in different problems when decision-makers feel a hesitant environment,e.g., disaccord and discrepancy situations. In almost all decision-making problems, decision-makers wish to get optimal solutions from real-life problematic data.

In this chapter, we discuss the hesitant interval-valued fuzzy linguistic element (HIFLE)-based environment and consider a new projection measure based ranking approach on HIFLE. The study of cotton production using mathematically considered uncertain fuzzy variables with hesitance characteristics in linguistic form is rare in use. A new methodology is proposed in this chapter to handle the optimal production evaluation of cotton with special reference to the Sundarban area of West Bengal in India.

The main highlights of this chapter can be characterized as:

- This chapter describes the optimal production evaluation of cotton in different soil and water conditions.
- Optimality is considered here under hesitant interval fuzzy environment.
- A new projection measure is treated under a hesitant interval fuzzy linguistic-based environment.

4.2 LITERATURE REVIEW

Zadeh (1965) first defined the FS and considered fuzzy concept to tackle real-life problems. Several articles have been published using fuzzy sets and concept (Bhaumik et al., 2017, 2021,?, 2020; Bhaumik & Roy, 2021, 2022; Jana & Roy, 2021, 2018, 2019; Roy & Bhaumik, 2018; Roy & Jana, 2021). Torra (2010) defined the HFS and Xia & Xu (2011) described some advanced operations on HFS. Chen & Xu (2014) defined the properties of interval-valued hesitant fuzzy sets. Chen et al. (2013) considered interval-valued hesitant preference relations with related applications into group decision-making problems. Herrera & Herrera-Viedma (2000) made an analysis on stepwise solving decision problems under linguistic information. A fuzzy linguistic methodology was proposed by Herrera et al. (2008) to deal with unbalanced linguistic term sets. Herrera et al. (1995) considered a sequential selection process in group decision-making with linguistic assessment. Jiang & Wei (2014) depicted some Bonferroni mean operators with 2-tuple linguistic information and applied these operators to multiple attribute decision-making. Kacprzyk & Zadrozny (2005) considered linguistic database summaries and their protoforms towards natural language based knowledge discovery tools. Several articles (Martinez et al., 2005; Xu, 2005, 2004) have been published on fuzzy linguistic information, hesitant fuzzy linguistic information. Yager (1977) considered multiple objective decision-making problems using fuzzy sets. Yu et al. (2013) introduced a hesitant fuzzy group decision-based solution to personnel evaluation. Rodriguez et al. (2012) extensively studied hesitant fuzzy linguistic term sets for decision-making. Rodriguez & Martinez (2013) analyzed symbolic linguistic computing models in decision-making. Özkan et al. (2019) successfully introduced a hesitant fuzzy linguistic MCDM approach in a real-life problem.

With respect to cotton production and its history, species, varieties, morphology, breeding, culture with related environments, diseases with recovery strategies, marketing and uses, etc., several articles (Ali et al., 2014; Anapalli et al., 2016; Brown, 2002; El Titi, 2003; Khan et al., 2004) have been published. Ahmad & Raza (2014) obtained some optimization of management practices to improve cotton fiber quality in an irrigated arid environment. Lee & Fang (2015), in their study, elaborately described cotton as a world crop with its origin, history, and current status. Sawan (2018) described the effects, like, evaporation, sunshine, relative humidity, soil and air temperature on cotton production. Aggarwal et al. (2017), in their research, formed models to tackle the problem of the soil water balance and root water uptake in cotton grown under different soil conservation practices in the Indo Gangetic Plain. Numerous research papers (Hillocks, 1995; Hira et al., 2004; Luo et al., 2016; Rahman et al., 2018; Williams et al., 2015; Yazar et al., 2002) have been published based on various issues of cotton production and uses in different territories and continents. Buttar et al. (2017) scrutinized the effect of saline-sodic water in cotton-wheat cropping system. Singh (2015) investigated soil salinization and water logging as a threat to environment and agricultural sustainability. Dong (2012) considered technological studies and management for controlling soil salinity effects on cotton in field. Foster et al. (2018), in their paper, compared two tillage practices in a semi-arid cotton. Karmakar et al. (2016) reviewed the potential effects of climate change on soil properties. Loka & Oosterhuis (2012) confined their study in water stress and reproductive development in cotton, whereas Lokhande & Reddy (2014) analyzed reproduction and fiber quality responses of upland cotton to moisture deficiency. Ma et al. (2002) studied the water deficit of cotton. Reddy et al. (1991) discussed the effect of temperature on the growth and development of cotton. Sun et al. (2017) described, in their works, the impacts of ridge-furrow planting on salt stress and cotton yield under drip irrigation. Torbert et al. (2015) analyzed high residue conservation tillage system for cotton production through a farmer's perspective. Verma et al. (2018) discussed the effects of inorganic and organic fertilizers in soils under a cotton-wheat cropping system. Paz et al. (2012) explored the optimal different planting dates and spatial aggregation levels of cotton in their research works. Iftikhar et al. (2010) and Zwart & Bastiaanssen (2004), in their studies, discussed the impact of land patterns and the hydrological properties of soil on cotton cultivation. Constable & Bange (2015) analyzed the potential for cotton cultivation. Grantz (2003) discussed ozone induction on cotton.

Numerous articles (Keshavarzi et al., 2010; Ashraf et al., 2014; Baral, 2013; Biswas & Pal, 2005; Burrough, 1989; Sharma, 2007; Kurtener et al., 2008) have been published in the farming sector considering fuzzy techniques. Hitherto, a lot of research articles have been published on cotton production based on irrigation systems and the salinity of the land, but our consideration of hesitant interval-valued linguistic fuzzy elements to obtain the optimal production of cotton in different soil and water conditions using projection measure may be claimed as rare.

4.3 MATHEMATICAL PRELIMINARIES

In this section we introduce some preliminary definitions of fuzzy set, intuitionistic fuzzy set, hesitant fuzzy set, interval number with examples, properties, etc.

Definition 4.3.1. (Zadeh, 1965) Consider χ to be a universe of discourse. A fuzzy set (FS) $\check{F}(\in \chi)$ is defined by a membership function $\mu_{\check{F}} : \chi \to [0,1]$. A FS $\check{F}(\in \chi)$ can be represented as $\check{F} = \{(x, \mu_{\check{F}}(x)) : \mu_{\check{F}}(x) \in [0,1],\ x \in \chi\}$. Obviously, the membership degrees $\mu_{\check{F}}(x)$ of $\check{F}$ are crisp numbers.

Definition 4.3.2. (Atanassov, 1986) Let χ denote a universe of discourse, then an intuitionistic fuzzy set (IFS), $\hat{F}$ in χ is given by a set of ordered triplets, $\hat{F} = \{\langle x, \mu_{\hat{F}}(x), \gamma_{\hat{F}}(x)\rangle : x \in \chi\}$, where both $\mu_{\hat{F}}$ and $\gamma_{\hat{F}}$ map the elements of χ to $[0,1]$, in such a way that $0 \leq \mu_{\hat{F}}(x) + \gamma_{\hat{F}}(x) \leq 1,\ \forall\, x \in \chi$. For each x, $\mu_{\hat{F}}(x)$ and $\gamma_{\hat{F}}(x)$ represent the degree of membership and the degree of non-membership, respectively. Again $\pi_{\hat{F}}(x)(= 1 - \mu_{\hat{F}}(x) - \gamma_{\hat{F}}(x))$ is the *"degree of hesitation"* of the element x in $\hat{F}$. If $\pi_{\hat{F}}(x) = 0,\ \forall\, x \in \chi$, then the IFS represents a FS.

Definition 4.3.3. (Torra, 2010) A hesitant fuzzy set (HFS) A_{HF} on a reference set χ is expressed in terms of a function $h_A(x)$, when it is applied to χ it gives a subset of $[0,1]$. Then, $A_{HF} = \{\langle x, h_A(x)\rangle : x \in \chi\}$, where $h_A(x)$ is termed as hesitant fuzzy element (HFE). HFE is a basic unit of HFS. A HFE represents the possible membership degrees of the corresponding element $x \in \chi$.

Example 4.3.4. $A_{HF} = \{\langle x_1, 0.1, 0.45\rangle, \langle x_2, 0.32, 0.35\rangle, \langle x_3, 0.2, 0.3, 0.4, 0.69, 0.81\rangle\}$ is a HFS. Here $\{x_1, x_2, x_3\} \in \chi$, a reference set and $h_A(x_1) = \{0.1, 0.45\}$, $h_A(x_2) = \{0.32, 0.35\}$, $h_A(x_3) = \{0.2, 0.3, 0.4, 0.69, 0.81\}$ are HFEs.

Property 4.3.5. (Torra, 2010) Assume h, h_1 and h_2 be three HFEs. Then,

(i) $h^c = \{1-\gamma : \gamma \in h\}$, complement of h;
(ii) $h_1 \cup h_2 = \{\gamma_1 \vee \gamma_2 : \gamma_1 \in h_1, \gamma_2 \in h_2\}$;
(iii) $h_1 \cap h_2 = \{\gamma_1 \wedge \gamma_2 : \gamma_1 \in h_1, \gamma_2 \in h_2\}$;
Furthermore, in order to aggregate hesitant fuzzy information, some new operations on h, h_1 and h_2 with $\lambda > 0$ are as below:
(iv) $h_1 \oplus h_2 = \{\gamma_1 + \gamma_2 - \gamma_1\gamma_2 : \gamma_1 \in h_1, \gamma_2 \in h_2\}$;
(v) $h_1 \otimes h_2 = \{\gamma_1\gamma_2 : \gamma_1 \in h_1, \gamma_2 \in h_2\}$;
(vi) $h^\lambda = \{\gamma^\lambda : \gamma \in h\}$;
(vii) $\lambda h = \{1-(1-\gamma)^\lambda : \gamma \in h\}$.

The score function (Farhadinia, 2014) $S(h)$ for any HFE h is: $S(h) = \sum_{\gamma \in h}^{e_h} \frac{\gamma}{|e_h|}$, where e_h indicates the set of all elements in h and $|e_h|$ for the cardinality of e_h.

Now, $S(h_1) > S(h_2)$ implies that $h_1 > h_2$; $S(h_1) < S(h_2)$ implies that $h_2 > h_1$; otherwise, $S(h_1) = S(h_2)$ implies that $h_1 = h_2$ for HFEs h_1, h_2.

Definition 4.3.6. (Chen & Xu, 2014) Specific membership degrees to an element are not appropriate to describe the acceptance of the element. To overcome this complexity, IVHFS was defined as: $B_{IVHFS} = \{\langle x_i, \tilde{h}_B(x_i)\rangle : x_i \in X\}$, and $\tilde{h}_B(x_i)$ denotes the fuzzy interval in $[0,1]$, i.e., $B_{IVHFS} = \{\langle x_i, \tilde{\gamma}_i\rangle : x_i \in X, \tilde{\gamma}_i = [\tilde{\gamma^l}, \tilde{\gamma^u}] \in \tilde{h}_B(x_i)\}$, and $\tilde{h}_B(x_i)$ is called an interval-valued hesitant fuzzy element (IVHFE), where $\tilde{\gamma^l} = \inf \tilde{\gamma}, \tilde{\gamma^u} = \sup \tilde{\gamma}$ represent the lower and upper limits of $\tilde{\gamma}$, respectively.

Example 4.3.7. We consider an IVHFS as: $A = \{\langle x_1; [0.1, 0.3]\rangle, \langle x_2; [0.1, 0.4], [0.3, 0.8]\rangle, \langle x_3; [0.4, 0.6]\rangle, \langle x_4; [0.1, 0.4], [0.5, 0.7], [0.75, 0.9]\rangle\}$.

Property 4.3.8. (Chen & Xu, 2014) Considering $\tilde{h} = \{\cup_{\tilde{\gamma} \in \tilde{h}} [\tilde{\gamma}^l, \tilde{\gamma}^u]\}$, $\tilde{h}_1 = \{\cup_{\tilde{\gamma}_1 \in \tilde{h}_1} [\tilde{\gamma}_1^l, \tilde{\gamma}_1^u]\}$ and $\tilde{h}_2 = \{\cup_{\tilde{\gamma}_2 \in \tilde{h}_2} [\tilde{\gamma}_2^l, \tilde{\gamma}_2^u]\}$ as IVHFEs, some operations are as follows:

(i) $\tilde{h}^c = \{[1-\tilde{\gamma}^u, 1-\tilde{\gamma}^l] : \tilde{\gamma} \in \tilde{h}\}$, complement of $\tilde{h}$;
(ii) $\tilde{h}_1 \cup \tilde{h}_2 = \{[\max(\tilde{\gamma}_1^l, \tilde{\gamma}_2^l), \max(\tilde{\gamma}_1^u, \tilde{\gamma}_2^u)] : \tilde{\gamma}_1 \in \tilde{h}_1, \tilde{\gamma}_2 \in \tilde{h}_2\}$;
(iii) $\tilde{h}_1 \cap \tilde{h}_2 = \{[\min(\tilde{\gamma}_1^l, \tilde{\gamma}_2^l), \min(\tilde{\gamma}_1^u, \tilde{\gamma}_2^u)] : \tilde{\gamma}_1 \in \tilde{h}_1, \tilde{\gamma}_2 \in \tilde{h}_2\}$;
(iv) $\tilde{h}_1 \oplus \tilde{h}_2 = \{\cup_{\tilde{\gamma}_1 \in \tilde{h}_1, \tilde{\gamma}_2 \in \tilde{h}_2} [\tilde{\gamma}_1^l + \tilde{\gamma}_2^l - \tilde{\gamma}_1^l\tilde{\gamma}_2^l, \tilde{\gamma}_1^u + \tilde{\gamma}_2^u - \tilde{\gamma}_1^u\tilde{\gamma}_2^u]\}$;

(v) $\tilde{h}_1 \otimes \tilde{h}_2 = \{\cup_{\tilde{\gamma}_1 \in \tilde{h}_1, \tilde{\gamma}_2 \in \tilde{h}_2}[\tilde{\gamma}_1^l \tilde{\gamma}_2^l, \tilde{\gamma}_1^u \tilde{\gamma}_2^u]\}$;
(vi) $\tilde{h}^k = \{\cup_{\tilde{\gamma} \in \tilde{h}}[(\tilde{\gamma}^l)^k, (\tilde{\gamma}^u)^k]\}$, where $k \in [0, 1]$;
(vii) $k\tilde{h} = \{\cup_{\tilde{\gamma} \in \tilde{h}}[1 - (1 - \tilde{\gamma}^l)^k, 1 - (1 - \tilde{\gamma}^u)^k]\}$, with $k \in [0, 1]$.

Definition 4.3.9. (Moore et al., 2009) Let $a = [a^l, a^u] = \{x : a^l \leq x \leq a^u\}$, then a is termed as an interval number. For positive interval number a, $a^l \geq 0$.

For any two positive interval numbers $a = [a^l, a^u]$ and $b = [b^l, b^u]$ and $\lambda \in [0, 1]$, the following operations are defined.

(i) $a + b = [a^l + b^l, a^u + b^u]$;
(ii) $a^\lambda = ([a^l, a^u])^\lambda = [(a^l)^\lambda, (a^u)^\lambda]$;
(iii) $\lambda a = \lambda[a^l, a^u] = [\lambda a^l, \lambda a^u]$;
(iv) $a.b = [a^l, a^u].[b^l, b^u] = [a^l.b^l, a^u.b^u]$;
(v) $a \cup b = [\max\{a^l, b^l\}, \max\{a^u, b^u\}]$;
(vi) $a \cap b = [\min\{a^l, b^l\}, \min\{a^u, b^u\}]$.

Definition 4.3.10. Consider $a = [a^l, a^u] = \{x : 0 \leq a^l \leq x \leq a^u\}$ as an interval number. The expected value of a is $E(a) = (1/2)(a^l + a^u)$.

Definition 4.3.11. Assume $\tilde{a} = [\tilde{a}^l, \tilde{a}^u] = \{x : \tilde{a}^l \leq x \leq \tilde{a}^u,\ \tilde{a}^l = \inf \tilde{a},\ \tilde{a}^u = \sup \tilde{a}\}$ as an interval fuzzy number. The expected value (Helipern, 1992) of $\tilde{a}$ is: $E(\tilde{a}) = (1/2)(\tilde{a}^l + \tilde{a}^u)$.

Definition 4.3.12. (Chanas, 2001) Let $\tilde{a}_1 = [\tilde{a}_1^l, \tilde{a}_1^u]$ and $\tilde{a}_2 = [\tilde{a}_2^l, \tilde{a}_2^u]$ be two interval fuzzy numbers and let $l(\tilde{a}_1) = \tilde{a}_1^u - \tilde{a}_1^l$ and $l(\tilde{a}_2) = \tilde{a}_2^u - \tilde{a}_2^l$, then the possibility degree of $\tilde{a}_1 \geq \tilde{a}_2$ is given by:

$$Pos(\tilde{a}_1 \geq \tilde{a}_2) = \max\left\{1 - \max\left\{\frac{\tilde{a}_2^u - \tilde{a}_1^l}{l(\tilde{a}_1) + l(\tilde{a}_2)}, 0\right\}, 0\right\}. \tag{4.1}$$

Similarly, the possibility degree of $\tilde{a}_2 \geq \tilde{a}_1$ is defined as:

$$Pos(\tilde{a}_2 \geq \tilde{a}_1) = \max\left\{1 - \max\left\{\frac{\tilde{a}_1^u - \tilde{a}_2^l}{l(\tilde{a}_1) + l(\tilde{a}_2)}, 0\right\}, 0\right\}. \tag{4.2}$$

From Definition 3.12, we get:

(i) $0 \leq Pos(\tilde{a}_1 \geq \tilde{a}_2) \leq 1$;
(ii) $0 \leq Pos(\tilde{a}_2 \geq \tilde{a}_1) \leq 1$;
(iii) $Pos(\tilde{a}_1 \geq \tilde{a}_2) + Pos(\tilde{a}_2 \geq \tilde{a}_1) = 1$; and specially,
(iv) $Pos(\tilde{a}_1 \geq \tilde{a}_1) = Pos(\tilde{a}_2 \geq \tilde{a}_2) = 0.5$.

In many real-life problems, problem-variables are expressed by languages, words, or sentences rather than numbers. Variables of this type are called *linguistic variables* (Zadeh, 1975).

Definition 4.3.13. Linguistic Term Set (Zadeh, 1975): We consider,

(a) H, the name of the variable;
(b) $T(H)$, the term set of H, i.e., the set of its linguistic values;
(c) U, a universe of discourse;
(d) G, the way by which the terms of $T(H)$ are generated;
(e) M, a semantic rule for associating each linguistic value X with its meaning, i.e., $M(X)$, a fuzzy subset of U;

Then, a linguistic variable is characterized by a quintuple $(H, T(H), U, G, M)$. A linguistic variable is linguistically described by its semantics.

Several expressions of linguistic descriptors and the corresponding semantics exist in literature (Herrera & Herrera-Viedma, 2000; Herrera et al., 2008; Rodriguez et al., 2012; Rodriguez & Martinez, 2013). Among these, seven scales of linguistic term-based semantics are used frequently:

$S_1 = \{s_0 =$ nothing, $s_1 =$ extremely low, $s_2 =$ low, $s_3 =$ medium, $s_4 =$ high, $s_5 =$ very high, $s_6 =$ perfectly high$\}$ or, $S_2 = \{s_0 =$ very very poor, $s_1 =$ very poor, $s_2 =$ slightly poor, $s_3 =$ poor, $s_4 =$ slightly good, $s_5 =$ good, $s_6 =$ very good$\}$.

Property 4.3.14. The properties of the linguistic term set $S = \{s_i : i = 1, 2, \ldots, t\}$ are:

(i) The set is ordered, i.e., $s_i > s_j$, if $i > j$;
(ii) There is a negation operator, i.e., $\neg(s_i) = s_j$, for $i + j = t + 1$;
(iii) There is a maximizing operator, i.e., $\max(s_i, s_j) = s_j$, if $s_j \geq s_i$;
(iv) There is a minimizing operator, i.e., $\min(s_i, s_j) = s_j$, if $s_j \leq s_i$.

This discrete term set can be converted into a continuous term set as:

$\overline{S} = \{s_i : s_1 \leq s_i \leq s_q, i \in [1, q]\}$, where q is a sufficiently large positive number. s_i is called the original linguistic term if $s_i \in S$, otherwise, the virtual linguistic term.

Property 4.3.15. Assume s_γ and s_δ be two linguistic variables; $s_\gamma, s_\delta \in \overline{S}; \lambda, \kappa \in [0, 1]$. The operational laws are defined as (Xu, 2004):

(i) $s_\gamma \oplus s_\delta = s_{\gamma+\delta}$;
(ii) $s_\gamma \otimes s_\delta = s_{\gamma\delta}$;
(iii) $\lambda s_\gamma = s_{\lambda\gamma}$;
(iv) $(\lambda + \kappa)s_\gamma = \lambda s_\gamma \oplus \kappa s_\gamma$;
(v) $(s_\gamma)^\lambda = s_{\gamma\lambda}$

Human judgements and perception always flow in a hesitant environment and basically these environments are nurtured with linguistic characters of responses having fuzziness sense. We, in the next section, demonstrate the hesitant fuzzy linguistic term set, first defined by Rodriguez et al. (2012).

Definition 4.3.16. Hesitant Fuzzy Linguistic Term Set(HFLTS) (Rodriguez et al., 2012): Assume $S = \{s_i : i = 1, 2, \dots, t\}$ to be a linguistic term set. An HFLTS H_S is defined as an ordered finite subset of the consecutive linguistic terms of the set S. Null HFLTS and full HFLTS are described by $H_{S_{null}}(z) = \{\}$ and $H_{S_{full}}(z) = S$ respectively, where z indicates linguistic variable.

Example 4.3.17. Let $S_2 = \{s_0$=very poor, s_1=poor, s_2=slightly poor, s_3=fair, s_4=slightly good, s_5=good, s_6=very good$\}$ be a linguistic term set. Different HFLTSs can be depicted as $H^1_{S_2}(z) = \{s_0, s_2\}$, $H^2_{S_2}(z) = \{s_2, s_3, s_5\}$, $H^3_{S_2}(z) = \{s_0, s_4, s_6\}$.

4.4 HESITANT INTERVAL-VALUED FUZZY LINGUISTIC ELEMENT (HIFLE)

We express here the hesitant interval-valued fuzzy linguistic numbers through definition.

Definition 4.4.1. A hesitant interval-valued fuzzy linguistic set (HIFLS) $\mathcal{A}$ on a reference set χ is characterized in terms of a function $\tilde{h}_{\mathcal{A}}(x_i)$. When $\tilde{h}_{\mathcal{A}}(x_i)$ is applied to χ we get a subset of $[s_0, s_1]$ (unit linguistic interval), i.e., $\mathcal{A} = \{\langle x_i, \tilde{\gamma}_i\rangle : x_i \in \chi\}$, where $\tilde{\gamma}_i = \{\cup_i[s_{\tilde{\gamma}l}, s_{\tilde{\gamma}u}]\} \in h_{\mathcal{A}}(x_i)$. Here, $\tilde{h}_{\mathcal{A}}(x_i)$ is the hesitant interval-valued fuzzy linguistic element. For the sake of simplicity, we can define HIFLS as: $\mathcal{A} = \{\langle\tilde{\gamma}_i\rangle\}$.

Example 4.4.2. Consider a HIFLS $\mathcal{A} = \{\langle[s_{0.1}, s_{0.3}]\rangle, \langle[s_{0.1}, s_{0.4}], [s_{0.3}, s_{0.8}]\rangle, \langle[s_{0.4}, s_{0.6}]\rangle, \langle[s_{0.1}, s_{0.4}], [s_{0.5}, s_{0.7}], [s_{0.75}, s_{0.9}]\rangle\}$. Here, $\langle[s_{0.1}, s_{0.4}], [s_{0.5}, s_{0.7}], [s_{0.75}, s_{0.9}]\rangle\}$ or $\langle[s_{0.1}, s_{0.3}]\rangle$ are hesitant fuzzy elements of the HIFLS $\mathcal{A}$. Similarly, the other elements are expressed.

Property 4.4.3. The operations and the comparison among HIFLEs are expressed as follows: Considering $\tilde{h} = \{\cup_{s_{\tilde{\gamma}}\in\tilde{h}}[s_{\tilde{\gamma}l}, s_{\tilde{\gamma}u}]\}$, $\tilde{h}_1 = \{\cup_{s_{\tilde{\gamma}_1}\in\tilde{h}_1}[s_{\tilde{\gamma}^l_1}, s_{\tilde{\gamma}^u_1}]\}$ and $\tilde{h}_2 = \{\cup_{s_{\tilde{\gamma}_2}\in\tilde{h}_2}[s_{\tilde{\gamma}^l_2}, s_{\tilde{\gamma}^u_2}]\}$,

(i) $\tilde{h}^c = \{[s_1 - s_{\tilde{\gamma}u}, s_1 - s_{\tilde{\gamma}l}] : s_{\tilde{\gamma}} \in \tilde{h}\}$, complement of $\tilde{h}$;
(ii) $\tilde{h}_1 \cup \tilde{h}_2 = \{[\max(s_{\tilde{\gamma}^l_1}, s_{\tilde{\gamma}^l_2}), \max(s_{\tilde{\gamma}^u_1}, s_{\tilde{\gamma}^u_2})] : s_{\tilde{\gamma}_1} \in \tilde{h}_1, s_{\tilde{\gamma}_2} \in \tilde{h}_2\}$;
(iii) $\tilde{h}_1 \cap \tilde{h}_2 = \{[\min(s_{\tilde{\gamma}^l_1}, s_{\tilde{\gamma}^l_2}), \min(s_{\tilde{\gamma}^u_1}, s_{\tilde{\gamma}^u_2})] : s_{\tilde{\gamma}_1} \in \tilde{h}_1, s_{\tilde{\gamma}_2} \in \tilde{h}_2\}$;
(iv) $\tilde{h}_1 \oplus \tilde{h}_2 = \{\cup_{s_{\tilde{\gamma}_1}\in\tilde{h}_1, s_{\tilde{\gamma}_2}\in\tilde{h}_2}[s_{\tilde{\gamma}^l_1} + s_{\tilde{\gamma}^l_2} - s_{\tilde{\gamma}^l_1}s_{\tilde{\gamma}^l_2}, s_{\tilde{\gamma}^u_1} + s_{\tilde{\gamma}^u_2} - s_{\tilde{\gamma}^u_1}s_{\tilde{\gamma}^u_2}]\}$;
(v) $\tilde{h}_1 \otimes \tilde{h}_2 = \{\cup_{s_{\tilde{\gamma}_1}\in\tilde{h}_1, s_{\tilde{\gamma}_2}\in\tilde{h}_2}[s_{\tilde{\gamma}^l_1}s_{\tilde{\gamma}^l_2}, s_{\tilde{\gamma}^u_1}s_{\tilde{\gamma}^u_2}]\}$;
(vi) $\tilde{h}^k = \{\cup_{s_{\tilde{\gamma}}\in\tilde{h}}[(s_{\tilde{\gamma}l})^k, (s_{\tilde{\gamma}u})^k]\}$, where $k \in [0, 1]$;
(vii) $k\tilde{h} = \{\cup_{s_{\tilde{\gamma}}\in\tilde{h}}[s_1 - (s_1 - s_{\tilde{\gamma}l})^k, s_1 - (s_1 - s_{\tilde{\gamma}u})^k]\}$, where $k \in [0, 1]$.

4.5 PROJECTION MEASURE IN HIFLE

In this section, we consider some definitions, and examples related to projection measure, besides our proposed projection measure.

Definition 4.5.1. Let $q = (q_1, q_2, \dots, q_m)$ and $r = (r_1, r_2, \dots, r_m)$ be two vectors, then

$$\cos(q, r) = \frac{\sum_{j=1}^{m} q_j r_j}{\sqrt{\sum_{j=1}^{m} q_j^2}\sqrt{\sum_{j=1}^{m} r_j^2}} \tag{4.3}$$

Definition 4.5.2. Consider the vector $q = (q_1, q_2, \dots, q_m)$. Then $|q| = \sqrt{\sum_{j=1}^{m} q_j^2}$ is called the modulus value of q.

Since the calculation of $\cos(q, r)$ only contemplates the similarity measures between the directions of the vectors, the projection of one vector on another reflects the degree of measure, globally.

Definition 4.5.3. Let $P_i, i = 1, 2$ be two hesitant interval-valued fuzzy linguistic elements. Then the cosine of included angle between P_1 and P_2 is defined as:

$$\cos(P_1, P_2) = \frac{a_1c_1 + b_1d_1 + a_2c_2 + b_2d_2 + \dots + a_nc_n + b_nd_n}{\sqrt{a_1^2 + b_1^2 + a_2^2 + b_2^2 + \dots + a_n^2 + b_n^2}\sqrt{c_1^2 + d_1^2 + c_2^2 + d_2^2 + \dots + c_n^2 + d_n^2}} \tag{4.4}$$

where,

$$P_1 = \langle [s_{a_1}, s_{b_1}], [s_{a_2}, s_{b_2}], \dots, [s_{a_n}, s_{b_n}] \rangle; P_2 = \langle [s_{c_1}, s_{d_1}], [s_{c_2}, s_{d_2}], \dots, [s_{c_n}, s_{d_n}] \rangle$$

Definition 4.5.4. Consider $P_i, i = 1, 2$ be two hesitant interval-valued fuzzy linguistic elements. Then the projection of P_2 on P_1 is defined as:

$$Proj_{P_1}(P_2) = ||P_2|| \cos(P_1, P_2) \tag{4.5}$$

$$= \frac{1}{||P_1||}(P_1P_2) \tag{4.6}$$

Here we consider two examples corresponding to Definition 5.4.

Example 4.5.5. Assuming $P_1 = \langle [s_{0.3}, s_{0.5}], [s_{0.7}, s_{0.9}], [s_{0.2}, s_{0.3}] \rangle$ as a hesitant interval-valued fuzzy linguistic element, the projection of P_1 on $\mathcal{U}$ is:

$$Proj_{\mathcal{U}}(P_1) = \frac{0.5 + 0.9 + 0.3}{\sqrt{1 + 1 + 1}} = \frac{1.7}{\sqrt{3}} \approx 0.98152, \tag{4.7}$$

since the calculations on linguistic terms are based on corresponding semantics. Here, the unit positive ideal hesitant interval-valued fuzzy linguistic element is defined as:

$$\mathcal{U} = \langle [s_0, s_1], [s_0, s_1], [s_0, s_1], \dots, [s_0, s_1] \rangle. \tag{4.8}$$

Example 4.5.6. Assume $P_1 = \langle [s_{0.3}, s_{0.5}], [s_{0.7}, s_{0.9}], [s_{0.2}, s_{0.3}] \rangle$ and $P_2 = \langle [s_{0.2}, s_{0.3}], [s_{0.4}, s_{0.5}], [s_{0.3}, s_{0.4}] \rangle$ be two hesitant interval-valued fuzzy linguistic elements. Then the projection of P_1 on P_2 is:

$$Proj_{P_2}(P_1) = \frac{0.06 + 0.15 + 0.28 + 0.45 + 0.06 + 0.12}{\sqrt{0.04 + 0.09 + 0.16 + 0.25 + 0.09 + 0.16}} \approx 1.26012, \tag{4.9}$$

since the calculations on linguistic terms are based on corresponding semantics.

4.6 FUZZY MCDM PROBLEM

Since fuzziness helps to express and evaluate uncertainty in different forms incorporated in a multi-criteria decision-making problem, fuzzy MCDM problems are characterized in various ways. Mathematically, a MCDM problem in a crisp environment has the form:

$$\begin{cases} \text{Optimize } \mathcal{A}_i,\ i = 1(1)m; \\ \text{Set of criteria (set of constraints): } \mathcal{C}_j;\ j = 1(1)n. \end{cases} \tag{4.10}$$

where $\mathcal{A}_i$ denotes the set of m alternatives and $\mathcal{C}_j$ denotes the set of n criteria. This can be written in matrix format as:

$$\begin{array}{c} \\ \mathcal{A}_1 \\ \mathcal{A}_2 \\ \vdots \\ \mathcal{A}_m \end{array} \begin{array}{c} \begin{array}{ccccc} \mathcal{C}_1 & \mathcal{C}_2 & \mathcal{C}_3 & \cdots & \mathcal{C}_n \end{array} \\ \begin{pmatrix} z_{11} & z_{12} & z_{13} & \cdots & z_{1n} \\ z_{21} & z_{22} & z_{23} & \cdots & z_{2n} \\ \vdots & \vdots & \vdots & \ddots & \vdots \\ z_{m1} & z_{m2} & z_{m3} & \cdots & z_{mn} \end{pmatrix} \end{array}$$

with weight vector $\omega = [\omega_1, \omega_2, \omega_3, \ldots, \omega_n]$.

Here, $\mathcal{A}_i$s is the set of alternatives from which decision-makers can choose the alternatives based on the set of criteria $\mathcal{C}_j$s.

In fuzzy sense, when the rating of alternatives with reference to criteria is measured by decision-makers' judgements or by preferences, these judgemental values are measured by fuzzy variables. Similarly, weights associated with criteria can be assumed in fuzzy sense.

However, in literature, numerous works have been established with different types of multi-criteria decision-making problematic models with different forms of fuzzy numbers and uncertainty.

In this chapter, we discuss multi-criteria decision-making models with the help of projection measure in hesitant interval-valued fuzzy linguistic element oriented environment.

4.7 MCDM METHOD BASED ON HIFLE USING PROJECTION

In this section we consider the stepwise solution procedure of multi-attribute decision-making problems using projection measure. The elements are expressed here as hesitant interval-valued fuzzy linguistic numbers. The stepwise algorithm structure is given as Algorithm 4.1.

Algorithm 4.1 Solving MCDM using projection measure in HIFLE environment.

Step 1: Consider the MCDM problem with weights associated to criteria. The decision-maker evaluates all the alternative ($\mathcal{A}_i$) values with respect to the criteria ($\mathcal{C}_j$), where $i = 1(1)m,\ j = 1(1)n$

Step 2: Construct the decision matrix corresponding to the set of alternatives and set of criteria as:

$$\begin{array}{c} \\ \mathcal{A}_1 \\ \mathcal{A}_2 \\ \vdots \\ \mathcal{A}_m \end{array} \begin{array}{c} \begin{array}{ccccc} \mathcal{C}_1 & \mathcal{C}_2 & \mathcal{C}_3 & \cdots & \mathcal{C}_n \end{array} \\ \begin{pmatrix} \tilde{h}_{z_{11}} & \tilde{h}_{z_{12}} & \tilde{h}_{z_{13}} & \cdots & \tilde{h}_{z_{1n}} \\ \tilde{h}_{z_{21}} & \tilde{h}_{z_{22}} & \tilde{h}_{z_{23}} & \cdots & \tilde{h}_{z_{2n}} \\ \vdots & \vdots & \vdots & \ddots & \vdots \\ \tilde{h}_{z_{m1}} & \tilde{h}_{z_{m2}} & \tilde{h}_{z_{m3}} & \cdots & \tilde{h}_{z_{mn}} \end{pmatrix} \end{array}.$$

Step 3: We transform the decision matrix into the normalized decision matrix

Step 4: On the normalized decision matrix, we perform the weight information and get the weighted normalized decision matrix

Step 5: Utilizing Eq.(8), we define the positive ideal hesitant interval-valued fuzzy linguistic element, $\mathcal{U} = \langle [s_0, s_1], [s_0, s_1], [s_0, s_1], \ldots, [s_0, s_1] \rangle$

Step 6: Considering Eqs.(4-6), we derive the projection measure of $Proj_{\mathcal{U}}(\tilde{h}_{z_{ij}})$ for all i, j on the positive ideal hesitant interval-valued fuzzy linguistic element $\mathcal{U}$

Step 7: Ranking and selecting of the alternatives $\mathcal{A}_i$ are done according to $Proj_{\mathcal{U}}(\tilde{h}_{z_{ij}})$

Step 8: Optimal results are obtained.

4.7.1 Case Studies: Cotton Production in Sundarban

In West Bengal, India, North and South 24 Parganas comprise the coastal part, mainly. Sundarban (N21°30′ E88° to N23° E89°) lies within it. The coastal part of West Bengal can be classified by the mainland and islands. Both the mainland and islands are under agricultural systems through different complex and risk prone criteria. Most of the lands of Sundarban are saline. In the summer season, salinity increases. Most of these lands are gone under water in monsoon with poor drainage system. Sundarban area can be classified into upland, medium-upland, and lowland areas. Cultivation of cotton is done in all these three areas (Saha & Ghosh, 1999), but mainly in the lowland and medium-land areas. We consider the three areas as three alternatives: low-land area ($\mathcal{A}_1$), medium-upland area ($\mathcal{A}_2$), and up-land area ($\mathcal{A}_3$). Choosing between these alternatives, i.e., the optimal choices of alternatives, are done based upon the considered criteria.

Varieties of criteria can figure the optimal choices of alternatives in Sundarban (Mandal, 2019). The reasonable criteria considered in Sundarban for choosing the cultivation area are irrigation system, rainfed land, salinity of land, temperature, water drainage system, use of fertilizers and pesticides, etc. In this chapter, we consider the cultivation of cotton in summer (Rabi cotton) (Saha & Ghosh, 1999) under different reasonable criteria. These criteria can be assumed as different attributes used to evaluate the production of cotton. Different criteria with sub-criteria and corresponding linguistic information are shown in Table 4.1.

TABLE 4.1
Criteria Definition of Land–Water Characteristics in the Decision-Making Process

Criteria	Sub-Criteria	Definition	Linguistic Interval
$\mathcal{C}_1$(Land)	$\mathcal{C}_{11}$	Very highly saline	$[s_{0.90}, s_{1.00}]$
	$\mathcal{C}_{12}$	Highly saline	$[s_{0.85}, s_{0.95}]$
	$\mathcal{C}_{13}$	Moderately saline	$[s_{0.70}, s_{0.80}]$
	$\mathcal{C}_{14}$	Medium saline	$[s_{0.50}, s_{0.75}]$
	$\mathcal{C}_{15}$	Slightly saline	$[s_{0.30}, s_{0.60}]$
	$\mathcal{C}_{16}$	Very slightly saline	$[s_{0.10}, s_{0.35}]$
$\mathcal{C}_2$(Water)	$\mathcal{C}_{21}$	Very highly rainfed	$[s_{0.90}, s_{0.98}]$
	$\mathcal{C}_{22}$	Highly rainfed	$[s_{0.80}, s_{0.95}]$
	$\mathcal{C}_{23}$	Medium rainfed	$[s_{0.60}, s_{0.85}]$
	$\mathcal{C}_{24}$	Partially medium irrigated	$[s_{0.40}, s_{0.65}]$
	$\mathcal{C}_{25}$	Partially low irrigated	$[s_{0.30}, s_{0.50}]$
	$\mathcal{C}_{26}$	Very low irrigated	$[s_{0.10}, s_{0.35}]$

If we consider the highly saline and medium rainfed area, we consider this as an linguistic interval $\langle \mathcal{C}_{12}, \mathcal{C}_{23} \rangle$, i.e., $\langle [s_{0.85}, s_{0.95}], [s_{0.60}, s_{0.85}] \rangle$. In many cases, due to the hesitant characteristics of the decision-maker's variables, one can choose the linguistic interval as $\langle [s_{0.80}, s_{0.90}], [s_{0.60}, s_{0.85}] \rangle$. So, finally, we can get the hesitant interval-valued fuzzy linguistic interval as $\langle [s_{0.80}, s_{0.90}], [s_{0.85}, s_{0.95}], [s_{0.60}, s_{0.85}] \rangle$. Now, we discuss our proposed decision-making problem in Algorithm 4.2, using the steps of Algorithm 4.1.

Algorithm 4.2

Step 1: When we consider the decision-making problematic matrix $\mathcal{D}_M$, we consider the criteria set as: $\langle \mathcal{C}^1 \rangle$ for $\langle \mathcal{C}_{13}, \mathcal{C}_{23} \rangle$. Similarly, we define, $\langle \mathcal{C}^2 \rangle := \langle \mathcal{C}_{12}, \mathcal{C}_{23} \rangle$, $\langle \mathcal{C}^3 \rangle := \langle \mathcal{C}_{15}, \mathcal{C}_{23} \rangle$, $\langle \mathcal{C}^4 \rangle := \langle \mathcal{C}_{13}, \mathcal{C}_{25} \rangle$, $\langle \mathcal{C}^5 \rangle := \langle \mathcal{C}_{13}, \mathcal{C}_{24} \rangle$, $\langle \mathcal{C}^6 \rangle := \langle \mathcal{C}_{11}, \mathcal{C}_{25} \rangle$, $\langle \mathcal{C}^7 \rangle := \langle \mathcal{C}_{15}, \mathcal{C}_{22} \rangle$, $\langle \mathcal{C}^8 \rangle := \langle \mathcal{C}_{15}, \mathcal{C}_{24} \rangle$, and $\langle \mathcal{C}^9 \rangle := \langle \mathcal{C}_{16}, \mathcal{C}_{21} \rangle$. We acknowledge the alternatives as $\mathcal{A}_1, \mathcal{A}_2$ and $\mathcal{A}_3$.
Step 2: The uncertain decision matrix $\mathcal{D}_M$ is considered for normalization. If needed, the normalization is done by equating the number of elements (cardinality) in every hesitant entity by adding the interval $[s_{0.5}, s_{0.5}]$.
Step 3: Since, both the characteristics of land-salinity and water-irrigation are equally important in time of cotton cultivation, we assign the weight vectors to these criteria equally distributed, and for this reason the weighted decision matrix remains same as $\mathcal{D}_M$.
Step 4: Considering Eqs.(4-6), we derive the projection measure of $Proj_{\mathcal{U}}(\tilde{h}_{z_{ij}})$ for all i, j on the positive ideal hesitant interval-valued fuzzy linguistic element $\mathcal{U}$. Thus, we get, $Proj_{\mathcal{U}}(\tilde{h}_{z_{11}}) = 1.4433, Proj_{\mathcal{U}}(\tilde{h}_{z_{12}}) = 1.3279$. Similarly, the other projection values are achieved, and we get the projection-value based decision matrix $\mathcal{D}'_M$.

$$
\mathcal{D}_M = \begin{array}{c} \\ \langle \mathcal{C}^1 \rangle \\ \langle \mathcal{C}^2 \rangle \\ \langle \mathcal{C}^3 \rangle \\ \langle \mathcal{C}^4 \rangle \\ \langle \mathcal{C}^5 \rangle \\ \langle \mathcal{C}^6 \rangle \\ \langle \mathcal{C}^7 \rangle \\ \langle \mathcal{C}^8 \rangle \\ \langle \mathcal{C}^9 \rangle \end{array}
\begin{array}{c} \begin{array}{ccc} \qquad\qquad\mathcal{A}_1\qquad\qquad & \qquad\qquad\mathcal{A}_2\qquad\qquad & \qquad\qquad\mathcal{A}_3\qquad\qquad \end{array} \\
\left(\begin{array}{ccc}
\langle [s_{0.70}, s_{0.80}], [s_{0.75}, s_{0.85}], [s_{0.60}, s_{0.85}] \rangle & \langle [s_{0.70}, s_{0.75}], [s_{0.60}, s_{0.82}] \rangle & \langle [s_{0.60}, s_{0.80}], [s_{0.60}, s_{0.85}] \rangle \\
\langle [s_{0.85}, s_{0.95}], [s_{0.60}, s_{0.85}] \rangle & \langle [s_{0.85}, s_{0.95}], [s_{0.60}, s_{0.85}], [s_{0.65}, s_{0.90}] \rangle & \langle [s_{0.80}, s_{0.97}], [s_{0.80}, s_{0.95}], [s_{0.50}, s_{0.75}] \rangle \\
\langle [s_{0.30}, s_{0.65}], [s_{0.70}, s_{0.85}] \rangle & \langle [s_{0.30}, s_{0.60}], [s_{0.60}, s_{0.85}] \rangle & \langle [s_{0.40}, s_{0.60}], [s_{0.65}, s_{0.70}] \rangle \\
\langle [s_{0.70}, s_{0.80}], [s_{0.40}, s_{0.60}] \rangle & \langle [s_{0.70}, s_{0.80}], [s_{0.40}, s_{0.65}] \rangle & \langle [s_{0.70}, s_{0.80}], [s_{0.45}, s_{0.65}] \rangle \\
\langle [s_{0.70}, s_{0.75}], [s_{0.40}, s_{0.60}] \rangle & \langle [s_{0.70}, s_{0.85}], [s_{0.45}, s_{0.65}], [s_{0.50}, s_{0.60}] \rangle & \langle [s_{0.70}, s_{0.80}], [s_{0.40}, s_{0.67}] \rangle \\
\langle [s_{0.90}, s_{1.00}], [s_{0.80}, s_{0.95}], [s_{0.30}, s_{0.45}] \rangle & \langle [s_{0.90}, s_{0.95}], [s_{0.30}, s_{0.50}] \rangle & \langle [s_{0.90}, s_{1.00}], [s_{0.30}, s_{0.40}] \rangle \\
\langle [s_{0.30}, s_{0.60}], [s_{0.80}, s_{0.95}] \rangle & \langle [s_{0.35}, s_{0.60}], [s_{0.80}, s_{0.95}] \rangle & \langle [s_{0.40}, s_{0.65}], [s_{0.80}, s_{0.90}] \rangle \\
\langle [s_{0.30}, s_{0.50}], [s_{0.40}, s_{0.65}] \rangle & \langle [s_{0.30}, s_{0.60}], [s_{0.40}, s_{0.60}], [s_{0.45}, s_{0.65}] \rangle & \langle [s_{0.30}, s_{0.60}], [s_{0.40}, s_{0.65}] \rangle \\
\langle [s_{0.10}, s_{0.30}], [s_{0.20}, s_{0.35}], [s_{0.90}, s_{0.96}] \rangle & \langle [s_{0.20}, s_{0.30}], [s_{0.90}, s_{0.98}] \rangle & \langle [s_{0.10}, s_{0.35}], [s_{0.95}, s_{0.98}] \rangle
\end{array} \right) \end{array}.
$$

$$
\mathcal{D}'_M = \begin{array}{c} \\ \langle \mathcal{C}^1 \rangle \\ \langle \mathcal{C}^2 \rangle \\ \langle \mathcal{C}^3 \rangle \\ \langle \mathcal{C}^4 \rangle \\ \langle \mathcal{C}^5 \rangle \\ \langle \mathcal{C}^6 \rangle \\ \langle \mathcal{C}^7 \rangle \\ \langle \mathcal{C}^8 \rangle \\ \langle \mathcal{C}^9 \rangle \end{array}
\begin{array}{c} \begin{array}{ccc} \mathcal{A}_1 & \mathcal{A}_2 & \mathcal{A}_3 \end{array} \\
\left(\begin{array}{ccc}
1.4433 & 1.1951 & 1.2413 \\
1.3279 & 1.5588 & 1.5415 \\
1.1547 & 1.1258 & 1.0392 \\
1.0969 & 1.1258 & 1.1258 \\
1.0680 & 1.2124 & 1.1373 \\
1.3856 & 1.1258 & 1.0969 \\
1.1835 & 1.1835 & 1.1835 \\
0.9526 & 1.0680 & 1.0103 \\
0.9295 & 1.0276 & 1.0565
\end{array} \right) \end{array}.
$$

TABLE 4.2
Projection Measure-Based Average Ranking of the Alternatives

Alternatives	Average Ranking	Ranking Order
$\mathcal{A}_1$	1.17133	2
$\mathcal{A}_2$	1.18031	1
$\mathcal{A}_3$	1.15914	3

Step 5: Now, we consider the projection measure-based average ranking of the alternatives, and obtain Table 4.2. Calculations are done upon the semantics of the linguistic elements.

4.8 RESULTS AND DISCUSSION

Ranking the alternatives $\mathcal{A}_i$s according to projection measure, we conclude that

$$\mathcal{A}_2 \succ \mathcal{A}_1 \succ \mathcal{A}_3.$$

Therefore, $\mathcal{A}_2$ can be considered as a better alternative than $\mathcal{A}_1$ or $\mathcal{A}_3$. This consideration depends upon the characteristics: land salinity and tillage system at time of cultivation of cotton in Sundarban of West Bengal, India. Simply to say, medium-upland areas are more suitable for cotton production than lowland and upland areas in Sundarban. Considering different weightage to different criteria, we get different results. Again, the consideration of other factors, like the uses of fertilizers and pesticides, soil temperature, erosion of land, etc., may provide a new direction for choosing the alternatives. The main advantage of our proposed method is its simple calculation and the assumption of the linguistic format of considered variables, whereas the main disadvantage of this proposed method is its time consumption during calculation.

4.9 CONCLUSION

Cotton is termed as "white gold" in different parts of world due to its importance in terms of foreign revenues. Cotton fiber is used from industrial sectors to medical sectors, and obviously it plays an important role in the socio-economic factors, like the livelihood of cotton-farmers to laborers involved in the cotton industry, etc. But the production of cotton depends upon numerous constraints, e.g., salinity of land, planting time, temperature of air and soil, soil erosion, fertilizers and pesticides, drought and flood, greenhouse gas-emission, irrigation and tillage practices, etc.

With the present work, the application of a proposed projection measure is shown as a measurement of choosing alternatives in a hesitant interval-valued fuzzy linguistic environment. We have simply applied our proposed methodology to a case study in the production-evaluation of cotton in Sundarban, West Bengal, India. At the time of considering the problem of cotton cultivation in Sundarban, we accept the salinity of land and the water-irrigation systems as criteria. We have done our calculations to pick out the optimal alternative from others.

Several new alternatives with a new set of criteria can be considered in the future to validate the proposed projection measure.

CONFLICT OF INTEREST

The authors would like to declare that there is no conflict of interest. Authors did not receive any funding for this work.

REFERENCES

Aggarwal, P., Bhattacharyya, R., Mishra A. K., Das, T. K., Simunek, J., Pramanik, P., Sudhishri, S., Vashisth, A., Krishnan, P., Chakraborty, D., Kamble, K. H. (2017) Modelling soil water balance and root water uptake in cotton grown under different soil conservation practices in the Indo Gangetic Plain. *Agriculture, Ecosystems and Environment*, *240*, 287–299.

Ahmad, S., Raza, I. (2014) Optimization of management practices to improve cotton fibre quality under irrigated arid environment. *Journal of Food, Agriculture & Environment*, *12*(2), 609–613.

Ali, H., Hameed, R. A., Ahmad, S., Shahzad, A. N., Sarwar, N. (2014) Efficacy of different techniques of nitrogen application on American cotton under semi-arid conditions. *Journal of Food, Agriculture & Environment*, *12*(1), 157–160.

Anapalli, S. S., Pettigrew, W. T., Reddy, K. N., Ma, L., Fisher, D. K., Sui, R. (2016) Climate-optimized planting windows for cotton in the lower Mississippi delta region. *Agronomy*, *6*, 46, 1–15.

Anjum, R., Ahmed, A., Ullah, R., Jahangir, M., Yousaf, M. (2005) Effect of soil salinity/sodicity on the growth and yield of different varieties of cotton. *International Journal of Agriculture & Biology*, *7*(4), 606–608.

Ashraf, A., Akram, M., Sarwar, M. (2014) Type-II fuzzy decision support system for fertilizer. *The Scientific World Journal*, Volume 2014, Article ID 695815, 1–9.

Atanassov, K. T. (1986) Intuitionistic fuzzy sets. *Fuzzy Sets and Systems*, *20*(1), 87–96.

Baral, A. (2013) An application of fuzzy concept of agricultural farm for decision-making. *International Journal of Computer Applications*, *71*(21), 18–23.

Bhaumik, A., Roy, S. K. (2021) Intuitionistic interval-valued hesitant fuzzy matrix games with a new aggregation operator for solving management problem. *Granular Computing*, *6*, 359–375.

Bhaumik, A., Roy, S. K. (2022) Evaluations for medical diagnoses phenomena through $2 * 2$ linguistic neutrosophic environment-based game situation. *Soft Computing*, *26*, 4883–4893.

Bhaumik, A., Roy, S. K., Li, D. F. (2017) Analysis of triangular intuitionistic fuzzy matrix games using robust ranking. *Journal of Intelligent and Fuzzy Systems*, *33*, 327–336.

Bhaumik, A., Roy, S. K., Li, D. F. (2021) (α, β, γ)-cut set based ranking approach to solving bi-matrix games in neutrosophic environment. *Soft Computing*, *25*, 2729–2739.

Bhaumik, A., Roy, S. K., Weber, G. W. (2020) Hesitant interval-valued intuitionistic fuzzy linguistic term set approach in Prisoners' dilemma game theory using TOPSIS: a case study on Human-trafficking. *Central European Journal of Opertions Research*, *28*, 797–816.

Bhaumik, A., Roy, S. K., Weber, G. W. (2021) Multi-objective linguistic-neutrosophic matrix game and its applications to tourism management. *Journal of Dynamics & Games*, *8*, 101–118.

Biswas, A., Pal, B. B. (2005) Application of fuzzy goal programming technique to land use planning in agriculture system. *Omega*, *33*, 391–398.

Brown, H. B. (2002) *Cotton history, species, varieties, morphology, breeding, culture, diseases, marketing and uses*. 2nd Edition, Biotech Books, Delhi, 27–40.

Burrough, P.A. (1989) Fuzzy mathematical methods for soil survey and land evaluation. *Journal of Soil Science*, *40*(8), 477–492.

Buttar, G. S., Thind, H. S., Sekhon, K. S., Kaur, A., Gill, R. S., Sidhu, B. S., Aujla, M. S. (2017) Management of saline-sodic water in cotton-wheat cropping system. *Journal of Agricultural Science and Technology*, *19*, 465–474.

Chanas, S. (2001) On the interval approximation of a fuzzy number. *Fuzzy Sets and Systems*, *122*(2), 353–356.

Chen, N., Xu, Z. S. (2014) Properties of interval-valued hesitant fuzzy sets. *Journal of Intelligent and Fuzzy Systems*, *27*, 143–158.

Chen, N., Xu, Z. S., Xia, M. M. (2013) Interval-valued hesitant preference relations and their applications to group decision-making. *Knowledge-Based Systems*, *37*, 528–540.

Constable, G. A., Bange, M. P. (2015) The yield potential of cotton (*Gossypium hirsutum* L.). *Field Crops Research*, *182*, 98–106.

Dong, H. (2012) Technology and field management for controlling soil salinity effects on cotton. *Australian Journal of Crop Science*, *6*, 333–341.

El Titi, A. (2003) *Soil tillage in agroecosystems*. CRC Press, Boca Raton, 383–386.

Farhadinia, B. (2014) A series of score functions for hesitant fuzzy sets. *Information Sciences*, *277*, 102–110.

Foster J. L., Bean M. E., Morgan, C., Morgan, G., Mohtar, R., Landivar, J., Young, M. (2018) Comparison of two tillage practices in a semi-arid cotton. *Agronomy Journal*, *110*(4), 1572–1579.

Grantz, D. A. (2003) Ozone impacts on cotton: towards an integrated mechanism. *Environmental Pollution*, *126*(3), 331–344.

Gregory, P. J., George, T. S. (2011) Feeding nine billion: the challenge to sustainable crop production. *Journal of Experimental Botany*, *62*(15), 5233–5239.

Heilpern, S. (1992) The expected value of a fuzzy number. *Fuzzy sets and Systems*, *47*(1), 81–86.

Herrera, F., Herrera-Viedma, E. (2000) Linguistic decision analysis: steps for solving decision problems under linguistic information. *Fuzzy Sets and Systems*, *115*, 67–82.

Herrera, F., Herrera-Viedma, E., Martinez, L. (2008) A fuzzy linguistic methodology to deal with unbalanced linguistic term sets. *IEEE Transactions on Fuzzy Systems*, *16*, 354–370.

Herrera, F., Herrera-Viedma, E., Verdegay, J. L. (1995) A sequential selection process in group decision-making with linguistic assessment. *Information Sciences*, *85*, 223–239.

Hillocks, R. J. (1995) Integrated management of insects pests, diseases and weeds of cotton in Africa. *Integrated Pest Management Reviews*, *1*, 31–47.

Hira, G. S., Jalota, S. K., Arora, V. K. (2004) Efficient management of water resources for sustainable cropping in Punjab. *Research Bulletin, Department of Soils, Punjab Agricultural University, Ludhiana*, 20.

Iftikhar, T., Babar, L. K., Zahoor, S., Khan, N. G. (2010) Impact of land pattern and hydrological properties of soil on cotton yield. *Pakistan Journal of Botany*, *42*(5), 3023–3028.

ITC (2011) Trade Map. International Trade Centre, Switzerland. http://www.trademap.org

Jana, J., Roy, S. K. (2018) Solution of Matrix Games with Generalised Trapezoidal Fuzzy Payoffs. *Fuzzy Information and Engineering*, *10*, 213–224.

Jana, J., Roy, S. K. (2019) Dual hesitant fuzzy matrix games: based on new similarity measure. *Soft Computing*, *23*, 8873–8886.

Jana, J., Roy, S. K. (2021) Soft Matrix Game: A Hesitant Fuzzy MCDM Approach. *American Journal of Mathematical and Management Sciences*, *40*, 107–119.

Jana, J., Roy, S. K. (2022) Linguistic Pythagorean hesitant fuzzy matrix game and its application in multi-criteria decision-making. *Applied Intelligence*, *40*(20), https://doi.org/10.1007/s10489-022-03442-2.

Jiang, X. P., Wei, G. W. (2014) Some Bonferroni mean operators with 2-tuple linguistic infortaion and their application to multiple attribute decision-making. *Journal of Intelligent and Fuzzy Systems*, *27*, 2153–2162.

Kacprzyk, J., Zadrozny, S. (2005) Linguistic database summaries and their protoforms; towards natural language based knowledge discovery tools. *Information Sciences*, *173*(4), 281–304.

Karmakar, R., Das, I., Dutta, D., Rakshit, A. (2016) Potential effects of climate change on soil properties: a review. *Science International*, *4*(2), 51–73.

Keshavarzi, A., Sarmadian, F., Heidari, A., Omid, M. (2010) Land suitability evaluation using fuzzy continuous classification (A case study: Ziaran Region). *Modern Applied Science*, *4*(7), 72–81.

Khan, M. B., Khaliq, A., Ahmad, S. (2004) Performance of mashbean intercropped in cotton planted in different planting patterns. *Journal of Research (Science)*, *15*(2), 191–197.

Kurtener, D., Torbert, H. A., Krueger, E. (1989) *Evaluation of agricultural land suitability: application of fuzzy indicators*, Springer-Verlag, 475–490.

Lee, J. A., Fang, D. D. (2015) Cotton as a world crop: origin, history, and current status. *Cotton*, *57*, 1–23.

Loka, D. A., Oosterhuis, D. M. (2012) Water stress and reproductive development in cotton. *Department of Crop, Soil, and Environmental Sciences University of Arkansas, Fayetteville, AR, 72704*.

Lokhande, S., Reddy, K. R. (2014) Reproductive and fibre quality responses of upland cotton to moisture deficiency. *Agronomy Journal*, *106*, 1060–1069.

Luo, Q., Bange, M., Braunack, M., Johnston, D. (2016) Effectiveness of agronomic practices in dealing with climate change impacts in the Australian cotton industry– a simulation study. *Agricultural Systems*, *147*, 1–9.

Ma, F. Y., Li, M. C., Yang, J. R., Ji, X. J., Shentu, X. D., Tao, H. J. (2002) A study of effect of water deficit of three periods during cotton anthesis on canopy apparent photosynthesis and WUE. *Zhongguo Nongye Kexue (China)*.

Mandal, S. (2019) *Risks and profitability challenges of agriculture in Sundarbans India. in The Sundarbans: a disaster-prone eco-region*, Coastal Research Library, Sen, H. S. (Ed.) (2019) doi:10.1007/978-3-030-00680-8.

Martinez, L., Liu, J., Yang, J. B., Herrera, F. (2005) A multi-granular hierarchical linguistic model for design evaluation based on safety and cost analysis. *International Journal of Intelligent Systems*, *20*(12), 1161–1194.

Moore, R. E., Kearfott, R. B., & Cloud, M. J. (2009) Introduction to interval analysis. *Society for Industrial and Applied Mathematics*.

Özkan, B., Özceylan, E., Kabak, M., Dağdeviren, M. (2019) Evaluating the websites of academic departments through SEO criteria: a hesitant fuzzy linguistic MCDM approach. *Artifcial Intelligence Review*, https://doi.org/10.1007/s10462-019-09681-z.

Paz, J. O., Woli, P., Garcia, Y., Garcia, A., Hoogenboom, G. (2012) Cotton yields as influenced by ENSO at different planting dates and spatial aggregation levels. *Agricultural Systems*, *111*, 45–52.

Rahman, M. H., Ahmad, A., Wang, X., Wajid, A., Nasim, W., Hussain, M., Ahmad, B., Ahmad, I., Ali, Z., Ishaque, W., Awasis, M., Shelia, V., Ahmad, S., Fahad, S., Alam, M., Ullah, H.,

Hoogenboom, G. (2018) Multi-modal projections of future climate and climate change impacts uncertainty assessment for cotton production in Pakistan. *Agricultural and Forest Meteorology*, *253*, 94–113.
Reddy, V. R., Reddy, K. R., Baker, D. N. (1991) Temperature effect on growth and development of cotton during the fruiting period. *Agronomy Journal*, *83*(1), 211–217.
Rodriguez, R. M., Martinez, L. (2013) An analysis of symbolic linguistic computing models in decision-making. *International Journal of General Systems*, *42*, 121–136.
Rodriguez, R. M., Martinez, L., Herrera, F. (2012) Hesitant fuzzy linguistic term sets for decision-making. *IEEE Transactions on Fuzzy Systems*, *20*(1), 109–119.
Roy, S. K., Bhaumik, A. (2018) Intelligent Water Management: a Triangular Type-2 Intuitionistic Fuzzy Matrix Games Approach. *Water Resources Management*, *32*, 949–968.
Roy, S.K., Jana, J. (2021) The multi-objective linear production planning games in triangular hesitant fuzzy sets. *Sadhana*, *46*, 176.
Saha, D., Ghosh, S. S. (1999) Farmer-led extension, its strength and weakness with special reference to coastal Bengal. Paper presented at the VIII National Workshop of KVKs and TTCs, 15–17th January, 1999.
Sawan, Z. M. (2018) Climatic variables: Evaporation, sunshine, relative humidity, soil and air temperature and its adverse effects on cotton production. *Information Processing in Agriculture*, *5*(1), 134–148.
Sharma, D. K. (2007) Fuzzy goal programming for agricultural land allocation problems *Yugoslav Journal of Operation Research*, *17*(1), 31–42.
Singh, A. (2015) Soil salinization and water logging: a threat to environment and agricultural sustainability. *Ecological Indicators*, *57*, 128–130.
Sun, C., Feng, D., Mi, Z., Li, C., Zhang, J., Gao, Y., Sun, J. (2017) Impacts of ridge-furrow planting on salt stress and cotton yield under drip irrigation. *Water*, *9*, 49.
Torbert, H. A., Ingram, J. T., Prior, S. A. (2015) High residue conservation tillage system for cotton production: a farmer's perspective. *Making Conservation Tillage Conventional: Building a Future on 25 Years of Research*, 36.
Torra, V. (2010) Hesitant fuzzy sets. *International Journal of Intelligent Systems*, *25*(6), 529–539.
USDA (United States Department of Agriculture) (2018) Pakistan sugar annual report. Global Agricultural Information Network. https://gain.fas.usda.gov
Verma, N., Chaudhary, S., Goyal, S. (2018) Long term effects of inorganic fertilizers and organic amendments on ammonification and nitrification activity of soils under cotton-wheat cropping system. *International Journal of Current Microbiology and Applied Sciences*, *7*, 718–724.
Wang, R., Kang, Y., Wan, S. (2015) Effects of different drip irrigation regimes on saline-sodic soil nutrients and cotton yield in an arid region of Northwest China. *Agricultural Water Management*, *153*, 1–18.
Williams, A., White, N., Mushtaq, S., Cockfield, G., Power, B., Kouadio, L., (2015) Quantifying the response of cotton production in eastern Australia to climate change. *Climate Change*, *129*(1), 183–196.
Xia, M. M., Xu, Z. S. (2011) Hesitant fuzzy information aggregation in decision-making. *International Journal of Approximate Reasoning*, *52*, 395–407.
Xu, Z. S. (2005) Deviation measures of linguistic preference relations in group decision-making. *Omega*, *33*, 249–254.
Xu, Z. S. (2004) A method based on linguistic aggregation operators for group decision-making with linguistic preference relations. *Information Sciences*, *166*(1), 19–30.
Yager, R. R. (1977) Multiple objective decision-making using fuzzy sets. *International Journal of Man-Machine Studies*, *9*(4), 375–382.

Yazar, A., Sezen S. M., Sesveren, S. (2002) LEPA and trickle irrigation of cotton in the Southeast Anatolia Project (GAP) area in Turkey. *Agricultural Water Management*, *54*, 189–203.

Yu, D. J., Zhang, W. Y., Xu, Y. J. (2013) Group decision-making under hesitant fuzzy environment with application to personnel evaluation. *Knowledge-Based Systems*, *52*, 1–10.

Zadeh, L. A. (1965) Fuzzy sets. *Information and Control*, *8*(3), 338–353.

Zadeh, L. A. (1975) The concept of a linguistic variable and its application to approximate reasoning-1. *Information Sciences*, *8*, 199–249.

Zhang, H, Khan, A., Tan, D. K. Y., Luo, H. (2017) Rational water and management improves root growth, increases yield and maintains water use efficiency of cotton under mulch drip irrigation. *Frontiers in Plant Science*, *8*, 912.

Zwart, S. J., Bastiaanssen, W. G. M. (2004) Review of measured crop water productivity values for irrigated wheat, rice, cotton, and maize. *Agricultural water management*, *69*, 115–133.

5 A Novel Approach for Feature Detection in Vector Graphics

Karthik Jain[1], *Purvi Gujarathi*[2], *Priya Bannur*[3], *Pinak Wadilkar*[4] *and Pradnya V. Kulkarni*[5]
[1]IDeaS – A SAS Company, Pune, Maharashtra, India
[2]Persistent Systems, Pune, Maharashtra, India
[3]Viterbi School of Engineering, University of Southern California, Los Angeles, CA, United States
[4]Department of Electronics and Telecommunication Engineering, Pune Institute of Computer Technology, Pune, Maharashtra, India
[5]School of Computer Engineering & Technology, MIT World Peace University, Pune, Maharashtra, India

CONTENTS

5.1 INTRODUCTION

Feature detection is a low-level image processing operation, usually performed as the first step in a computer vision application. A feature is the primitive information of the image that helps to aid the computational task. It can be a point, edge or an object in the concerned image. The points at specific locations where the edges meet at some angle are known as corners. Corner detection is frequently used in 3D reconstruction [1], panorama stitching [2], image mosaicking [3], motion detection [4], etc.

DOI: 10.1201/9781003329039-5

The contemporary corner detection algorithms like Harris corner detector and Feature from Accelerated Segment Test (FAST) detector are designed to detect corners from raster images. The Harris corner detector calculates the locally averaged moment matrix computed from the image gradients, and then combines the eigenvalues of the moment matrix to compute a corner measure, from which maximum values indicate corners positions [5]. Some of the remarkable fields in which Harris corner detection was applied are Human eye detection [6], bone fracture detection from X-ray images [7]. The FAST detector is based on an accelerated segment test [8]. The machine learning approach in FAST makes it a high-speed feature detector. Simultaneous localization and mapping (SLAM) [9], augmented reality [10], and nutrient content measurement [11] are a few of the applications where the FAST corner detection algorithm was implemented.

Raster images are resolution dependent, where the intensity of the pixels is represented using bitmaps. The size of these bitmaps is directly dependent on the file size. As a result, raster images need more disk space. Image formats like Portable Network Graphics (PNG) and Joint Photographic Expert Group (JPEG) store the compressed version of images. Scaling up the bitmap makes it pixelated or blurred. Current implementations of real-time computer vision applications require resizing of images on multiple scales for accurate feature identification. These images of different resolutions are together known as an image pyramid [12]. The process of building image pyramids imposes a computational burden. In such cases raster images prove to be a disadvantage. A similar problem is highlighted in [13], where the authors intend to address the issue of computational complexity of Harris Algorithm on raster images for real-time frame-rate applications such as SLAM, here they have proposed a machine learning algorithm to remodel the feature detector in order to ameliorate the processing time.

On the other hand, vector graphics use sequential commands or mathematical statements to represent an image. This makes a vector image file comparatively smaller in size. Enlarging a vector image is equivalent to rendering a mathematical equation on a bigger dimension. Hence, vector images can be resized infinitely without affecting the image quality. In [14] the advantages of the vector graphics are accentuated, where they have targeted object localization and classification using a Graph Neural Network (GNN)-based pipeline that considers vector graphics as inputs. Vector graphics are now also being used for a variety application such as content-based image retrieval [15], image map tiles [16], SLAM [13]. Hence, there is a substantial scope for research considering this image format in computer vision applications. In this paper the authors have tried to address this need by leveraging the advantages of vector graphics and have proposed an approach to detect basic features like corners from scalable vector graphics (SVG) image with mathematical analysis, leading to improved accuracy and computational efficiency. SVG is an XML-based markup language for describing two-dimensional vector graphics.

In [17] the authors have come up with a novel model for detecting geometric shapes in a SVG image without converting it into its equivalent raster form. This model takes advantage of the mathematical characteristics of the SVG format and uses tree or graph cycle(s) for detecting closed-form objects in the image. Inspired by this, the algorithm proposed in this paper follows a mathematical approach for detection of corners in the SVG image. The XML file of SVG is thoroughly processed to extract the relevant information, which is further represented in the form of straight-line elements and curved elements. The intersections between these elements are found to determine the corners.

The remainder of this paper is organized as follows: Section 5.2 describes the mathematical formulation of the proposed algorithm; the results of testing the algorithm on the acquired dataset is presented in Section 5.3; and Section 5.4 concludes the research and presents a note on future directions of this paper.

5.2 MATHEMATICAL FORMULATION

This research work aims to reduce the computational complexity of traditional corner detection techniques by replacing raster images with vector images. The corner detection process is divided into five major steps, which are elaborated upon below, and a detailed example is illustrated in Appendix 5A.

Step 1. XML Preprocessing
The shape of an object is defined by the coordinate information stored in various SVG elements. SVG consists of predefined shape elements each having different attributes to store the shape properties. The elements can be broadly categorized as straight-line elements and curve elements. Rectangle <rect>, Line <line>, Polyline <polyline>and Polygon <polygon> are the basic straight-line elements, while Circle <circle> and Ellipse <ellipse> are the basic curve elements. The Path <path> element is a combination of both, the elements consisting of various straight-line commands like: M (moveto), L (lineto), H (horizontal lineto), V (vertical lineto) and Z (closepath) as well as curve commands like: C (curveto), S (smooth curveto), Q (quadratic Bézier curveto), T (smooth quadratic Bézier curveto) and A (elliptical Arc). The XML file is scanned for such elements and the relevant information is stored.

Step 2. Processing of path tag
The SVG paths allow different ways to represent coordinates on the image. The representation can be either Relative to other points or absolute in nature. The further steps of the algorithm require all coordinates to be of Absolute nature, hence prior conversion of the coordinate system is necessary. The SVG format is insensitive to inconsistent use of whitespaces and special characters. Therefore, the extracted data needs to be converted into a common standard for efficient processing. This is done using different text processing techniques based on regular expressions.

Step 3. Data extraction from SVG elements
The information from all straight-line elements is extracted and stored in a uniform format as a list of successive points. The center and radii information is stored from the basic curve elements. While, for the curve commands in path elements, the end points and control points are stored.

Step 4: Calculating intersection of lines and curves
Corners are the points at which two converging contour lines meet in their local neighborhood at some angle. Corners are points of the intersection of all straight and curve

line segments. The corners created due to the intersection of two straight line segments are calculated using the formula given in Equation (5.3).

Let the equations of the two lines be (written in the general form):

$$a_1x + b_1y + c_1 = 0$$
$$a_2x + b_2y + c_2 = 0$$
$$(x_0, y_0) = \left(\frac{b_1c_2 - b_2c_1}{a_1b_2 - a_2b_1}, \frac{c_1a_2 - c_2a_1}{a_1b_2 - a_2b_1}\right) \quad (5.1)$$

The corners created due to intersection of bezier curves with other straight lines as well as curves are found. This work uses the bezier Python package [18] for the computation. Similarly, the intersection of elliptical curves with other ellipses and straight lines is found. For this calculation, the authors used the SymPy Python package [19].

Step 5: Filtering the detected corners
The list of points found up to this step is filtered to remove duplicate points. Furthermore, the point of intersection of line segments is considered to be a corner if the converging lines make a significant angle with each other. This angle is calculated by using the slopes of the intersecting straight lines or tangents to intersecting curves as follows in Equation (5.2):

$$\theta = \left|\frac{m_2 - m_1}{1 + m_1m_2}\right| \quad (5.2)$$

The points making an angle equal to 0 degrees are discarded.

5.3 ANALYSIS OF RESULTS

5.3.1 Dataset Description

SVG images are primarily used for logo creation and graphical illustrations. The dataset was created considering these applications. The dataset is broadly classified into two types: manually created simple images and complex images taken from external sources. For every SVG image, the same image of PNG format was also stored. In total of 50 images were created for the following cases in SVG: images having a single element, images having a combination of different elements. For the complex type, 75 images were downloaded from [20].

5.3.2 Experimental Setup

The algorithm was coded in Python 3.7 on Anaconda 4.10.0 platform and was tested on Windows 10 operating system with Intel Core i5 2.5 GHz processor speed with 8 GB RAM. For efficient testing and comparative analysis of the algorithm, an automated testing platform was developed. The proposed algorithm as well as the Harris and FAST corner detection algorithms were executed on this platform. The results of the proposed algorithm were taken for a total of 125 SVG images and corresponding 125 PNG images for Harris and FAST corner detection algorithms.

The performance was analyzed using the following parameters: Accuracy and Execution time. The accuracy formula was devised considering Harris and FAST as benchmark algorithms. The objective of this research is to design an algorithm comparable in performance to these benchmark algorithms. The accuracy is based on the number of corners correctly matched with respect to the results of Harris and FAST algorithms, refer to Equation (5.3). The output corner coordinates of the proposed algorithm were compared to the output corner coordinates of Harris and FAST algorithm keeping a tolerance of 5 pixels. This threshold was decided empirically. PNG images of dimension 512 × 512 pixels having the corner of interest at a known coordinate were considered. Harris and FAST algorithms were run on these images. The deviation of the outputs of Harris and FAST as compared to the known coordinates was observed to be between 4 to 6 pixels. Therefore, the standard threshold was considered to be the mean of this deviation.

$$Accuracy = \frac{Number\ of\ corners\ exactly\ matched\ with\ Harris/FAST}{Total\ number\ of\ corners\ detected\ by\ Harris/FAST} \quad (5.3)$$

5.4 RESULTS

The following test cases were considered - Image having a single element, combination of different elements, complex images taken from external sources. On a total of 125 images the algorithm was tested, amongst it 5 images were of single element case, 45 images were of multiple element case, and the remaining 75 images were from external source. The result of execution for all the images for each of these cases is presented in Table 5.1. The results of accuracy with respect to Harris and FAST algorithms are summarized in terms of mean in Table 5.2.

TABLE 5.1
Accuracy and Execution Time for All Images

	Accuracy w.r.t. (%)		Execution Time (sec)		
Image No.	Harris	FAST	Proposed Algo.	Harris	FAST
			Single Element		
1	50.000	100.000	0.026	0.048	0.242
2	100.000	100.000	0.030	0.038	0.027
3	28.125	33.333	0.015	0.033	0.046
4	100.000	100.000	0.017	0.028	0.020
5	100.000	100.000	0.029	0.023	0.020
			Multiple Elements		
6	66.667	20.000	0.124	0.029	0.018
7	66.667	53.333	0.326	0.029	0.027
8	100.000	100.000	0.279	0.050	0.026
9	100.000	100.000	0.913	0.041	0.066
10	100.000	100.000	0.258	0.027	0.025
11	100.000	100.000	0.101	0.028	0.031

(continued)

TABLE 5.1 (Continued)
Accuracy and Execution Time for All Images

Image No.	Accuracy w.r.t. (%)		Execution Time (sec)		
	Harris	FAST	Proposed Algo.	Harris	FAST
12	0.000	0.000	0.017	0.035	0.018
13	100.000	100.000	0.023	0.026	0.023
14	57.143	66.667	0.018	0.029	0.019
15	66.667	66.667	0.023	0.032	0.028
16	100.000	100.000	0.022	0.034	0.078
17	77.778	61.538	0.033	0.028	0.027
18	100.000	100.000	0.027	0.029	0.018
19	85.714	100.000	0.022	0.029	0.018
20	0.000	8.000	0.020	0.025	0.022
21	100.000	100.000	0.017	0.028	0.018
22	100.000	100.000	0.019	0.026	0.017
23	100.000	100.000	0.031	0.024	0.020
24	100.000	100.000	0.020	0.023	0.019
25	71.429	80.000	0.023	0.027	0.025
26	90.000	100.000	0.047	0.024	0.016
27	33.333	18.182	0.613	0.033	0.018
28	83.333	24.138	0.323	0.020	0.017
29	75.000	50.000	0.767	0.020	0.017
30	100.000	26.667	0.755	0.026	0.018
31	75.000	33.333	1.205	0.076	0.030
32	100.000	41.667	0.515	0.022	0.014
33	88.889	100.000	0.017	0.022	0.012
34	51.724	50.000	0.023	0.034	0.035
35	88.889	88.235	0.022	0.022	0.023
36	92.308	95.238	0.030	0.030	0.025
37	85.714	100.000	0.028	0.044	0.021
38	100.000	100.000	0.035	0.039	0.019
39	37.500	46.154	0.038	1.469	0.029
40	43.333	41.176	0.029	0.041	0.031
41	100.000	100.000	0.031	0.038	0.019
42	45.455	48.889	0.026	0.039	0.023
43	69.231	52.941	0.034	0.078	0.022
44	60.000	37.500	0.269	0.030	0.028
45	100.000	96.774	0.093	0.048	0.025
46	76.190	70.370	0.049	0.037	0.029
47	84.848	72.973	0.079	0.035	0.025
48	100.000	100.000	0.082	0.041	0.025
49	100.000	100.000	0.048	0.037	0.023
50	76.786	69.474	0.141	1.785	0.047
External Source					
51	100.000	50.000	15.047	0.057	0.280
52	72.727	53.191	44.367	0.039	0.038

TABLE 5.1 (Continued)
Accuracy and Execution Time for All Images

	Accuracy w.r.t. (%)		Execution Time (sec)		
Image No.	Harris	FAST	Proposed Algo.	Harris	FAST
53	100.000	92.308	65.207	0.032	0.774
54	81.818	44.737	49.986	0.079	0.182
55	79.310	86.275	61.652	0.052	0.619
56	97.436	94.595	0.696	0.044	0.251
57	100.000	100.000	0.493	0.044	0.218
58	60.000	64.000	54.047	0.025	0.193
59	91.026	100.000	4.417	0.051	0.160
60	100.000	85.714	0.789	0.041	0.053
61	100.000	93.333	0.399	0.042	0.337
62	25.000	85.075	1.745	0.032	0.033
63	100.000	90.000	0.305	0.048	0.075
64	100.000	95.000	0.724	0.031	0.110
65	77.922	90.805	5.118	0.063	0.110
66	100.000	4.908	0.326	1.345	0.700
67	69.697	100.000	2.380	0.058	1.017
68	88.571	73.438	1.096	0.055	0.453
69	78.000	90.000	0.838	0.024	1.390
70	100.000	100.000	0.564	0.062	0.177
71	33.333	50.000	37.071	0.070	0.141
72	100.000	100.000	0.374	0.040	0.152
73	85.714	100.000	0.706	0.049	0.207
74	100.000	92.683	0.588	0.049	1.036
75	81.818	80.000	4.582	0.052	0.731
76	100.000	100.000	0.083	0.058	1.205
77	100.000	100.000	0.058	0.057	0.062
78	93.750	100.000	0.071	0.062	0.052
79	80.531	92.857	0.364	0.027	0.042
80	100.000	100.000	0.224	0.050	0.195
81	100.000	91.304	0.068	0.027	0.028
82	77.778	73.684	0.103	0.024	0.053
83	100.000	100.000	0.816	0.026	0.024
84	92.857	90.000	0.237	0.061	0.153
85	100.000	100.000	0.217	0.114	0.684
86	97.561	92.727	0.272	0.067	0.045
87	100.000	100.000	0.467	0.034	0.247
88	31.944	32.432	0.301	0.126	0.373
89	100.000	88.372	0.336	0.109	0.125
90	100.000	100.000	0.033	0.046	0.037
91	97.368	97.468	0.603	0.038	0.028
92	91.667	95.000	0.317	0.032	0.019
93	100.000	100.000	0.070	0.048	0.872
94	100.000	100.000	0.207	0.034	0.040

(continued)

TABLE 5.1 (Continued)
Accuracy and Execution Time for All Images

	Accuracy w.r.t. (%)		Execution Time (sec)		
Image No.	Harris	FAST	Proposed Algo.	Harris	FAST
95	92.982	78.125	0.460	0.028	0.029
96	90.244	92.308	0.352	0.045	0.044
97	100.000	100.000	0.354	0.065	0.056
98	78.873	96.000	0.282	1.285	0.136
99	87.500	88.889	0.345	0.040	0.041
100	75.862	87.500	0.183	0.047	0.054
101	20.000	100.000	0.052	0.046	0.185
102	52.941	91.667	6.897	0.034	0.034
103	16.667	58.065	5.289	0.993	0.088
104	100.000	100.000	3.694	0.042	0.181
105	71.429	67.568	2.456	0.039	0.162
106	89.474	84.848	18.274	0.044	0.064
107	37.500	25.000	4.637	0.039	0.276
108	33.333	42.188	13.687	0.034	0.130
109	60.870	50.000	0.077	0.036	0.103
110	47.222	88.235	1.437	0.047	0.089
111	100.000	79.310	7.121	0.082	0.254
112	69.231	93.333	3.629	0.038	0.223
113	82.222	100.000	3.092	0.055	0.917
114	0.000	0.000	0.233	0.037	0.286
115	33.333	100.000	2.442	0.053	0.767
116	48.485	96.875	4.775	0.043	0.273
117	37.037	89.474	1.877	0.043	0.287
118	34.615	100.000	2.516	0.058	0.295
119	100.000	90.909	7.045	0.046	0.209
120	100.000	76.471	3.384	0.044	0.485
121	100.000	100.000	8.088	0.156	0.421
122	66.667	74.419	1.448	0.051	0.033
123	20.000	84.000	6.251	0.025	0.006
124	38.462	100.000	6.248	0.059	0.011
125	25.000	88.889	0.322	0.123	0.007

TABLE 5.2
Mean of Accuracy

	Mean of Accuracy (%) w.r.t	
Case	Harris	FAST
Single Element	75.625	86.667
Multiple Elements	78.928	71.601
External Source	77.277	83.273

It is evident from Table 5.1 that in the Single Element case the accuracy for the majority of the images is 100% with respect to both the benchmark algorithms and the execution time of the proposed algorithm is comparable to both these algorithms. The mean of accuracy came out to be 75.63% for Harris algorithm and 86.67% for FAST algorithm. Hence, from this data it can be deduced that the SVG corner detection algorithm works exceptionally well for images having a single SVG element. As shown in Table 5.1, in the Multiple Element case the mean of accuracy for the Harris algorithm was 78.93% and the mean for FAST algorithm was 71.60%. For approximately 6-8 images the proposed algorithm took significantly more time than the benchmark algorithms. This was because of the complex nature of the images with a greater number of intersecting elements. For the rest of the images, the SVG corner detector completed the execution within comparable time with respect to Harris and FAST algorithms. Thus, the proposed algorithm proved to be akin to the benchmark algorithms in terms of performance. As presented in Table 5.1, in the External Source case the mean of accuracy was measured to be 77.23% for Harris algorithm and 83.27% for FAST algorithm. However, the execution of the proposed algorithm on the majority of the images took comparatively more time, since the images in this category were closer to real-world, making them extremely complex.

These results are also depicted in the form of visualization of Accuracy and Time taken for execution with respect to FAST and Harris Algorithm for all the cases in Figures 5.1 and 5.2. For visualizing the accuracy, histogram was used and execution time was visualized using scatter plot.

Figure 5.3 shows output examples of the SVG corner detector, Harris and FAST algorithm. From the first image it is evident that the proposed algorithm detected almost all corners correctly, while FAST could detect only a few corners in the low intensity areas and Harris could detect none. As the SVG corner detector works on mathematical equations, there is no effect of intensity change in the working of the algorithm. The output of the second image shows that the SVG corner detector identifies a single coordinate for each corner point. On the other hand, FAST and Harris algorithms detect multiple coordinates close to each other and are overlapped. The SVG corner detector could detect the corners on sharp curves in the third image, while FAST and Harris failed to do so.

Hence, by considering Tables 5.1, 5.2 and Figures 5.1, 5.2, 5.3, it can be concluded that the proposed algorithm works comparable to Harris and FAST in detecting corners accurately and efficiently. Also, it can be stated that there are some advantages of the proposed model over the traditional ones.

5.5 CONCLUSION AND FUTURE SCOPE

This paper proposes a novel algorithm for corner detection from vector graphics. Although almost all computer vision applications are done using raster images, many techniques require high computation and usually include complex processes. In such cases, this algorithm provides an advantage by reducing the computational power

Case	Accuracy Histogram
Single Element	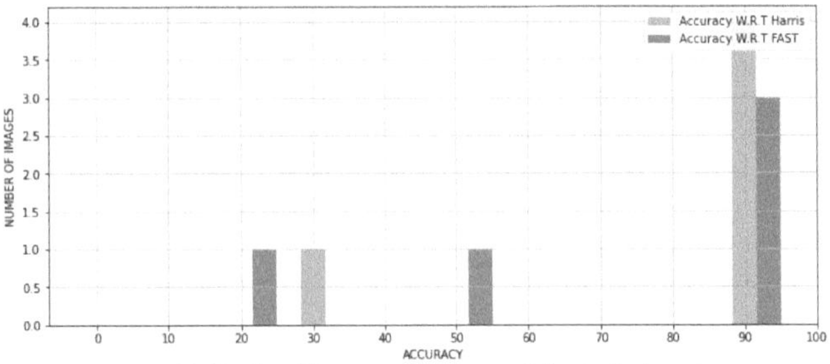 (a) Number of images vs Accuracy graph for single element case
Multiple Elements	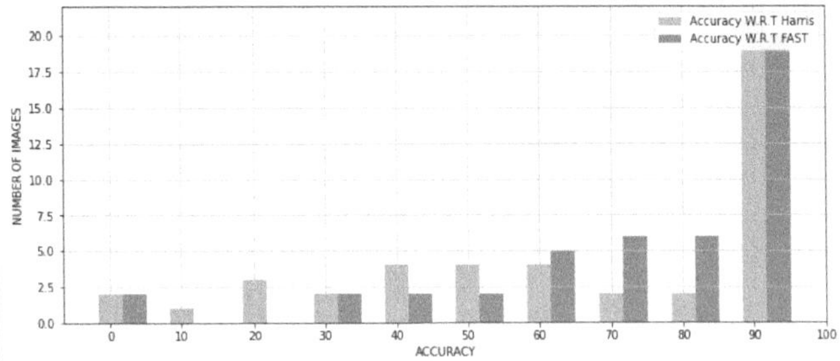 (b) Number of images vs Accuracy graph for multiple elements case
External Source	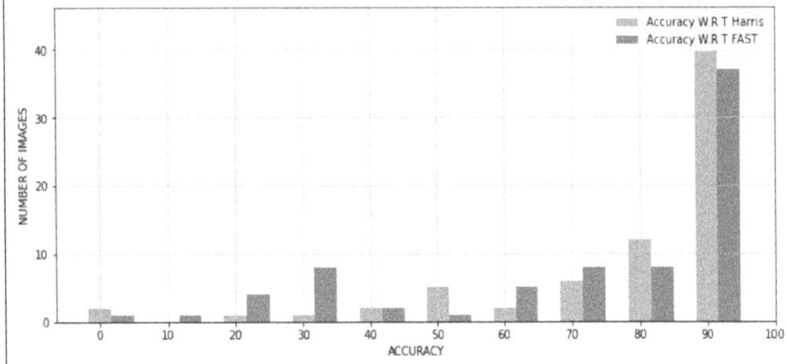 (c) Number of images vs Accuracy graph for external source case

FIGURE 5.1 Data visualization of number of images vs accuracy for different cases

through utilizing the advantages of vector graphics. It was compared with two traditional feature detection algorithms, FAST and Harris corner detector on the basis of accuracy and time taken. This implementation was successfully tested for the following test cases: image having a single element, images with a combination of different elements, and complex images taken from external sources. The results were such to conclude that the proposed algorithm worked exceptionally well on most of the simple and complex images closer to the real-world photos.

Feature detection in vector graphics is an underexplored domain in the field of computer vision. This research lays the foundation for this domain. The authors believe

<table>
<tr><th>Case</th><th>Execution Time Scatter Plot</th></tr>
<tr><td>Single Element</td><td>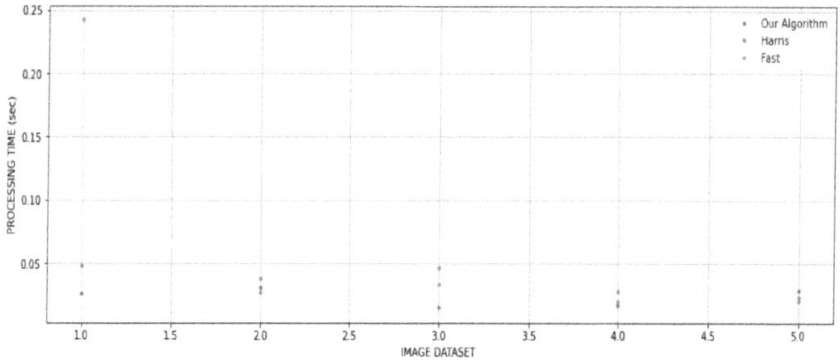

(a) Processing time vs number of images scatterplot for single element case</td></tr>
<tr><td>Multiple Elements</td><td>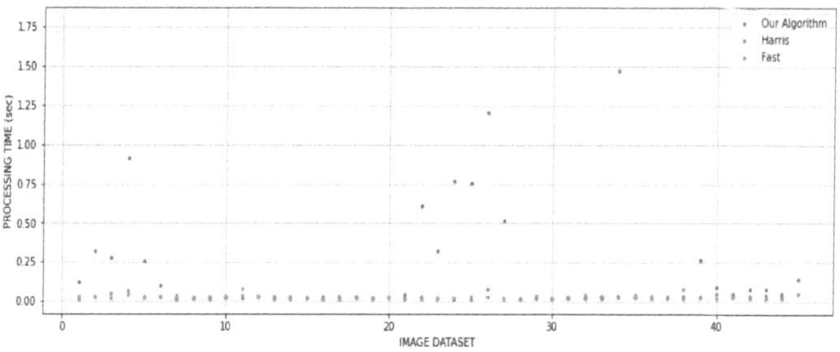

(b) Processing time vs number of images scatterplot for multiple elements case</td></tr>
<tr><td>External Source</td><td>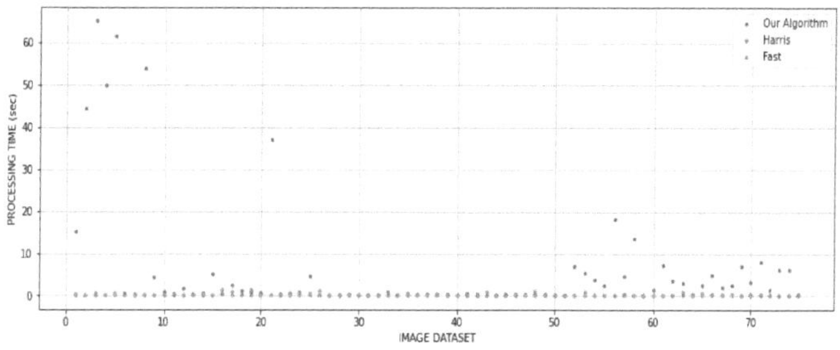

(c) Processing time vs number of images scatterplot for external source case</td></tr>
</table>

FIGURE 5.2 Data visualization of processing time vs number of images for different cases

that the proposed algorithm can be optimized further to increase its efficiency and accuracy.

ACKNOWLEDGEMENTS

This research was supported by Collaborative Research and Development Forum (CRDF). The authors would like to acknowledge them for providing insights, expertise, and guidance that greatly assisted the research.

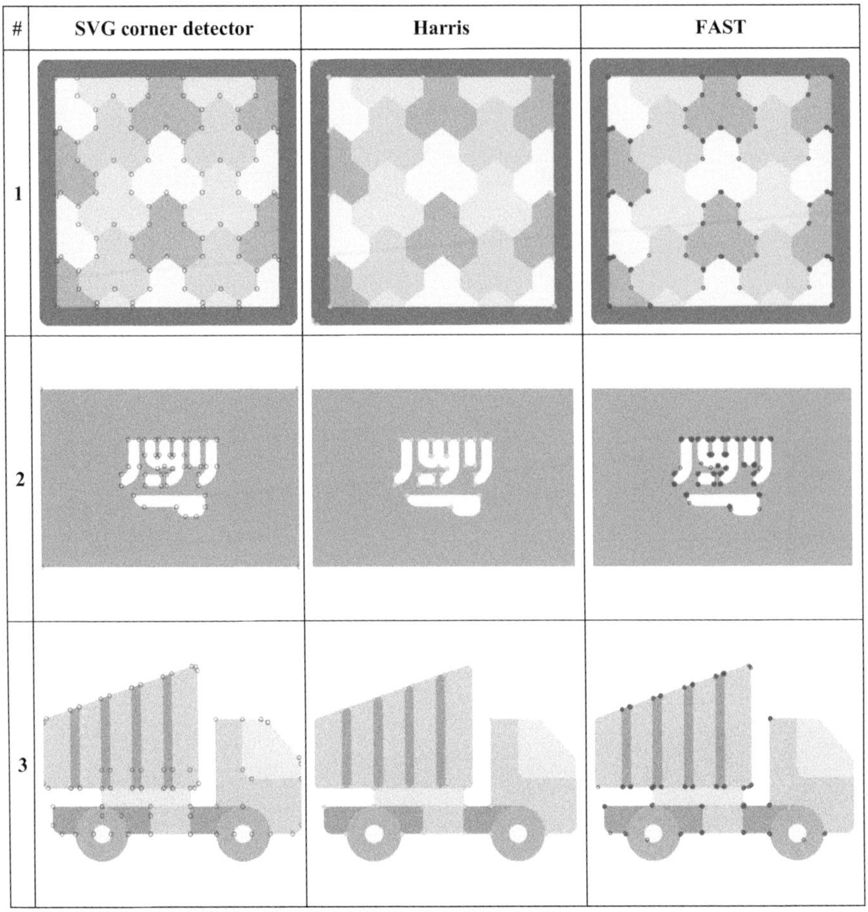

FIGURE 5.3 Output examples

APPENDIX 5A: AN EXAMPLE OF MATHEMATICAL PROCESSING OF XML OF A SVG IMAGE

The XML format of a SVG image consisting of a basic shape rectangle defined by "rect" element and a free form shape defined by "path" element is described below:

```
<svg>
<g>
<rect width="300" height="100" style="fill:rgb(0,0,255);stroke-width:3;stroke:rgb
(0,0,0)"/>
<path d="M 10 80 C 40 10, 65 10, 95 80 S 150 150, 180 80" stroke="black"
fill="None"/>
</g>
</svg>
```

Step 1: The above XML is processed to remove all the inconsequential information and only the shape defining tags are extracted.

Output after Step 1:

Preprocessed XML:

<rect width="300" height="100" style="fill:rgb(0,0,255);stroke-width:3;stroke: rgb(0,0,0)"/>

<path d="M 10 80 C 40 10, 65 10, 95 80 S 150 150, 180 80" stroke="black" fill="None"/>

Step 2: For path tags the extracted data is converted into a common standard using different text processing techniques.

Output after Step 2

Pre-processed path data:

[['m', '10', '80', 'c', '40', '10', '65', '10', '95', '80', 's', '150', '150', '180', '80']]

Step 3: The relevant information from these preprocessed data is extracted and stored as a list of points. For straight line element the successive points are stored, for curve elements the center and radii points are stored and for the curve commands in path elements, the end points and control points are stored.

Output after Step 3:

Straight line segments:

[[0, 0, 300.0, 0], [300.0, 0, 300.0, 100.0], [300.0, 100.0, 0, 100.0], [0, 100.0, 0, 0]]

Curved line segments:

[[10.0, 80.0, 40.0, 10.0, 65.0, 10.0, 95.0, 80.0], [95.0, 80.0, 125.0, 150.0, 150.0, 150.0, 180.0, 80.0]]

Step 4: The intersection points of these extracted lines and curves are calculated to determine the corner points.

Output after Step 4:

Detected Corners:

[(10.0, 80.0), (95.0, 80.0), (95.0, 80.0), (180.0, 80.0), (0, 0), (300.0, 0), (300.0, 0.0), (0.0, 0.0), (300.0, 0), (300.0, 100.0), (300.0, 100.0), (300.0, 100.0), (0, 100.0), (0.0, 100.0), (0, 100.0), (0, 0), (104.43584348312935, 100.0), (170.56415651687064, 100.0), (95.0, 80.0)]

Step 5: These calculated intersection points are further processed to find the true and exact corners. All the duplicate points are removed. Furthermore, intersection points which are making a significant angle are only considered.

Output after Step 5:

Filtered Corners:

[(10.0, 80.0), (95.0, 80.0), (180.0, 80.0), (0, 0), (300.0, 0), (300.0, 100.0), (0, 100.0), (104.43584348312935, 100.0), (170.56415651687064, 100.0)]

REFERENCES

1. Varkonyi-Koczy, A.R., 2008. Fuzzy logic supported corner detection. Journal of Intelligent & Fuzzy Systems, 19(1), pp. 41–50.

2. Mistry, S. and Patel, A., 2016. Image stitching using Harris feature detection. International Research Journal of Engineering and Technology (IRJET), 3(4), pp. 2220–6.
3. Ghosh, D. and Kaabouch, N., 2016. A survey on image mosaicing techniques. Journal of Visual Communication and Image Representation, 34, pp. 1–11.
4. Wang, H. and Brady, M., 1995. Real-time corner detection algorithm for motion estimation. Image and Vision Computing, 13(9), pp. 695–703.
5. Harris, C. and Stephens, M., 1988, August. A combined corner and edge detector. In Alvey Vision Conference (Vol. 15, No. 50, pp. 10–5244).
6. Das, A. and Ghoshal, D., 2012, December. Human eye detection of color images based on morphological segmentation using modified Harris corner detector. In 2012 International Conference on Emerging Trends in Electrical Engineering and Energy Management (ICETEEEM) (pp. 143–147). IEEE.
7. Basha, C.Z., Reddy, M.R.K., Nikhil, K.H.S., Venkatesh, P.S.M. and Asish, A.V., 2020, March. Enhanced computer aided bone fracture detection employing x-ray images by Harris Corner technique. In 2020 Fourth International Conference on Computing Methodologies and Communication (ICCMC) (pp. 991–995). IEEE.
8. Rosten, E. and Drummond, T., 2006, May. Machine learning for high-speed corner detection. In European Conference on Computer Vision (pp. 430–443). Springer, Berlin, Heidelberg.
9. Adam, N., Purnamasari, D. and Ibrahim, A., 2019, May. Implementation of Object Tracking Augmented Reality Markerless using FAST Corner Detection on User Defined-Extended Target Tracking in Multivarious Intensities. In Journal of Physics: Conference Series (Vol. 1201, No. 1, p. 012041). IOP Publishing.
10. Andriyandi, A.P., Darmalaksana, W., Adillah Maylawati, D.S., Irwansyah, F.S., Mantoro, T. and Ramdhani, M.A., 2020. Augmented reality using features accelerated segment test for learning tajweed. Telkomnika (Telecommunication Computing Electronics and Control), 18(1), pp. 208–216, doi: 10.12928/TELKOMNIKA. V18I1. 14750.
11. Sumiharto, R., Putra, R.G. and Demetouw, S., 2020. Methods for Determining Nitrogen, Phosphorus, and Potassium (NPK) Nutrient Content Using Features from Accelerated Segment Test (FAST). International Journal of Computer Science and Software Engineering, 9(1), pp. 1–5.
12. Adelson, E.H., Anderson, C.H., Bergen, J.R., Burt, P.J. and Ogden, J.M., 1984. Pyramid methods in image processing. RCA Engineer, 29(6), pp. 33–41.
13. Rosten, E. and Drummond, T., 2006, May. Machine learning for high-speed corner detection. In European conference on computer vision (pp. 430–443). Springer, Berlin, Heidelberg
14. Jiang, X., Liu, L., Shan, C., Shen, Y., Dong, X. and Li, D., 2021. Recognizing vector graphics without rasterization. Advances in Neural Information Processing Systems, 34, pp. 24569–24580.
15. Jiang, K., Fang, Z., Ge, Y. and Zhou, Y., 2007, October. Information retrieval through SVG-based vector images using an original method. In IEEE International Conference on e-Business Engineering (ICEBE'07) (pp. 183–188). IEEE
16. Netek, R., Masopust, J., Pavlicek, F. and Pechanec, V., 2020. Performance testing on vector vs. raster map tiles—comparative study on load metrics. ISPRS International Journal of Geo-Information, 9(2), p. 101.
17. Seel-audom, C., Naiyapo, W. and Chouvatut, V., 2017, February. A search for geometric-shape objects in a vector image: Scalable Vector Graphics (SVG) file format. In 2017 9th International Conference on Knowledge and Smart Technology (KST) (pp. 305–310). IEEE.

18. Hermes, D., 2017. Helper for Bézier curves, triangles, and higher order objects. Journal of Open Source Software, 2(16), p. 267.
19. Meurer, A., Smith, C.P., Paprocki, M., Čertík, O., Kirpichev, S.B., Rocklin, M., Kumar, A., Ivanov, S., Moore, J.K., Singh, S. and Rathnayake, T., 2017. SymPy: symbolic computing in Python. Peer Journal of Computer Science 3, p. e103.
20. Flaticon 2013, Europe, accessed on 1 February 2021, https://www.flaticon.com/

6 On Uncertain Matrix Games Involving Linguistic Pythagorean Fuzzy Sets

Deeba R. Naqvi[1,2] *and Geeta Sachdev*[1]
[1]Indira Gandhi Delhi Technical University for Women, Delhi, India
[2]Maharaja Surajmal Institute of Technology, C-4 Janakpuri, New Delhi, India

CONTENTS

6.1 INTRODUCTION

Zadeh (1965) employed membership degrees to describe fuzzy sets, which solely express measure of affiliation to the fuzzy set. Atanassov (1986) proposed the hypothesis of intuitionistic fuzzy (I-fuzzy) sets by incorporating non-membership degrees to complement membership degrees in fuzzy sets and reporting equivocal perspectives not only by the degree of belongingness but also by non-belongingness functions. The degree of hesitancy/unreliability for an I-fuzzy set is described together with the duo of degree of membership and non-membership of an estimated set. Such sets have been widely used in a variety of real-world problems by Szmidt and Kacprzyk (1996); Atanassov (1999); Bustince et al. (2000); De et al. (2001); Li and Chuntian (2002); Hung and Yang (2008); Li (2010b); Zhang et al. (2017), to name a few. Under certain circumstances the total value of belongingness (membership)

DOI: 10.1201/9781003329039-6

and non-belongingness (non-membership) degrees may be greater than one. For instance, the expert's priority for evaluating possibilities on a qualification in terms of belongingness is 0.65, whereas the grade for non-belongingness is 0.55. It may be clearly revealed that $0.65 + 0.55 \nleq 1$. Consequently, I-fuzzy sets are found to be incompetent for comprehending this scenario. Thus to overcome the shortcomings, Yager (2013, 2014) pioneered the Pythagorean fuzzy set (PFS), an augmentation of an I-fuzzy set, with the qualification that the sum of the square of its membership and non-membership degrees be less than or equivalent to one. Apparently, the predetermined sample demonstrates that $(0.65)^2 + (0.55)^2 \leq 1$. PFSs have been thoroughly investigated by various researchers, including Chen (2019); Garg (2020); Peng and Selvachandran (2019); Zeng et al. (2016), owing to its exemplary structure. Later, a variety of formulations were developed, including interval-valued PFS by Peng and Yang (2016); Garg (2016), hesitant PFS by Garg (2018a), and so forth.

Considering the fact that real-world decision-making circumstances are often more uncertain and complex, employing linguistic variables (LVs) instead of quantitative values further seems to be realistic. The paradigm of LVs were originally developed by Zadeh (1975a,b) to predominantly express the uncertain, indeterminate, or conflicting findings. Due to this reason, Geng et al. (2017) drafted the concept of Pythagorean fuzzy uncertain linguistic variables by associating uncertain linguistic entities with PFSs. Garg (2018b) suggested the concept of a linguistic Pythagorean fuzzy number (LPFN) and categorized a variety of aggregation operators on multi-criteria decision-making situations. For interval-valued 2-tuple LPFS scenarios, Wang et al. (2018) provided a strategy for handling multi-attribute assessment performing challenges. Deng et al. (2019) established a method for solving multi-attribute group decision-making problems utilizing the Heronian mean operator under the 2-tuple linguistic PFS environment.

Game theory is an influential assessment technique for resolving everyday issues. In some ways, this seems to be strategic planning engineering, or the best decision-making of autonomous yet competing participants/players in an estimated situation. The most influential designer for game theory was Neumann and Morgenstern (1944). Numerous academicians had already investigated game theory in depth, including Zhang and Zhang (2003); Leng and Parlar (2005); Collins and Hu (2008); Kalpiński and Tamošaitiene (2010); Madani (2010); Zhao et al. (2010); Barron (2013). Butnariu (1978); Aubin (1981) pioneered the extensive analysis of fuzzy game theory in real-life scenarios due to uncertainty or ambiguity in understanding of a situation. Thereafter, Campos (1989); Sakawa and Nishizaki (1994); Maeda (2003); Bector et al. (2004a,b); Vijay et al. (2007) incorporated fuzzy game theory into further research.

Atanassov (1995) was the first to provide a thorough assessment of matrix games having payoffs characterized using I-fuzzy sets. For finding the solutions of matrix games with I-fuzzy payoffs, Li and Nan (2009) proposed the non-linear programming methodology. Aggarwal et al. (2012a,b) applied I-fuzzy linear programming duality to zero-sum matrix games, henceforth advancing on fuzzy duality results from Bector et al. (2004b); Vijay et al. (2005). Later, Khan et al. (2017) devised a method for resolving indeterminacy throughout matrix games with I-fuzzy goals. Naqvi et al. (2021) recently described a new Tanaka and Asai (1984) approach for describing the numerical solution to I-fuzzy matrix games. So far, considerable attempts to secure strategies for matrix games characterized via I-fuzzy sets have been undertaken by Zhou et al. (2010); Li (2010a); Seikh et al. (2013, 2015); Khan et al. (2016); Bhaumik et al. (2017); Nan et al. (2010, 2017); Xia (2018); Xing and Qiu (2019); Bhaumik and Roy (2021), to list a few.

There is, however, very little research on matrix games incorporating Pythagorean fuzzy variables in the literature. Ding and Liu (2019) proposed an extended zero-sum game approach based on the best-worst method and Pythagorean fuzzy uncertain linguistic variable sets for solving emergency decision-making problems. Also, Han et al. (2020) applied interval-valued Pythagorean prioritized an operator-based game theoretical framework to study multi-criteria decision-making problems. In a recent study, Jana and Roy (2022) introduced a new concept of linguistic Pythagorean hesitant fuzzy set and employed it to study multi-criteria decision-making problems.

6.1.1 Inspiration and Significance for Suggested Methodology

However, in many realistic circumstances, rather than using numeric expressions, the payoffs of an uncertain matrix game are investigated using uncertain linguistic terms. Thus explaining the qualitative aspect of information. Many eminent researchers have significantly studied the uncertain entities including LVs and their applications. To manage real life situations, where the human assessment is based more on qualitative assessment than quantitative, Garg (2018b) introduced the concept of the linguistic Pythagorean fuzzy set (LPFS), which is described via linguistic membership degree and non-membership degree, so that their sum of squares is less than the cardinality of the linguistic set, while preserving the benefits of both PFS and LVs. Thus, it is employed as an economical device to survey the multi-attribute decision-taking challenges. This LPFS describes both the qualitative aspect of information together with its quantitative aspect. Furthermore, getting inspired by Li (2010a), the investigation on matrix games involving payoffs defined by linguistic intuitionistic fuzzy numbers was outlined by Verma and Aggarwal (2021). Thus, this tool has been extensively adopted to even resolve the solutions of matrix games, where uncertain and imprecise data are known. One of the greatest advantages of using Linguistic interval-valued Atanassov's intuitionistic fuzzy data is that while processing the information during the solution procedure there is no dissipation or deformation of the information.

Drawing inspiration from aforementioned studies, we attempt to implement this advancement to resolve uncertain matrix games with payoffs characterized via LPFSs, as this has yet to be investigated. Through employing weighted mean aggregation operators for LPFSs and the Li (2010a) approach on matrix games comprising LPFSs, we contribute to the reduction of the two-player problems into a couple of linear bi-objective programming problems. As a result, both players, player I and player II, are able to acquire ideal game values along with mixed strategies.

The significance of the proposed effort are:

1. The technique entails analyzing a solution procedure for an uncertain matrix game where the payoffs are characterized via LPFNs and thus describing the game's expectation in terms of an LPFN.

2. Through employing LPFSs, we can precisely define one's belongingness and non-belongingness. Therefore, during the process of resolving these uncertain matrix games involving LPFSs, there is marginal possibility of data leakage or deformation, enabling players to determine the best recourse in the end.
3. Well-defined comparative parameters and aggregation procedures are utilized to evaluate and consolidate the payoffs of the LPFSs, which aids in information processing.
4. The suggested framework feasibly acquires results to matrix games with LPFSs payoffs through addressing a pair of primal-dual linear programming problems. The outcomes imply that the suggested methodology may effectively illustrate the extent to which one player is preferable to the other in respect of competitive winning.

The remainder of this chapter is put together as follows: Within Section 6.2, we examine several fundamental LPFS terminologies and contemporary notions. The precise nature of matrix games incorporating LPFS data, as well as the resolution approach for the two players, are stated in Section 6.3. Section 6.4 offers a mathematical illustration to demonstrate the efficacy of the proposed methodology. Section 6.5 examines the current work in comparison to the literature. Section 6.6 in this chapter includes necessary findings and a discussion of the recommended approach's forthcoming implications.

6.2 DEFINITIONS AND PROPERTIES

This section inspects the definitions of intuitionistic, Pythagorean, and linguistic Pythagorean fuzzy sets along with their distinguishing characteristics. In the following discussion, the finite universe of discourse is represented through the set $X = \{x_1, x_2, \dots, x_n\}$.

Definition 6.2.1. Intuitionistic Fuzzy Set (I-fuzzy set) Atanassov (1986)
The I-fuzzy set $\mathfrak{I}$ in X is represented by

$$\mathfrak{I} = \{\langle x, \delta_{\mathfrak{I}}(x), \eta_{\mathfrak{I}}(x)\rangle |\ \delta_{\mathfrak{I}}(x) + \eta_{\mathfrak{I}}(x) \leq 1,\ x \in X\},$$

where $\delta_{\mathfrak{I}} : X \to [0, 1]$ and $\eta_{\mathfrak{I}} : X \to [0, 1]$ describes the degree of membership (DM) and degree of non-membership (DNM) functions, respectively.

In real-world circumstances, there are numerous decision-making challenges that exhibit qualitative aspects that are hard to conclude via numerical values. To overcome this complexity, Herrera and Martinez (2000, 2001) undertook the study of LVs.

Definition 6.2.2. Linguistic Term Set (LTS)
The linguistic term set $\mathbb{S} = \{\mathfrak{s}_0, \mathfrak{s}_1, \dots, \mathfrak{s}_t\}$ should be totally ordered discrete linguistic term set (LTS) with an odd cardinality, where t should be an even positive integer. The LTS should fulfill the below mentioned properties:

1. $\mathfrak{s}_{\mathfrak{a}} \leq \mathfrak{s}_{\mathfrak{b}} \Leftrightarrow \mathfrak{a} \leq \mathfrak{b}$;

2. $neg(\mathfrak{s}_c) = \mathfrak{s}_{t-c}$;
3. $\max(\mathfrak{s}_{\mathfrak{a}}, \mathfrak{s}_{\mathfrak{b}}) = \mathfrak{s}_{\mathfrak{a}} \Leftrightarrow \mathfrak{a} \geq \mathfrak{b}$;
4. $\min(\mathfrak{s}_{\mathfrak{a}}, \mathfrak{s}_{\mathfrak{b}}) = \mathfrak{s}_{\mathfrak{a}} \Leftrightarrow \mathfrak{a} \leq \mathfrak{b}$.

For instance, $\mathbb{S}$ may be described as:

$\mathbb{S} = \{\mathfrak{s}_0$: ignorant, $\mathfrak{s}_1$: slightly confused, $\mathfrak{s}_2$: confused, $\mathfrak{s}_3$: unintelligent, $\mathfrak{s}_4$: moderately intelligent, $\mathfrak{s}_5$: intelligent, $\mathfrak{s}_6$: more intelligent, $\mathfrak{s}_7$: most intelligent, $\mathfrak{s}_8$: genius.

Xu et al. (2017) extended the discontinuous linguistic term set $\mathbb{S} = \{\mathfrak{s}_0, \mathfrak{s}_1, \dots, \mathfrak{s}_t\}$ to a continuous term set $\mathbb{S}_{[0,t]} = \{\mathfrak{s}_{\mathfrak{p}} | \ \mathfrak{s}_0 \leq \mathfrak{s}_{\mathfrak{p}} \leq \mathfrak{s}_t, \mathfrak{p} \in [0,t]\}$ to preserve the required information. Hence, all the entities of the set $\mathbb{S}_{[0,t]}$ display each characteristics of the set $\mathbb{S}$. If $\mathfrak{s}_{\mathfrak{p}} \in \mathbb{S}$, then $\mathfrak{s}_{\mathfrak{p}}$ is the real linguistic term, otherwise $\mathfrak{s}_{\mathfrak{p}}$ is the virtualized linguistic term.

Definition 6.2.3. Pythagorean Fuzzy Set (PFS) Yager (2013, 2014)
The PFS $\mathcal{A}$ in X is described as

$$\mathcal{A} = \{\langle x, \zeta_{\mathcal{A}}(x), \nu_{\mathcal{A}}(x)\rangle | \ \zeta_{\mathcal{A}}^2(x) + \nu_{\mathcal{A}}^2(x) \leq 1, \ x \in X\},$$

for each $x \in X$, $\zeta_{\mathcal{A}} : X \to [0,1]$ and $\nu_{\mathcal{A}} : X \to [0,1]$ defines the DM and DNM functions, respectively. The measure of indeterminacy can be characterized by $\pi_{\mathcal{A}}(x) = 1 - \zeta_{\mathcal{A}}^2(x) - \nu_{\mathcal{A}}^2(x)$. For notational convenience, we denote Pythagorean fuzzy number by $\beta = \langle \zeta_{\mathcal{A}}, \nu_{\mathcal{A}}\rangle$, such that $\zeta_{\mathcal{A}}^2 + \nu_{\mathcal{A}}^2 \leq 1$, $\zeta_{\mathcal{A}}$, $\eta_{\mathcal{A}} \in [0,1]$.

6.2.1 Linguistic Pythagorean Fuzzy Set (LPFS)

The interval-valued or 2-tuple LVs reveal qualitative evaluation of information presented by the decision-makers. However, LPFSs established by Garg (2018b) describe not only the decision-maker's quantitative but also qualitative evaluation of information. Due to this reason, we tend to express the payoff matrix of the game with the linguistic Pythagorean fuzzy set.

Definition 6.2.1.4. Let $\mathbb{S}_{[0,t]}$ be a continuous LTS, then LPFS $\tilde{\mathrm{A}}$ in X is expressed as

$$\tilde{\mathrm{A}} = \{\langle x, \mathfrak{s}_{\phi_{\tilde{\mathrm{A}}}}(x), \ \mathfrak{s}_{\psi_{\tilde{\mathrm{A}}}}(x)\rangle \ | \ x \in X\},$$

where $\mathfrak{s}_{\phi_{\tilde{\mathrm{A}}}}(x), \mathfrak{s}_{\psi_{\tilde{\mathrm{A}}}}(x) \in \mathbb{S}_{[0,t]}$ illustrates DM and DNM, respectively. Specifically, $\phi_{\tilde{\mathrm{A}}}^2 + \psi_{\tilde{\mathrm{A}}}^2 \leq t^2$ always exists. In general, the LPFS is expressed as $\tilde{\mathrm{A}} = \langle \mathfrak{s}_{\phi_{\tilde{\mathrm{A}}}}, \mathfrak{s}_{\psi_{\tilde{\mathrm{A}}}}\rangle$ and is also called the linguistic Pythagorean fuzzy value (LPFV) or number.

Properties of Linguistic Pythagorean Fuzzy Sets
Garg (2018b) has defined the notion of LPFS in detail. We present some of the properties of LPFVs below. For LPFVs $\tilde{\mathrm{A}} = \langle \mathfrak{s}_{\phi_{\tilde{\mathrm{A}}}}, \ \mathfrak{s}_{\psi_{\tilde{\mathrm{A}}}}\rangle$ and $\tilde{\mathrm{C}} = \langle \mathfrak{s}_{\phi_{\tilde{\mathrm{C}}}}, \ \mathfrak{s}_{\psi_{\tilde{\mathrm{C}}}}\rangle$

1. $\tilde{\mathrm{A}} = \tilde{\mathrm{C}}$ iff $\mathfrak{s}_{\phi_{\tilde{\mathrm{A}}}} = \mathfrak{s}_{\phi_{\tilde{\mathrm{C}}}}$ and $\mathfrak{s}_{\psi_{\tilde{\mathrm{A}}}} = \mathfrak{s}_{\psi_{\tilde{\mathrm{C}}}}$;

2. $\tilde{\mathbf{A}} < \tilde{\mathbf{C}}$ iff $\mathfrak{s}_{\phi_{\tilde{A}}} < \mathfrak{s}_{\phi_{\tilde{C}}}$ and $\mathfrak{s}_{\psi_{\tilde{A}}} > \mathfrak{s}_{\psi_{\tilde{C}}}$;
3. $\tilde{\mathbf{A}}^{c} = \left\langle \mathfrak{s}_{\psi_{\tilde{A}}}, \mathfrak{s}_{\phi_{\tilde{A}}} \right\rangle$, where $\tilde{\mathbf{A}}^{c}$ denotes the complement of $\tilde{\mathbf{A}}$;
4. $\tilde{\mathbf{A}} \cup \tilde{\mathbf{C}} = \left\langle \max\left\{\mathfrak{s}_{\phi_{\tilde{A}}}, \mathfrak{s}_{\phi_{\tilde{C}}}\right\},\ \min\left\{\mathfrak{s}_{\psi_{\tilde{A}}}, \mathfrak{s}_{\psi_{\tilde{C}}}\right\}\right\rangle$;
5. $\tilde{\mathbf{A}} \cap \tilde{\mathbf{C}} = \left\langle \min\left\{\mathfrak{s}_{\phi_{\tilde{A}}}, \mathfrak{s}_{\phi_{\tilde{C}}}\right\},\ \max\left\{\mathfrak{s}_{\psi_{\tilde{A}}}, \mathfrak{s}_{\psi_{\tilde{C}}}\right\}\right\rangle$;

Further, the fundamental algebraic operations are as follows:

1. $$\tilde{\mathbf{A}} \oplus \tilde{\mathbf{C}} = \left\langle \mathfrak{s}_{t\sqrt{\left(\frac{\phi_{\tilde{A}}^2}{t^2} + \frac{\phi_{\tilde{C}}^2}{t^2} - \frac{\phi_{\tilde{A}}^2 \phi_{\tilde{C}}^2}{t^4}\right)}}, \mathfrak{s}_{t\left(\frac{\psi_{\tilde{A}}\psi_{\tilde{C}}}{t^2}\right)} \right\rangle;$$

2. $$\tilde{\mathbf{A}} \otimes \tilde{\mathbf{C}} = \left\langle \mathfrak{s}_{t\left(\frac{\phi_{\tilde{A}}\phi_{\tilde{C}}}{t^2}\right)}, \mathfrak{s}_{t\sqrt{\left(\frac{\psi_{\tilde{A}}^2}{t^2} + \frac{\psi_{\tilde{C}}^2}{t^2} - \frac{\psi_{\tilde{A}}^2 \psi_{\tilde{C}}^2}{t^4}\right)}} \right\rangle;$$

3. $$\lambda\tilde{\mathbf{A}} = \left\langle \mathfrak{s}_{t\sqrt{1-\left(1-\frac{\phi_{\tilde{A}}^2}{t^2}\right)^{\lambda}}}, \mathfrak{s}_{t\left(\frac{\psi_{\tilde{A}}}{t}\right)^{\lambda}} \right\rangle, \text{ for a real number } \lambda > 0;$$

4. $$\tilde{\mathbf{A}}^{\lambda} = \left\langle \mathfrak{s}_{t\left(\frac{\phi_{\tilde{A}}}{t}\right)^{\lambda}}, \mathfrak{s}_{t\sqrt{1-\left(1-\frac{\psi_{\tilde{A}}^2}{t}\right)^{\lambda}}} \right\rangle, \text{ for a real number } \lambda > 0.$$

As observed by Garg and Kumar (2019), for any LPFVs $\tilde{\mathbf{A}}$ and $\tilde{\mathbf{C}}$, the algebraic operations defined above are also LPFVs. In addition, for $\tilde{\mathbf{A}} = \left\langle \mathfrak{s}_{\phi_{\tilde{A}}}, \mathfrak{s}_{\psi_{\tilde{A}}} \right\rangle$, $\tilde{\mathbf{A}}_i = \left\langle \mathfrak{s}_{\phi_{\tilde{A}_i}}, \mathfrak{s}_{\psi_{\tilde{A}_i}} \right\rangle$, $i = 1, 2$, and for real values $\lambda, \lambda_1, \lambda_2 > 0$, the following holds:

1. $\tilde{\mathbf{A}}_1 \oplus \tilde{\mathbf{A}}_2 = \tilde{\mathbf{A}}_2 \oplus \tilde{\mathbf{A}}_1$;
2. $\lambda(\tilde{\mathbf{A}}_1 \oplus \tilde{\mathbf{A}}_2) = \lambda\tilde{\mathbf{A}}_1 \oplus \lambda\tilde{\mathbf{A}}_2$;
3. $\lambda_1\tilde{\mathbf{A}} \oplus \lambda_2\tilde{\mathbf{A}} = (\lambda_1 + \lambda_2)\tilde{\mathbf{A}}$;
4. $\tilde{\mathbf{A}}_1 \otimes \tilde{\mathbf{A}}_2 = \tilde{\mathbf{A}}_2 \otimes \tilde{\mathbf{A}}_1$;
5. $\tilde{\mathbf{A}}^{\lambda_1} \otimes \tilde{\mathbf{A}}^{\lambda_2} = \tilde{\mathbf{A}}^{\lambda_1+\lambda_2}$;
6. $\tilde{\mathbf{A}}_1^{\lambda} \otimes \tilde{\mathbf{A}}_2^{\lambda} = (\tilde{\mathbf{A}}_1 \otimes \tilde{\mathbf{A}}_2)^{\lambda}$;

Definition 6.2.1.5. Garg (2018b) The score and accuracy values of the LPFV $\tilde{\mathbf{A}} = \left\langle \mathfrak{s}_{\phi_{\tilde{A}}}, \mathfrak{s}_{\psi_{\tilde{A}}} \right\rangle$ are defined as

$$\mathcal{S}(\tilde{\mathbf{A}}) = \mathfrak{s}_{\left(\sqrt{\frac{t^2+\phi_{\tilde{A}}^2-\psi_{\tilde{A}}^2}{2}}\right)}$$

and

$$\mathcal{H}(\tilde{\mathbf{A}}) = \mathfrak{s}_{\left(\sqrt{\phi_{\tilde{A}}^2+\psi_{\tilde{A}}^2}\right)},$$

respectively.

It is clear from Garg (2018b) that $0 \leq \frac{t^2+\phi_{\tilde{A}}^2-\psi_{\tilde{A}}^2}{2} \leq t^2$ and $\phi_{\tilde{A}}{}^2 + \psi_{\tilde{A}}{}^2 \leq t^2$. Thus, $\mathfrak{s}_{\left(\sqrt{\frac{t^2+\phi_{\tilde{A}}^2-\psi_{\tilde{A}}^2}{2}}\right)}, \mathfrak{s}_{\left(\sqrt{\phi_{\tilde{A}}^2+\psi_{\tilde{A}}^2}\right)} \in \mathbb{S}_{[0,t]}$. Based on these values, the order relation amongst any two LPFVs $\tilde{A} = \left\langle \mathfrak{s}_{\phi_{\tilde{A}}}, \mathfrak{s}_{\psi_{\tilde{A}}} \right\rangle$ and $\tilde{C} = \left\langle \mathfrak{s}_{\phi_{\tilde{C}}}, \mathfrak{s}_{\psi_{\tilde{C}}} \right\rangle$ can be explained as

1. if $\mathcal{S}(\tilde{A}) < \mathcal{S}(\tilde{C})$ then $\tilde{A} \prec \tilde{C}$.
2. if $\mathcal{S}(\tilde{A}) = \mathcal{S}(\tilde{C})$ and
   a. if $\mathcal{H}(\tilde{A}) < \mathcal{H}(\tilde{C})$ then $\tilde{A} \prec \tilde{C}$.
   b. if $\mathcal{H}(\tilde{A}) = \mathcal{H}(\tilde{C})$ then $\tilde{A} = \tilde{C}$.

Definition 6.2.1.6. Linguistic Pythagorean Fuzzy Weighted Average Operator for LPFVs Garg (2018b)
Let $\tilde{A}_r = \left\langle \mathfrak{s}_{\phi_{\tilde{A}_r}}, \mathfrak{s}_{\psi_{\tilde{A}_r}} \right\rangle$, $r = 1,2,\ldots,l$ be the collection of LPFVs and $\varpi = (\varpi_1, \varpi_2 \ldots, \varpi_l)^T$ be the reasonable weight vectors such that $0 \leq \varpi_r \leq 1, r = 1,2,\ldots,l$, $\sum_{r=1}^{l} \varpi_r = 1$, then the weighted average operator for LPFVs is defined as

$$LPFWA\left(\tilde{A}_1, \tilde{A}_2, \ldots, \tilde{A}_l\right) = \bigoplus_{r=1}^{l} \varpi_r \tilde{A}_r = \bigoplus_{r=1}^{l} \varpi_r \left\langle \mathfrak{s}_{\phi_{\tilde{A}_r}}, \mathfrak{s}_{\psi_{\tilde{A}_r}} \right\rangle$$

$$= \left\langle \mathfrak{s}_{t\sqrt{1-\prod_{r=1}^{l}\left(1-\frac{\phi_{\tilde{A}_r}^2}{t^2}\right)^{\varpi_r}}}, \mathfrak{s}_{t\prod_{r=1}^{l}\left(\frac{\psi_{\tilde{A}_r}}{t}\right)^{\varpi_r}} \right\rangle$$

6.3 MATRIX GAMES WITH UNCERTAIN INFORMATION

The fact that linguistic variable framework expresses the granularity of information motivated Singh et al. (2018) to utilize the linguistic scheme in defining the opinions of players in constant-sum game. Xia (2018) proposed a generalized study of interval-valued intuitionistic fuzzy matrix games originated from Archimedian t-conorm and t-norm. Here, we propose an application of Xia (2018) results for two-person matrix games under uncertainty involving LPFVs. The matrix games having payoff matrix with uncertain information involving LPFVs can be explained as follows:

Consider the set of pure strategies for player I (maximin player) as $\chi = \{\chi_1, \chi_2, \cdots, \chi_m\}$ and the set of pure strategies for player II (minimax player) be $\tau = \{\tau_1, \tau_2, \ldots, \tau_n\}$. Assuming that player I opts for a pure strategy $\chi_i \in \chi$ and player II opts for a pure strategy $\tau_j \in \tau$, subsequently player I gains a payoff $\tilde{a}_{ij} = \left\langle \mathfrak{s}_{\phi_{a_{ij}}}, \mathfrak{s}_{\psi_{a_{ij}}} \right\rangle$ that is again a LPFV, where $\mathfrak{s}_{\phi_{a_{ij}}}, \mathfrak{s}_{\psi_{a_{ij}}} \in \mathbb{S}_{[0,t]} = \{\mathfrak{s}_w | \mathfrak{s}_0 \leq \mathfrak{s}_w \leq \mathfrak{s}_t, w \in [0,t]\}$. Also, $0 \leq \phi_{a_{ij}}^2 + \psi_{a_{ij}}^2 \leq t^2$ is always true. Hence, an uncertain matrix game having payoffs comprising LPFVs is designated via $\tilde{\mathcal{A}} = [\tilde{a}_{ij}]_{m \times n}$. Nevertheless, if $\tilde{\mathcal{A}}$ denotes the payoff matrix of player I, then

$\tilde{\mathcal{A}}^c = \left\langle \mathfrak{s}_{\psi_{a_{ij}}}, \mathfrak{s}_{\phi_{a_{ij}}} \right\rangle$ represents the payoff for player II. $\mathbb{R}_+^m$ and $\mathbb{R}_+^n$ depicts the $m-$dimensional and $n-$dimensional non-negative Euclidean spaces, respectively.

Let player I select the pure strategy $\chi_i \in \chi$ with the probability x_i $(i = 1, 2, \dots, m)$ and player II select the pure strategy τ_j with probability y_j $(j = 1, 2, \dots, n)$. The probabilities $x = (x_1, x_2, \dots, x_m)^T \in \mathbb{R}_+^m$ and $y = (y_1, y_2, \dots, y_m)^T \in \mathbb{R}_+^n$ are such that $e^T x = 1$ and $e^T y = 1$, where e^T is a vector of ones having appropriate dimensions. Then x (y) is known as a mixed strategy for player I (player II) and the mixed strategy spaces for the two players are $S^m = \{x \mid x \in \mathbb{R}_+^m,\ e^T x = 1\}$ and $S^n = \{y \mid y \in \mathbb{R}_+^n,\ e^T y = 1\}$, respectively.

The linguistic Pythagorean fuzzy two person matrix game ($LPFG$) is defined by

$$LPFG = \left(S^m, S^n, \tilde{\mathcal{A}}, \mathbb{S}_{[0,t]}\right).$$

Player I is considered to be a maximizing player, while player II is presumed to be a minimizing player.

For a linguistic matrix game having uncertain information of the payoff matrix, the expected value of the game is attainable. The expected payoff for $LPFG$, when player I selects a mixed strategy $x \in S^m$ and player II selects a mixed strategy $y \in S^n$, for solving the game can be explained using Definition 6.2.1.6 in the following way:

$$\begin{aligned}\mathrm{Exp}_{\tilde{\mathcal{A}}}(x, y) &= x^T \tilde{\mathcal{A}} y \\ &= \left\langle \mathfrak{s}_{t\sqrt{1-\prod_{j=1}^{n}\prod_{i=1}^{m}\left(1-\frac{\phi_{a_{ij}}^2}{t^2}\right)^{x_i y_j}}}, \mathfrak{s}_{t\prod_{j=1}^{n}\prod_{i=1}^{m}\left(\frac{\psi_{a_{ij}}}{t}\right)^{x_i y_j}} \right\rangle\end{aligned} \tag{6.1}$$

Therefore,

$$\begin{aligned}\mathrm{Exp}_{\tilde{\mathcal{A}}}(x, y) &= x^T \tilde{\mathcal{A}} y \\ &= \begin{bmatrix} x_1 & x_2 & \dots x_m \end{bmatrix} \otimes \begin{bmatrix} \tilde{a}_{11} & \tilde{a}_{12} & \dots & \tilde{a}_{1n} \\ \tilde{a}_{21} & \tilde{a}_{22} & \dots & \tilde{a}_{2n} \\ \vdots & \vdots & & \vdots \\ \tilde{a}_{m1} & \tilde{a}_{m2} & \dots & \tilde{a}_{mn} \end{bmatrix} \otimes \begin{bmatrix} y_1 \\ y_2 \\ \vdots \\ y_n \end{bmatrix} \\ &= \begin{bmatrix} x_1 & x_2 & \dots x_m \end{bmatrix} \\ &\otimes \begin{bmatrix} \left\langle \mathfrak{s}_{\phi_{a_{11}}}, \mathfrak{s}_{\psi_{a_{11}}} \right\rangle & \left\langle \mathfrak{s}_{\phi_{a_{12}}}, \mathfrak{s}_{\psi_{a_{12}}} \right\rangle & \dots & \left\langle \mathfrak{s}_{\phi_{a_{1n}}}, \mathfrak{s}_{\psi_{a_{1n}}} \right\rangle \\ \left\langle \mathfrak{s}_{\phi_{a_{21}}}, \mathfrak{s}_{\psi_{a_{21}}} \right\rangle & \left\langle \mathfrak{s}_{\phi_{a_{22}}}, \mathfrak{s}_{\psi_{a_{22}}} \right\rangle & \dots & \left\langle \mathfrak{s}_{\phi_{a_{2n}}}, \mathfrak{s}_{\psi_{a_{2n}}} \right\rangle \\ \vdots & \vdots & & \vdots \\ \left\langle \mathfrak{s}_{\phi_{a_{m1}}}, \mathfrak{s}_{\psi_{a_{m1}}} \right\rangle & \left\langle \mathfrak{s}_{\phi_{a_{m2}}}, \mathfrak{s}_{\psi_{a_{m2}}} \right\rangle & \dots & \left\langle \mathfrak{s}_{\phi_{a_{mn}}}, \mathfrak{s}_{\psi_{a_{mn}}} \right\rangle \end{bmatrix} \\ &\otimes \begin{bmatrix} y_1 \\ y_2 \\ \vdots \\ y_n \end{bmatrix}\end{aligned}$$

The $\mathrm{Exp}_{\tilde{\mathcal{A}}}(x, y)$ so attained is again a LPFV. Clearly, as player I is a maximin player and player II is a minimax player, they will select the mixed strategies according to maximin and minimax principles, respectively. Supposing the duo of mixed strategies $(x^\star \in S^m,\ y^\star \in S^n)$, for the two players are in such a way that

$$x^{\star T}\tilde{A}y^\star = \max_{x\in S^m}\min_{y\in S^n}\{x^T\tilde{A}y\} = \min_{y\in S^n}\max_{x\in S^m}\{x^T\tilde{A}y\},$$

then, $x^\star$ ($y^\star$) is called the optimal strategy for player I (player II). Moreover, $x^{\star T}\tilde{A}y^\star$ expresses the value of $LPFG$ having payoff matrix $\tilde{A}$. The notion of solutions for the uncertain matrix game $LPFG$ is described in accordance with the Pareto optimal solution, as suggested by Bector and Chandra (2005).

Definition 6.3.7. Reasonable Solution of LPFG Suppose $\tilde{\mu}_1$ and $\tilde{\mu}_2$ are LPFVs. Then, $(x^\star,\ y^\star) \in S^m \times S^n$ is defined as the reasonable solution for the game LPFG if

1. $x^{\star T}\tilde{A}y \geq \tilde{\mu}_1$ for all $y \in S^n$ and
2. $x^T\tilde{A}y^\star \leq \tilde{\mu}_2$ for all $x \in S^m$

If $(x^\star,\ y^\star) \in S^m \times S^n$ is the reasonable solution of the matrix game $LPFG$, then $\tilde{\mu}_1$ ($\tilde{\mu}_2$) is known as the reasonable value of player I (player II). The set of all reasonable values for player I is represented by $\mathfrak{M}_1$ and for player II is represented by $\mathfrak{M}_2$.

Definition 6.3.8. Solution of LPFG Assume that there exists $\tilde{\mu}_1^\star \in \mathfrak{M}_1$ and $\tilde{\mu}_2^\star \in \mathfrak{M}_2$, then $(x^\star,\ y^\star) \in S^m \times S^n$ is known as the solution of $LPFG$ if there do not exist any $\tilde{\mu}_1 \in \mathfrak{M}_1$ and $\tilde{\mu}_2 \in \mathfrak{M}_2$ in such a way

1. $\tilde{\mu}_1 > \tilde{\mu}_1^\star$, where $\tilde{\mu}_1 \neq \tilde{\mu}_1^\star$ and
2. $\tilde{\mu}_2 < \tilde{\mu}_2^\star$, where $\tilde{\mu}_2 \neq \tilde{\mu}_2^\star$.

Hence, $\tilde{\mu}_1^\star$ ($\tilde{\mu}_2^\star$) is known as an optimal value for player I (player II), whereas $x^\star$ ($y^\star$) is known as the optimal solution for player I (player II). Moreover, $x^\star$ ($y^\star$) is termed as a maximin (minimax) strategy for player I (player II). Hence, $(x^\star, y^\star, x^{\star T}\tilde{A}y^\star)$ is termed as the complete solution of the matrix game $LPFG$ having payoffs matrix $\tilde{A}$ incorporated with LPFVs. Also

$$\tilde{\mathfrak{V}}^\star = \tilde{\mu}_1^\star \cap \tilde{\mu}_2^\star.$$

where $\tilde{\mathfrak{V}}^\star$ is a LPFV and is the value of matrix game LPFG. Consequently, the Pareto optimal solution proposed by Steuer (1986) will be used to capture the essence of solutions for the matrix game $LPFG$ under uncertainty.

6.3.1 Solution Technique for Matrix Games Characterized by LPFVs

The minimum expected gain $\aleph$ for maximin player I is defined as:

$$
\begin{aligned}
\aleph(x) &= \min_{y \in S^n} \mathrm{Exp}_{\tilde{A}}(x,y) = \left\langle \mathfrak{s}_{\phi_{\aleph}}, \mathfrak{s}_{\psi_{\aleph}} \right\rangle \\
&= \left\langle \min_{y \in S^n} \left\{ \mathfrak{s}_{\phi_{axy}} \right\}, \max_{y \in S^n} \left\{ \mathfrak{s}_{\psi_{axy}} \right\} \right\rangle \\
&= \left\langle \min_{y \in S^n} \left\{ \mathfrak{s}_{t\sqrt{1-\prod_{j=1}^{n}\prod_{i=1}^{m}\left(1-\frac{\phi^2_{a_{ij}}}{t^2}\right)^{x_i y_j}}} \right\}, \max_{y \in S^n} \left\{ \mathfrak{s}_{t\prod_{j=1}^{n}\prod_{i=1}^{m}\left(\frac{\psi_{a_{ij}}}{t}\right)^{x_i y_j}} \right\} \right\rangle
\end{aligned}
$$

Thus, player I selects the mixed strategy $x^* \in S^m$ to maximize $\aleph$, i.e.

$$
\begin{aligned}
\aleph^* &= \left\langle \mathfrak{s}_{\phi_{\aleph^*}}, \mathfrak{s}_{\psi_{\aleph^*}} \right\rangle = \left\langle \max_{x \in S^m} \min_{y \in S^n} \left\{ \mathfrak{s}_{\phi_{axy}} \right\}, \min_{x \in S^m} \max_{y \in S^n} \left\{ \mathfrak{s}_{\psi_{axy}} \right\} \right\rangle \\
&= \Bigg\langle \max_{x \in S^m} \min_{y \in S^n} \left\{ \mathfrak{s}_{t\sqrt{1-\prod_{j=1}^{n}\prod_{i=1}^{m}\left(1-\frac{\phi^2_{a_{ij}}}{t^2}\right)^{x_i y_j}}} \right\}, \\
&\qquad \min_{x \in S^m} \max_{y \in S^n} \left\{ \mathfrak{s}_{t\prod_{j=1}^{n}\prod_{i=1}^{m}\left(\frac{\psi_{a_{ij}}}{t}\right)^{x_i y_j}} \right\} \Bigg\rangle
\end{aligned}
\tag{6.2}
$$

where $\aleph^*$ is termed as the minimum gain for player I and its corresponding mixed strategy x^* is known as a maximin strategy.

Equivalently, the maximum expected loss for player II is displayed by:

$$
\begin{aligned}
\mathrm{Y}(y) &= \max_{x \in S^m} \mathrm{Exp}_{\tilde{A}}(x,y) = \left\langle \mathfrak{s}_{\phi_{\mathrm{Y}}}, \mathfrak{s}_{\psi_{\mathrm{Y}}} \right\rangle \\
&= \left\langle \max_{x \in S^m} \left\{ \mathfrak{s}_{\phi_{axy}} \right\}, \min_{x \in S^m} \left\{ \mathfrak{s}_{\psi_{axy}} \right\} \right\rangle \\
&= \left\langle \max_{x \in S^m} \left\{ \mathfrak{s}_{t\sqrt{1-\prod_{j=1}^{n}\prod_{i=1}^{m}\left(1-\frac{\phi^2_{a_{ij}}}{t^2}\right)^{x_i y_j}}} \right\}, \min_{x \in S^m} \left\{ \mathfrak{s}_{t\prod_{j=1}^{n}\prod_{i=1}^{m}\left(\frac{\psi_{a_{ij}}}{t}\right)^{x_i y_j}} \right\} \right\rangle
\end{aligned}
\tag{6.3}
$$

Player II opts for a mixed strategy $y^* \in S^n$ to minimize Y, such that

$$\mathrm{Y}^* = \left\langle \mathfrak{s}_{\phi_{\mathrm{Y}^*}}, \mathfrak{s}_{\psi_{\mathrm{Y}^*}} \right\rangle = \left\langle \min_{y \in S^n} \max_{x \in S^m} \left\{ \mathfrak{s}_{\phi_{axy}} \right\}, \max_{y \in S^n} \min_{x \in S^m} \left\{ \mathfrak{s}_{\psi_{axy}} \right\} \right\rangle$$

$$= \left\langle \min_{y \in S^n} \max_{x \in S^m} \left\{ \mathfrak{s}_{t\sqrt{1-\prod_{j=1}^{n}\prod_{i=1}^{m}\left(1-\frac{\phi_{a_{ij}}^2}{t^2}\right)^{x_i y_j}}} \right\}, \right.$$

$$\left. \max_{y \in S^n} \min_{x \in S^m} \left\{ \mathfrak{s}_{t\prod_{j=1}^{n}\prod_{i=1}^{m}\left(\frac{\psi_{a_{ij}}}{t}\right)^{x_i y_j}} \right\} \right\rangle \tag{6.4}$$

where Y^* is called the maximum loss for player II, while the mixed strategy y^* is known as the minimax strategy for player II.

Theorem: 6.3.1.1. *Suppose $\aleph^*$ denotes the minimum gain for player I and Y^* denotes the maximum loss for player II, then both $\aleph^*$ and Y^* are LPFVs such that $\aleph^* \precsim \mathrm{Y}^*$.*

Proof. Using (6.1), certainly $\mathrm{Exp}_{\tilde{A}}(x, y)$ is a LPFV set. Now, $\mathfrak{s}_{\phi_{axy}}, \mathfrak{s}_{\psi_{axy}} \in \mathbb{S}_{[0,t]}$ therefore it is observed that

$$\phi_{axy}^2 + \psi_{axy}^2 \le t^2. \tag{6.5}$$

Then, using (6.2), we get

$$\max_{x \in S^m} \min_{y \in S^n} \left\{ \mathfrak{s}_{\phi_{axy}} \right\} = \mathfrak{s}_{\phi_{\aleph^*}} \quad \text{and} \quad \min_{x \in S^m} \max_{y \in S^n} \left\{ \mathfrak{s}_{\psi_{axy}} \right\} = \mathfrak{s}_{\psi_{\aleph^*}}$$

such that $\mathfrak{s}_{\phi_{\aleph^*}}, \mathfrak{s}_{\psi_{\aleph^*}} \in \mathbb{S}_{[0,t]}$.

Next, to show that $\left\langle \mathfrak{s}_{\phi_{\aleph^*}}, \mathfrak{s}_{\psi_{\aleph^*}} \right\rangle$ is a LPFV, we need to prove that $\phi_{\aleph^*}^2 + \psi_{\aleph^*}^2 \le t^2$ or $\max_{x \in S^m} \min_{y \in S^n} \left\{ \phi_{axy}^2 \right\} + \min_{x \in S^m} \max_{y \in S^n} \left\{ \psi_{axy}^2 \right\} \le t^2$.

Using (6.5), it can be seen

$$0 \le \phi_{axy}^2 + \min_{y \in S^n} \left\{ \psi_{axy}^2 \right\} \le \phi_{axy}^2 + \max_{y \in S^n} \left\{ \psi_{axy}^2 \right\} \le t^2.$$

Furthermore,

$$0 \le \min_{y \in S^n} \left\{ \phi_{axy}^2 \right\} + \min_{y \in S^n} \left\{ \psi_{axy}^2 \right\} \le \phi_{axy}^2 + \max_{y \in S^n} \left\{ \psi_{axy}^2 \right\} \le t^2.$$

Therefore,

$$0 \le \min_{y \in S^n} \left\{ \phi_{axy}^2 \right\} + \min_{x \in S^m} \min_{y \in S^n} \left\{ \psi_{axy}^2 \right\} \le \min_{y \in S^n} \left\{ \phi_{axy}^2 \right\} + \min_{x \in S^m} \max_{y \in S^n} \left\{ \psi_{axy}^2 \right\}$$

$$\le \phi_{axy}^2 + \min_{x \in S^m} \max_{y \in S^n} \left\{ \psi_{axy}^2 \right\} \le t^2,$$

which results in

$$0 \leq \max_{x\in S^m} \min_{y\in S^n} \left\{\phi^2_{a_{xy}}\right\} + \min_{x\in S^m} \min_{y\in S^n} \left\{\psi^2_{a_{xy}}\right\} \leq \max_{x\in S^m} \min_{y\in S^n} \left\{\phi^2_{a_{xy}}\right\} + \min_{x\in S^m} \max_{y\in S^n} \left\{\psi^2_{a_{xy}}\right\}$$

$$\leq \max_{x\in S^m} \phi^2_{a_{xy}} + \min_{x\in S^m} \max_{y\in S^n} \left\{\psi^2_{a_{xy}}\right\} \leq t^2.$$

Hence,

$$0 \leq \max_{x\in S^m} \min_{y\in S^n} \left\{\phi^2_{a_{xy}}\right\} + \min_{x\in S^m} \max_{y\in S^n} \left\{\psi^2_{a_{xy}}\right\} \leq t^2.$$

Thus, $\aleph^*$ is a LPFV.

On similar lines, Υ^* can also be proved to be a LPFV.

Also, for any $x \in S^m$ and $y \in S^n$

$$\min_{y\in S^n} \left\{\phi_{a_{xy}}\right\} \leq \phi_{a_{xy}} \leq \max_{x\in S^m} \left\{\phi_{a_{xy}}\right\}$$

and

$$\min_{y\in S^n} \left\{\phi_{a_{xy}}\right\} \leq \min_{y\in S^n} \max_{x\in S^m} \left\{\phi_{a_{xy}}\right\}.$$

Hence, we obtain

$$\max_{x\in S^m} \min_{y\in S^n} \left\{\phi_{a_{xy}}\right\} \leq \min_{y\in S^n} \max_{x\in S^m} \left\{\phi_{a_{xy}}\right\}.$$

Therefore

$$\phi_{\aleph^*} \leq \phi_{\Upsilon^*}. \tag{6.6}$$

Analogously, for $x \in S^m$ and $y \in S^n$

$$\max_{y\in S^n} \left\{\psi_{a_{xy}}\right\} \geq \psi_{a_{xy}} \geq \min_{x\in S^m} \left\{\psi_{a_{xy}}\right\}$$

which implies

$$\max_{y\in S^n} \left\{\psi_{a_{xy}}\right\} \geq \max_{y\in S^n} \min_{x\in S^m} \left\{\psi_{a_{xy}}\right\}$$

Hence, it is derived that

$$\min_{x\in S^m} \max_{y\in S^n} \left\{\psi_{a_{xy}}\right\} \geq \max_{y\in S^n} \min_{x\in S^m} \left\{\psi_{a_{xy}}\right\}.$$

As a consequence

$$\psi_{\Upsilon^*} \geq \psi_{\aleph^*} \tag{6.7}$$

Utilizing (6.6) and (6.7) and Section 6.2.1

$$\aleph^* \precsim \Upsilon^*. \qquad \square$$

Thus, from (6.2) we obtain the following non-linear bi-objective programming problem which on solving yields the maximin strategy x^* and the maximum gain $\aleph^*$ for player I:

$$\textbf{(LPFPI-1)} \quad \max\{\mathfrak{s}_{\phi_\aleph}\}, \ \min\{\mathfrak{s}_{\psi_\aleph}\}$$

$$\text{s.t.} \quad \mathfrak{s}_{t\sqrt{1-\prod_{j=1}^{n}\prod_{i=1}^{m}\left(1-\frac{\phi^2_{a_{ij}}}{t^2}\right)^{x_i y_j}}} \geq \mathfrak{s}_{\phi_\aleph}, \qquad \text{for any } y \in S^n$$

$$\mathfrak{s}_{t\prod_{j=1}^{n}\prod_{i=1}^{m}\left(\frac{\psi_{a_{ij}}}{t}\right)^{x_i y_j}} \leq \mathfrak{s}_{\psi_\aleph}, \qquad \text{for any } y \in S^n$$

$$\mathfrak{s}_{\phi_\aleph}, \mathfrak{s}_{\psi_\aleph} \in \mathbb{S}_{[0,t]}$$

$$0 \leq \phi^2_\aleph + \psi^2_\aleph \leq t^2,$$

$$\sum_{i=1}^{m} x_i = 1, \ x_i \geq 0, \ i = 1, 2, \ldots, m.$$

where

$$\mathfrak{s}_{\phi_\aleph} = \min_{y \in S^n}\left\{\mathfrak{s}_{t\sqrt{1-\prod_{j=1}^{n}\prod_{i=1}^{m}\left(1-\frac{\phi^2_{a_{ij}}}{t^2}\right)^{x_i y_j}}}\right\}$$

and

$$\mathfrak{s}_{\psi_\aleph} = \max_{y \in S^n}\left\{\mathfrak{s}_{t\prod_{j=1}^{n}\prod_{i=1}^{m}\left(\frac{\psi_{a_{ij}}}{t}\right)^{x_i y_j}}\right\}.$$

Therefore, we have the following relationships

$$\mathfrak{s}_{t\sqrt{1-\prod_{j=1}^{n}\prod_{i=1}^{m}\left(1-\frac{\phi^2_{a_{ij}}}{t^2}\right)^{x_i y_j}}} \geq \mathfrak{s}_{\phi_\aleph},$$

$$\mathfrak{s}_{t\prod_{j=1}^{n}\prod_{i=1}^{m}\left(\frac{\psi_{a_{ij}}}{t}\right)^{x_i y_j}} \leq \mathfrak{s}_{\psi_\aleph}.$$

Here, the solution of bi-objective non-linear programming problem **(LPFPI-1)** exists in the sense of Pareto optimality.

Also, the following is true for $t > 0$

$$\max\{\mathfrak{s}_{\phi_\aleph}\} \Leftrightarrow \min\{\mathfrak{s}_{t^2-\phi^2_\aleph}\} \Leftrightarrow \min\left\{\mathfrak{s}_{\ln\left(1-\frac{\phi^2_\aleph}{t^2}\right)}\right\},$$

$$\min\{\mathfrak{s}_{\psi_\aleph}\} \Leftrightarrow \min\left\{\mathfrak{s}_{\ln\left(\frac{\psi_\aleph}{t}\right)}\right\}.$$

Constraints in **(LPFPI-1)** holds if

$$\mathfrak{s}_{\prod_{j=1}^{n}\prod_{i=1}^{m}\left(1-\frac{\phi^2_{a_{ij}}}{t^2}\right)^{x_i y_j}} \leq \mathfrak{s}_{1-\frac{\phi^2_{\aleph}}{t^2}} \Leftrightarrow \mathfrak{s}_{\sum_{j=1}^{n}\sum_{i=1}^{m} x_i y_j\left[\ln\left(1-\frac{\phi^2_{a_{ij}}}{t^2}\right)\right]} \leq \mathfrak{s}_{\ln\left(1-\frac{\phi^2_{\aleph}}{t^2}\right)},$$

$$\mathfrak{s}_{\prod_{j=1}^{n}\prod_{i=1}^{m}\left(\frac{\psi_{a_{ij}}}{t}\right)^{x_i y_j}} \leq \mathfrak{s}_{\frac{\psi_{\aleph}}{t}} \Leftrightarrow \mathfrak{s}_{\sum_{j=1}^{n}\sum_{i=1}^{m} x_i y_j\left[\ln\left(\frac{\psi_{a_{ij}}}{t}\right)\right]} \leq \mathfrak{s}_{\ln\left(\frac{\psi_{\aleph}}{t}\right)}. \tag{6.8}$$

Utilizing the weighted average approach, the objective function of **(LPFPI-1)** reduces to

$$\min\left\{\mathfrak{s}_{\alpha\ln\left(1-\frac{\phi^2_{\aleph}}{t^2}\right)+(1-\alpha)\ln\left(\frac{\psi_{\aleph}}{t}\right)}\right\} \tag{6.9}$$

for $\alpha \in [0,1]$ is the amount of weight recommended mostly by two players of the game. Similarly, the constraints of **(LPFPI-1)** using (6.8) are reduced to

$$\mathfrak{s}_{\sum_{j=1}^{n}\sum_{i=1}^{m} x_i y_j\left(\alpha\ln\left(1-\frac{\phi^2_{a_{ij}}}{t^2}\right)+(1-\alpha)\ln\left(\frac{\psi_{a_{ij}}}{t}\right)\right)} \leq \mathfrak{s}_{\alpha\ln\left(1-\frac{\phi^2_{\aleph}}{t^2}\right)+(1-\alpha)\ln\left(\frac{\psi_{\aleph}}{t}\right)} \tag{6.10}$$

From (6.9) and (6.10), the **(LPFPI-1)** is transformed into:

(LPFPI-2) $\min\left\{\mathfrak{s}_{\alpha\ln\left(1-\frac{\phi^2_{\aleph}}{t^2}\right)+(1-\alpha)\ln\left(\frac{\psi_{\aleph}}{t}\right)}\right\}$

s.t. $\mathfrak{s}_{\sum_{j=1}^{n}\sum_{i=1}^{m} x_i y_j\left[\alpha\ln\left(1-\frac{\phi^2_{a_{ij}}}{t^2}\right)+(1-\alpha)\ln\left(\frac{\psi_{a_{ij}}}{t}\right)\right]}$

$\leq \mathfrak{s}_{\alpha\ln\left(1-\frac{\phi^2_{\aleph}}{t^2}\right)+(1-\alpha)\ln\left(\frac{\psi_{\aleph}}{t}\right)},$ for any $y \in S^n$

$\mathfrak{s}_{\phi_{\aleph}}, \mathfrak{s}_{\psi_{\aleph}} \in \mathbb{S}_{[0,t]}$

$0 \leq \phi^2_{\aleph} + \psi^2_{\aleph} \leq t^2,$

$\sum_{i=1}^{m} x_i = 1, \ x_i \geq 0, \ i = 1, 2, \ldots, m.$

The problem **(LPFPI-2)** is always true except for $\mathfrak{s}_{\phi_{\aleph}} = \mathfrak{s}_t$, $\mathfrak{s}_{\psi_{\aleph}} = \mathfrak{s}_0$, $\mathfrak{s}_{\phi_{a_{ij}}} = \mathfrak{s}_t$ and $\mathfrak{s}_{\psi_{a_{ij}}} = \mathfrak{s}_0$.

For computational convenience, put $\digamma = \alpha\ln\left(1-\frac{\phi^2_{\aleph}}{t^2}\right)+(1-\alpha)\ln\left(\frac{\psi_{\aleph}}{t}\right)$.

Here, $\digamma \leq 0$ as $\alpha \in [0,1]$ and $0 \leq \phi_{\aleph}, \psi_{\aleph} \leq t$. Consequently, $0 \leq 1-\frac{\phi^2_{\aleph}}{t^2} \leq 1$ and $0 \leq \frac{\psi_{\aleph}}{t} \leq 1$.

Thus, the objective function in **(LPFPI-2)** evolves as $\min\{\mathfrak{s}_{\digamma}\}$ which can be further established as a mathematical programming problem with objective function $\min\{\digamma\}$.

Specifically, adopting the features of linguistic terms in Definition 6.2.2, the constraints in **(LPFPI-2)** can be remodeled as

$$\sum_{j=1}^{n}\sum_{i=1}^{m} x_i y_j\left(\alpha \ln\left(1-\frac{\phi^2_{a_{ij}}}{t^2}\right)+(1-\alpha)\ln\left(\frac{\psi_{a_{ij}}}{t}\right)\right)$$
$$\leq \alpha \ln\left(1-\frac{\phi^2_{\aleph}}{t^2}\right)+(1-\alpha)\ln\left(\frac{\psi_{\aleph}}{t}\right)=\digamma.$$

Accordingly, **(LPFPI-2)** is remodeled as:

(LPFPI-3) $\min\{\digamma\}$

s.t. $\sum_{j=1}^{n}\sum_{i=1}^{m} x_i y_j\left(\alpha \ln\left(1-\frac{\phi^2_{a_{ij}}}{t^2}\right)+(1-\alpha)\ln\left(\frac{\psi_{a_{ij}}}{t}\right)\right)$

$\leq \digamma,$ for any $y \in S^n$

$\digamma \leq 0,$

$\sum_{i=1}^{m} x_i = 1,\ x_i \geq 0, \quad i = 1,2,\ldots,m.$

except for $\mathfrak{s}_{\phi_{a_{ij}}} = \mathfrak{s}_t$ and $\mathfrak{s}_{\psi_{a_{ij}}} = \mathfrak{s}_0$ i.e. $\frac{\phi_{a_{ij}}}{t} = 1$, $\frac{\psi_{a_{ij}}}{t} = 0$, respectively.

Nevertheless, with S^n being a closed compact bounded set, it is relevant to determine exclusively the extreme points of S^n in the constraints **(LPFPI-3)**. As a result, problem **(LPFPI-3)** is transformed into a mathematical programming problem:

(LPFPI-4) $\min\{\digamma\}$

s.t. $\sum_{i=1}^{m} x_i\left[\alpha \ln\left(1-\frac{\phi^2_{a_{ij}}}{t^2}\right)+(1-\alpha)\ln\left(\frac{\psi_{a_{ij}}}{t}\right)\right] \leq \digamma, \quad j = 1,2,\ldots,n$

$\digamma \leq 0,$

$\sum_{i=1}^{m} x_i = 1,\ x_i \geq 0, \quad i = 1,2,\ldots,m.$

The aforementioned linear programming problem can be resolved by employing the existing tools like on MATLAB® (2021b) software.

Equivalently, in the player II game, employing (6.3) and the Definitions 6.3.7 and 6.3.8, a minimax strategy y^* and the minimal loss Y^* in (6.4) for player II may be determined through resolving the non-linear bi-objective programming problem as follows:

(LPFPII-1) $\min\{\mathfrak{s}_{\phi_{\mathrm{Y}}}\},\ \max\{\mathfrak{s}_{\psi_{\mathrm{Y}}}\}$

s.t. $\mathfrak{s}_{t\sqrt{1-\prod_{j=1}^{n}\prod_{i=1}^{m}\left(1-\frac{\phi^2_{a_{ij}}}{t^2}\right)^{x_i y_j}}} \leq \mathfrak{s}_{\phi_{\mathrm{Y}}},$ for any $x \in S^m$

$$\mathfrak{s}_{t\prod_{j=1}^{n}\prod_{i=1}^{m}\left(\frac{\psi_{a_{ij}}}{t}\right)^{x_iy_j}} \geq \mathfrak{s}_{\psi_{Y}}, \qquad \text{for any } x \in S^m$$

$$\mathfrak{s}_{\phi_Y}, \mathfrak{s}_{\psi_Y} \in \mathbb{S}_{[0,t]}$$

$$0 \leq \phi_Y^2 + \psi_Y^2 \leq t^2,$$

$$\sum_{j=1}^{n} y_j = 1, \; y_j \geq 0, \; j = 1, 2, \ldots, n.$$

where

$$\mathfrak{s}_{\phi_Y} = \max_{x \in S^m}\left\{\mathfrak{s}_{t\sqrt{1-\prod_{j=1}^{n}\prod_{i=1}^{m}\left(1-\frac{\phi_{a_{ij}}^2}{t^2}\right)^{x_iy_j}}}\right\}$$

and

$$\mathfrak{s}_{\psi_Y} = \min_{x \in S^m}\left\{\mathfrak{s}_{t\prod_{j=1}^{n}\prod_{i=1}^{m}\left(\frac{\mathfrak{s}_{\psi_{a_{ij}}}}{t}\right)^{x_iy_j}}\right\}.$$

In this case too, the solution of bi-objective non-linear programming problem is based on the idea of Pareto optimality. Further for $t > 0$

$$\min\left\{\mathfrak{s}_{\phi_Y}\right\} \Leftrightarrow \max\left\{\mathfrak{s}_{1-\frac{\phi_Y^2}{t^2}}\right\} \Leftrightarrow \max\left\{\mathfrak{s}_{\ln\left(1-\frac{\phi_Y^2}{t^2}\right)}\right\} \tag{6.11}$$

$$\max\left\{\mathfrak{s}_{\psi_Y}\right\} \Leftrightarrow \max\left\{\mathfrak{s}_{\frac{\psi_Y}{t}}\right\} \Leftrightarrow \max\left\{\mathfrak{s}_{\ln\left(\frac{\psi_Y}{t}\right)}\right\} \tag{6.12}$$

whereas constraints are re-framed as

$$\left.\begin{array}{l}
\mathfrak{s}_{t\sqrt{1-\prod_{j=1}^{n}\prod_{i=1}^{m}\left(1-\frac{\phi_{a_{ij}}^2}{t^2}\right)^{x_iy_j}}} \leq \mathfrak{s}_{\phi_Y} \Leftrightarrow \mathfrak{s}_{\prod_{j=1}^{n}\prod_{i=1}^{m}\left(1-\frac{\phi_{a_{ij}}^2}{t^2}\right)^{x_iy_j}} \geq \mathfrak{s}_{1-\frac{\phi_Y^2}{t^2}} \\
\Leftrightarrow \mathfrak{s}_{\sum_{j=1}^{n}\sum_{i=1}^{m} x_iy_j \ln\left(1-\frac{\phi_{a_{ij}}^2}{t^2}\right)} \geq \mathfrak{s}_{\ln\left(1-\frac{\phi_Y^2}{t^2}\right)}, \\
\mathfrak{s}_{t\prod_{j=1}^{n}\prod_{i=1}^{m}\left(\frac{\psi_{a_{ij}}}{t}\right)^{x_iy_j}} \geq \mathfrak{s}_{\psi_Y} \Leftrightarrow \mathfrak{s}_{\prod_{j=1}^{n}\prod_{i=1}^{m}\left(\frac{\psi_{a_{ij}}}{t}\right)^{x_iy_j}} \geq \mathfrak{s}_{\frac{\psi_Y}{t}} \\
\Leftrightarrow \mathfrak{s}_{\sum_{j=1}^{n}\sum_{i=1}^{m} x_iy_j \ln\left(\frac{\psi_{a_{ij}}}{t}\right)} \geq \mathfrak{s}_{\ln\left(\frac{\psi_Y}{t}\right)}.
\end{array}\right\} \tag{6.13}$$

Employing the weighted average technique for the objective function, (6.11) and (6.12) are transformed to

$$\max\left\{\mathfrak{s}_{\alpha\ln\left(1-\frac{\phi_Y^2}{t^2}\right)+(1-\alpha)\ln\left(\frac{\psi_Y}{t}\right)}\right\} \tag{6.14}$$

the weight $\alpha \in [0,1]$ is provided by the players of the game. The constraints in **(LPFPII-1)** using (6.13) are reduced to

$$\mathfrak{s}_{\sum_{j=1}^{n}\sum_{i=1}^{m} x_i y_j \left[\alpha \ln\left(1-\frac{\phi_{a_{ij}}^2}{t^2}\right)+(1-\alpha)\ln\left(\frac{\psi_{a_{ij}}}{t}\right)\right]} \leq \mathfrak{s}_{\alpha \ln\left(1-\frac{\phi_Y^2}{t^2}\right)+(1-\alpha)\ln\left(\frac{\psi_Y}{t}\right)} \tag{6.15}$$

Consequently using (6.14) and (6.15), **(LPFPII-1)** is remodeled as:

(LPFPII-2) $\quad \max\left\{\mathfrak{s}_{\alpha \ln\left(1-\frac{\phi_Y^2}{t^2}\right)+(1-\alpha)\ln\left(\frac{\psi_Y}{t}\right)}\right\}$

s.t. $\quad \mathfrak{s}_{\sum_{j=1}^{n}\sum_{i=1}^{m} x_i y_j \left[\alpha \ln\left(1-\frac{\phi_{a_{ij}}^2}{t^2}\right)+(1-\alpha)\ln\left(\frac{\psi_{a_{ij}}}{t}\right)\right]}$

$\leq \mathfrak{s}_{\alpha \ln\left(1-\frac{\phi_Y^2}{t^2}\right)+(1-\alpha)\ln\left(\frac{\psi_Y}{t}\right)}$, $\quad$ for any $x \in S^m$

$$0 \leq \phi_Y^2 + \psi_Y^2 \leq \quad t^2$$

$$\sum_{j=1}^{n} y_j = 1,\ y_j \geq 0,\ j = 1, 2, \ldots, n.$$

The problem **(LPFPII-2)** always exists, except for $\mathfrak{s}_{\phi_Y} = \mathfrak{s}_t$, $\mathfrak{s}_{\psi_Y} = \mathfrak{s}_0$, $\mathfrak{s}_{\phi_{a_{ij}}} = \mathfrak{s}_t$, $\mathfrak{s}_{\psi_{a_{ij}}} = \mathfrak{s}_0$.

Taking $\wp = \alpha \ln\left(1 - \frac{\phi_{a_{ij}}^2}{t^2}\right) + (1-\alpha)\ln\left(\frac{\psi_{a_{ij}}}{t}\right)$. Then $\wp \leq 0$, since $\alpha \in [0,1]$ and $0 \leq \phi_Y,\ \psi_Y \leq t$.

Therefore, $0 \leq 1 - \frac{\phi_Y^2}{t^2} \leq 1,\ 0 \leq \frac{\psi_Y}{t} \leq 1.$

Consequently, the objective function in **(LPFPII-2)**, i.e., $\min\{\mathfrak{s}_{\wp}\}$ is established as $\min\{\wp\}$. In addition, utilizing the properties of linguistic terms in Definition 6.2.2, the constraints in **(LPFPII-2)** are evolved as:

$$\sum_{j=1}^{n}\sum_{i=1}^{m} x_i y_j \left(\alpha \ln\left(1 - \frac{\phi_{a_{ij}}^2}{t^2}\right) + (1-\alpha)\ln\left(\frac{\psi_{a_{ij}}}{t}\right)\right)$$
$$\geq \alpha \ln\left(1 - \frac{\phi_Y^2}{t^2}\right) + (1-\alpha)\ln\left(\frac{\psi_Y}{t}\right) = \wp,$$

Subsequently, **(LPFPII-2)** is re-framed as:

(LPFPII-3) $\quad \min\{\wp\}$

s.t. $\quad \sum_{j=1}^{n}\sum_{i=1}^{m} x_i y_j \left(\alpha \ln\left(1 - \frac{\phi_{a_{ij}}^2}{t^2}\right) + (1-\alpha)\ln\left(\frac{\psi_{a_{ij}}}{t}\right)\right)$

$\geq \wp$, $\quad$ for any $x \in S^m$

$\wp \leq 0$,

$$\sum_{j=1}^{n} y_j = 1,\ y_j \geq 0,\ j = 1, 2, \ldots, n.$$

which holds except for $\mathfrak{s}_{\phi_{a_{ij}}} = \mathfrak{s}_t$ and $\mathfrak{s}_{\psi_{a_{ij}}} = \mathfrak{s}_0$, i.e. $\frac{\phi_{a_{ij}}}{t} = 1$ and $\frac{\psi_{a_{ij}}}{t} = 0$, respectively.

Furthermore, with S^m being a closed compact bounded set, it is adequate to consider only the corner points of the set S^m in the constraints of **(LPFPII-3)**. Hence, **(LPFPII-3)** is transformed into:

(LPFPII-4) $\max\{\wp\}$

$$\text{s.t.} \quad \sum_{j=1}^{n} y_j \left(\alpha \ln\left(1 - \frac{\phi_{a_{ij}}^2}{t^2}\right) + (1-\alpha)\ln\left(\frac{\psi_{a_{ij}}}{t}\right)\right) \geq \wp,$$

for any $x \in S^m$

$$\wp \leq 0,$$

$$\sum_{j=1}^{n} y_j = 1,\ y_j \geq 0,\ j = 1, 2, \ldots, n.$$

The aforementioned mathematical problems can be resolved using existing tools, such as MATLAB (R2021b) software.

Theorem: 6.3.1.2. *The matrix game LPFG involving payoffs $\tilde{A}$ characterized by LPFV sets always acquire a solution $(x^*, y^*, x^{*T}\tilde{A}y^*)$ for every $\alpha \in [0,1]$,.*

Proof. Suppose the predetermined weight is $\alpha \in [0,1]$. Clearly, **(LPFPI-4)** and **(LPFPII-4)** establish primal-dual linear programming problems and is comparable to the matrix game involving payoffs:

$$\left(\alpha \ln\left(1 - \frac{\phi_{a_{ij}}^2}{t^2}\right) + (1-\alpha)\ln\left(\frac{\psi_{a_{ij}}}{t}\right)\right)_{m \times n}.$$

Using the minimax theorem for matrix games given by Owen (1982), **(LPFPI-4)** and **(LPFPII-4)** have optimal solutions $(x^*, \digamma^*)$ and $(y^*, \wp^*)$, respectively, where $\digamma^* = \wp^*$. As a result, whatever is the supplied weight $\alpha \in [0,1]$, the matrix game *LPFG* with payoffs matrix $\tilde{A}$ characterized via LPFV sets constantly possess a solution $(x^*, y^*, x^{*T}\tilde{A}y^*)$. □

$\digamma$ and $\wp$ are defined in terms of each $\alpha \in [0,1]$, and they preserve the succeeding property:

Theorem: 6.3.1.3. *$\digamma$ and $\wp$ are monotonically non-decreasing functions of $\alpha \in [0,1]$.*

Proof. Following **(LPFPI-3)**, $\digamma = \alpha \ln\left(1 - \frac{\phi_{\aleph}^2}{t^2}\right) + (1-\alpha)\ln\left(\frac{\psi_{\aleph}}{t}\right)$, where $0 \leq 1 - \frac{\phi_{\aleph}^2}{t^2} \leq 1$ and $0 \leq \frac{\psi_{\aleph}}{t} \leq 1$ for $t > 0$. Moreover,

$$\ln\left(1 - \frac{\phi_{\aleph}^2}{t^2}\right) \leq 0 \text{ and } \ln\left(\frac{\psi_{\aleph}}{t}\right) \leq 0.$$

Now,

$$\frac{\partial \digamma}{\partial \alpha} = \ln\left(1 - \frac{\phi_{\aleph}^2}{t^2}\right) - \ln\left(\frac{\psi_{\aleph}}{t}\right) = \ln\left(\frac{1 - \frac{\phi_{\aleph}^2}{t^2}}{\frac{\psi_{\aleph}}{t}}\right).$$

Also, $0 \leq \phi_{\aleph}^2 + \psi_{\aleph}^2 \leq t^2$ or $0 \leq \frac{\phi_{\aleph}^2}{t^2} + \frac{\psi_{\aleph}^2}{t^2} \leq 1$, i.e., $1 - \frac{\phi_{\aleph}^2}{t^2} \geq \frac{\psi_{\aleph}^2}{t^2}$ which is further expressed as $\sqrt{1 - \frac{\phi_{\aleph}^2}{t^2}} \geq \frac{\psi_{\aleph}}{t}$. Therefore,

$$\ln\left(1 - \frac{\phi_{\aleph}^2}{t^2}\right) \geq 2\ln\left(\frac{\psi_{\aleph}}{t}\right) \geq \ln\left(\frac{\psi_{\aleph}}{t}\right)$$

i.e.,

$$\ln\left(\frac{1 - \frac{\phi_{\aleph}^2}{t^2}}{\frac{\psi_{\aleph}}{t}}\right) \geq 0.$$

Thus, $\frac{\partial \digamma}{\partial \alpha} \geq 0$ except for $\frac{\psi_{\aleph}}{t} = 0$.

Hence, $\digamma$ is a monotonically non-decreasing function for each $\alpha \in [0, 1]$.

Similarly, given $\alpha \in [0, 1]$, $\wp$ as defined in **(LPFPII-3)** is a monotonically non-decreasing function. □

The optimal solutions of **(LPFPI-4)** and **(LPFPII-4)** and the solution of the matrix game *LPFG* with payoffs $\tilde{A}$ determined by LPFV sets eventually form a straightforward association:

Theorem: 6.3.1.4. *Assume that* $(x^*, \digamma^*)$ *and* $(y^*, \wp^*)$ *are optimal solutions of* ***(LPFPI-4)*** *and* ***(LPFPII-4)****, respectively, with* $0 < \alpha < 1$. *Then the Pareto optimal solutions of* ***(LPFPI-1)*** *and* ***(LPFPII-1)*** *are presented by* $(x^*, \aleph^*)$ *and* (y^*, Y^*)*, where*

$$\aleph^* = \left\langle \max_{x \in S^m} \min_{y \in S^n} \left\{ \mathfrak{s}_{t\left(1 - \prod_{j=1}^{n} \prod_{i=1}^{m} \left(1 - \frac{\phi_{a_{ij}}^2}{t^2}\right)^{x_i y_j}\right)} \right\}, \right.$$
$$\left. \min_{x \in S^m} \max_{y \in S^n} \left\{ \mathfrak{s}_{t \prod_{j=1}^{n} \prod_{i=1}^{m} \left(\frac{\psi_{a_{ij}}}{t}\right)^{x_i y_j}} \right\} \right\rangle$$

and

$$Y^* = \left\langle \min_{y\in S^n} \max_{x\in S^m} \left\{ \mathfrak{s}_{t\left(1-\prod_{j=1}^{n}\prod_{i=1}^{m}\left(1-\frac{\phi_{a_{ij}}^2}{t^2}\right)^{x_i y_j}\right)} \right\}, \right.$$
$$\left. \max_{y\in S^n} \min_{x\in S^m} \left\{ \mathfrak{s}_{t\prod_{j=1}^{n}\prod_{i=1}^{m}\left(\frac{\psi_{a_{ij}}}{t}\right)^{x_i y_j}} \right\} \right\rangle.$$

Proof. Suppose $(x^*, \aleph^*)$ is not a Pareto optimal solution of **(LPFPI-1)**. Then, there exists a feasible solution $(\hat{x}, \hat{\aleph})$ for $\hat{x} \in S^m$ and $\hat{\aleph} = \left\langle \mathfrak{s}_{\phi_{\hat{\aleph}}}, \mathfrak{s}_{\psi_{\hat{\aleph}}} \right\rangle$ such that

$$\left.\begin{array}{ll}
\mathfrak{s}_{t\sqrt{1-\prod_{j=1}^{n}\prod_{i=1}^{m}\left(1-\frac{\phi_{a_{ij}}^2}{t^2}\right)^{\hat{x}_i y_j}}} \geq \mathfrak{s}_{\phi_{\hat{\aleph}}}, & \text{for any } y \in S^n \\
\mathfrak{s}_{t\prod_{j=1}^{n}\prod_{i=1}^{m}\left(\frac{\psi_{a_{ij}}}{t}\right)^{\hat{x}_i y_j}} \leq \mathfrak{s}_{\psi_{\hat{\aleph}}}, & \text{for any } y \in S^n \\
\mathfrak{s}_{\phi_{\hat{\aleph}}}, \mathfrak{s}_{\psi_{\hat{\aleph}}} \in \mathbb{S}_{[0,t]}, & \\
0 \leq \phi_{\hat{\aleph}}^2 + \psi_{\hat{\aleph}}^2 \leq t^2 & \\
\sum_{i=1}^{m} \hat{x}_i = 1,\ \hat{x}_i \geq 0, & i = 1, 2, \ldots, m.
\end{array}\right\} \quad (6.16)$$

Here, $\mathfrak{s}_{\phi_{\hat{\aleph}}} \geq \mathfrak{s}_{\phi_{\aleph^*}}$ and $\mathfrak{s}_{\psi_{\hat{\aleph}}} \leq \mathfrak{s}_{\psi_{\aleph^*}}$ and one of these can be a strict inequality.

Specifically, for $0 < \alpha < 1$, (6.16) can be rewritten as:

$$\left.\begin{array}{l}
\sum_{j=1}^{n}\sum_{i=1}^{m} \hat{x}_i y_j \left(\alpha \ln\left(1 - \frac{\phi_{a_{ij}}^2}{t^2}\right) + (1-\alpha)\ln\left(\frac{\psi_{a_{ij}}}{t}\right)\right) \\
\quad \leq \alpha \ln\left(1 - \frac{\phi_{\hat{\aleph}}^2}{t^2}\right) + (1-\alpha)\ln\left(\frac{\psi_{\hat{\aleph}}}{t}\right), \quad \text{for any } y \in S^n \\
\quad 0 \leq \phi_{\hat{\aleph}}^2 + \psi_{\hat{\aleph}}^2 \leq t^2 \\
\quad \sum_{i=1}^{m} \hat{x}_i = 1,\ \hat{x}_i \geq 0, \quad i = 1, 2, \ldots, m.
\end{array}\right\} \quad (6.17)$$

This gives,

$$\alpha \ln\left(1 - \frac{\phi_{\hat{\aleph}}^2}{t^2}\right) + (1-\alpha)\ln\left(\frac{\psi_{\hat{\aleph}}}{t}\right) \leq \alpha \ln\left(1 - \frac{\phi_{\aleph^*}^2}{t^2}\right) + (1-\alpha)\ln\left(\frac{\psi_{\aleph^*}}{t}\right) \quad (6.18)$$

Let $\hat{\digamma} = \alpha \ln\left(1 - \frac{\phi_{\hat{\aleph}}^2}{t^2}\right) + (1-\alpha)\ln\left(\frac{\psi_{\hat{\aleph}}}{t}\right)$. Using Equations (6.18) in (6.17), it is obtained as

$$\hat{\digamma} \leq \digamma^*.$$

As S^n is a definite compact convex set, we need to determine only the set of extreme points. Thus, (6.16) is remodeled as:

$$\sum_{i=1}^{m} \hat{x}_i \left(\alpha \ln\left(1 - \frac{\phi_{a_{ij}}^2}{t^2}\right) + (1-\alpha)\ln\left(\frac{\psi_{a_{ij}}}{t}\right)\right) \leq \hat{\digamma}, \quad \text{for any } j = 1, 2, \ldots, n,$$

$$0 \leq \phi_{\aleph}^2 + \psi_{\aleph}^2 \leq t^2$$

$$\sum_{i=1}^{m} \hat{x}_i = 1,\ \hat{x}_i \geq 0, \quad i = 1, 2, \ldots, m.$$

i.e., $(\hat{x}, \hat{\digamma})$ is also a feasible solution of **(LPFPI-4)** such that $\hat{\digamma} \leq \digamma^*$.

This is contrary to the statement $(x^*, \digamma^*)$ exists as the optimal solution for **(LPFPI-4)**. Thus, $(x^*, \aleph^*)$ is the Pareto optimal solution of **(LPFPI-1)**.

In an analogous way $(y^*,\ \mathrm{Y}^*)$ can be shown to be the Pareto optimal solution of **(LPFPII-1)**. □

Further on applying Theorem 6.3.1.4, Definitions 6.3.7, and 6.3.8 for each $\alpha \in [0,1]$, from **(LPFPI-4)** the maximin strategy for player I is obtained as x^* and from **(LPFPII-4)** the minimax strategy for player II is obtained as y^*. Thus, $(x^*, y^*, x^{*T}\tilde{A}y^*)$ is obtained as a solution of the uncertain game $LPFG$ involving LP-FVs in the payoffs matrix. Whenever $\phi_{a_{ij}} = t,\ \psi_{a_{ij}} = 0$, then $\ln\left(1 - \frac{\phi_{a_{ij}}^2}{t^2}\right) \to -\infty$ and $\ln\left(\frac{\psi_{a_{ij}}}{t}\right) \to -\infty$. Therefore, the problems **(LPFPI-4)** and **(LPFPII-4)** become invalid. Thus, **(LPFPI-4)** for player I and **(LPFPII-4)** for player II are reformulated as the subsequent non-linear programming problems:

(LPFPI-5)
$$\min\left\{\left(1 - \frac{\phi_{\aleph}^2}{t^2}\right)^{\alpha} \left(\frac{\psi_{\aleph}}{t}\right)^{(1-\alpha)}\right\}$$

s.t.
$$\prod_{i=1}^{m}\left[\left(1 - \frac{\phi_{a_{ij}}^2}{t^2}\right)^{\alpha}\left(\frac{\psi_{a_{ij}}}{t}\right)^{(1-\alpha)}\right]^{x_i}$$

$$\leq \left(1 - \frac{\phi_{\aleph}^2}{t^2}\right)^{\alpha}\left(\frac{\psi_{\aleph}}{t}\right)^{(1-\alpha)}, \qquad j = 1, 2, \ldots, n,$$

$$0 \leq \phi_{\aleph}^2 + \psi_{\aleph}^2 \leq t^2$$

$$\sum_{i=1}^{m} x_i = 1,\ x_i \geq 0,\ i = 1, 2, \ldots, m.$$

and

(LPFPII-5)
$$\min\left\{\left(1 - \frac{\phi_{\mathrm{Y}}^2}{t^2}\right)^{\alpha}\left(\frac{\psi_{\mathrm{Y}}}{t}\right)^{(1-\alpha)}\right\}$$

s.t.
$$\prod_{j=1}^{n}\left[\left(1 - \frac{\phi_{a_{ij}}^2}{t^2}\right)^{\alpha}\left(\frac{\psi_{a_{ij}}}{t}\right)^{(1-\alpha)}\right]^{y_j}$$

$$\leq \left(1-\frac{\phi_{\mathbb{Y}}^2}{t^2}\right)^{\alpha}\left(\frac{\psi_{\mathbb{Y}}}{t}\right)^{(1-\alpha)}, \qquad i=1,2,\ldots,m,$$
$$0 \leq \phi_{\mathbb{Y}}^2+\psi_{\mathbb{Y}}^2 \leq t^2$$
$$\sum_{j=1}^{n} y_j = 1,\ y_j \geq 0,\ j=1,2,\ldots,n.$$

respectively.

On substituting $\mathbb{U}=\left(1-\frac{\phi_{\mathbb{N}}^2}{t^2}\right)^{\alpha}\left(\frac{\psi_{\mathbb{N}}}{t}\right)^{(1-\alpha)}$ in **(LPFPI-5)**, then $0 \leq \mathbb{U} \leq 1$ because $\alpha \in [0,1]$, $0 \leq 1-\frac{\phi_{\mathbb{N}}^2}{t^2} \leq 1$, and $0 \leq 1-\frac{\psi_{\mathbb{N}}}{t} \leq 1$. Thus, **(LPFPI-5)** reduces to the following non-linear programming problem:

(LPFPI-6) $\min\{\mathbb{U}\}$

$$\text{s.t.} \quad \prod_{i=1}^{m}\left[\left(1-\frac{\phi_{a_{ij}}^2}{t^2}\right)^{\alpha}\left(\frac{\psi_{a_{ij}}}{t}\right)^{(1-\alpha)}\right]^{x_i} \leq \mathbb{U}, \qquad j=1,2,\ldots,n,$$
$$0 \leq \mathbb{U} \leq 1,$$
$$\sum_{i=1}^{m} x_i = 1,\ x_i \geq 0,\ i=1,2,\ldots,m.$$

Similarly, taking $\mathbb{V}=\left(1-\frac{\phi_{\mathbb{Y}}^2}{t^2}\right)^{\alpha}\left(\frac{\psi_{\mathbb{Y}}}{t}\right)^{(1-\alpha)}$ in **(LPFPII-5)**. Using the aforementioned assertion we get $0 \leq \mathbb{V} \leq 1$.

(LPFPII-6) $\max\{\mathbb{V}\}$

$$\text{s.t.} \quad \prod_{j=1}^{n}\left[\left(1-\frac{\phi_{a_{ij}}^2}{t^2}\right)^{\alpha}\left(\frac{\psi_{a_{ij}}}{t}\right)^{(1-\alpha)}\right]^{y_j} \geq \mathbb{V}, \qquad i=1,2,\ldots,m,$$
$$0 \leq \mathbb{V} \leq 1,$$
$$\sum_{j=1}^{n} y_j = 1,\ y_j \geq 0,\ j=1,2,\ldots,n.$$

From **(LPFPI-4)**, **(LPFPII-4)**, **(LPFPI-6)**, and **(LPFPII-6)** it is evident that $\mathbb{U}^* = \mathbb{V}^*$, such that $\mathbb{U}^* = e^{\digamma}$ and $\mathbb{V}^* = e^{\wp}$, while $(x^*, \digamma^*)$ and $(y^*, \wp^*)$ exist as the optimal solutions of **(LPFPI-4)** and **(LPFPII-4)**, respectively. Also, $(x^*, \mathbb{U}^*)$ and $(y^*, \mathbb{V}^*)$ are the optimal solutions of (**LPFPI-6**) and (**LPFPII-6**), respectively.

Figure 6.1 explains the entire algorithm for the proposed technique.

6.4 NUMERICAL ILLUSTRATION

This section demonstrates the application of the solution procedure explained in Section 6.3 for a matrix game having LPFVs in the payoffs using a mathematical formulation.

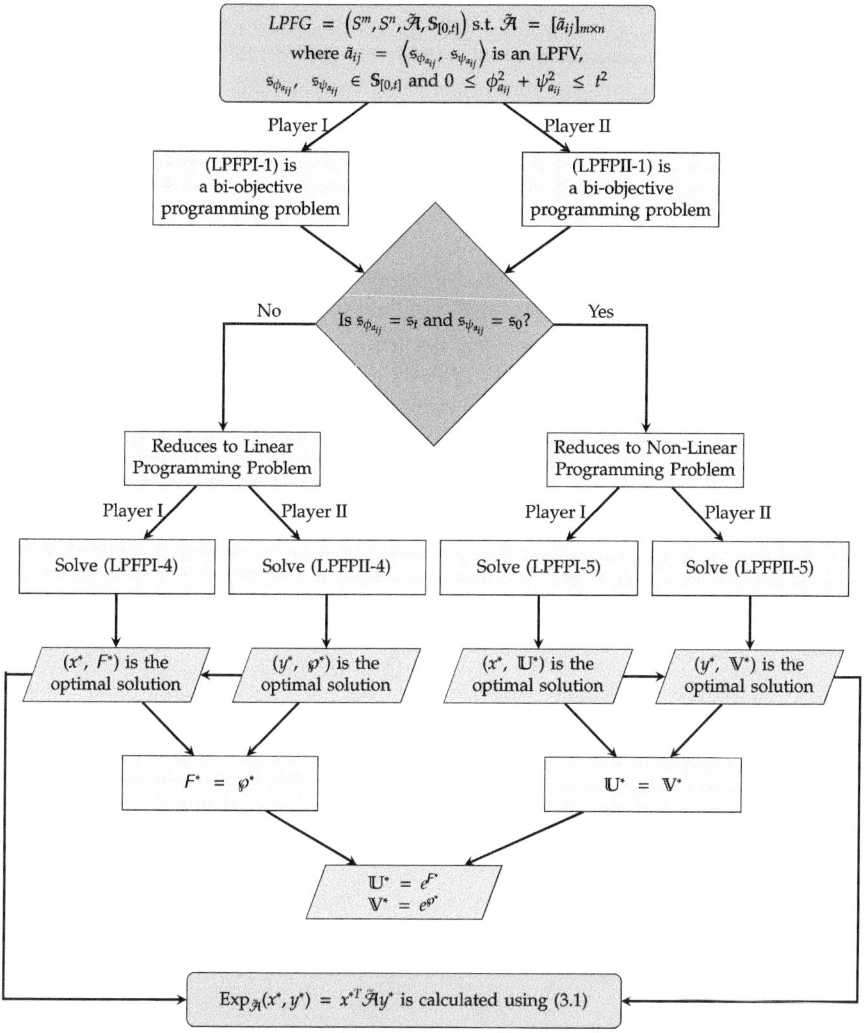

FIGURE 6.1 Algorithm for obtaining the solution for LPFG

Example 6.4.1. Market Sales Problem Suppose that the two ventures, W_1 and W_2 produces laptops. The ventures presume that the demand for laptops is primarily stable. This way, when the return of one venture is increased, at the same time the return of other venture declines. With the aim of accomplishing their goals, the two ventures desire to select their marketing strategies wisely. There seem to be plenty of limitations, but the venture W_1 prefers three fundamental possibilities:

(i) to cut down the cost of the laptops (σ_1); (ii) to upgrade the latest technology (σ_2); (iii) to publicize the laptops in print/e-media and promote their sale at an appropriate price (σ_3).

On the contrary, venture W_2 has three options based on priority: (i) to deliver laptops at a minimum affordable price with home delivery as complimentary (ζ_1); (ii) to publicize the product on e-media and offer a modest price (ζ_2); (iii) to upgrade the laptops with the latest technology (ζ_3).

Both the ventures are struggling with limited inventory simultaneously, leaving them with only one option to choose from. Therefore, the aforementioned ambitious decision problem is considered a matrix game. The ventures W_1 and W_2 can be considered as two players, "player I" and "player II", respectively. The market return that a venture awaits to acquire, depending upon the selection of a strategy by the individual venture, constitutes the payoffs of the game. Moreover, the economic fundamentals are rarely recognized or unrecognized, and they are generally uncertain. As a consequence, a single quantitative value for the return cannot accurately depict the payoff. Therefore, a matrix game involving linguistic Pythagorean fuzzy terms may adequately depict the game from the viewpoints of both players (ventures).

Considering the linguistic terms set $\mathbb{S} = \{\mathfrak{s}_0 = \text{no return}, \mathfrak{s}_1 = \text{very low return}, \mathfrak{s}_2 = \text{low return}, \mathfrak{s}_3 = \text{moderate return}, \mathfrak{s}_4 = \text{slightly good return}, \mathfrak{s}_5 = \text{good return}, \mathfrak{s}_6 = \text{very good return}, \mathfrak{s}_7 = \text{best return}, \mathfrak{s}_8 = \text{excellent returns}\}$, let the payoff matrix involving LPFVs $\tilde{\mathcal{A}}$, exhibiting the given set of circumstances, be

$$\tilde{\mathcal{A}} = \begin{bmatrix} \langle \mathfrak{s}_5, \mathfrak{s}_1 \rangle & \langle \mathfrak{s}_4, \mathfrak{s}_1 \rangle & \langle \mathfrak{s}_1, \mathfrak{s}_4 \rangle \\ \langle \mathfrak{s}_0, \mathfrak{s}_3 \rangle & \langle \mathfrak{s}_7, \mathfrak{s}_1 \rangle & \langle \mathfrak{s}_3, \mathfrak{s}_1 \rangle \\ \langle \mathfrak{s}_0, \mathfrak{s}_2 \rangle & \langle \mathfrak{s}_1, \mathfrak{s}_4 \rangle & \langle \mathfrak{s}_2, \mathfrak{s}_1 \rangle \end{bmatrix}$$

It is significant here that the first element in the payoffs $\langle \mathfrak{s}_5, \mathfrak{s}_1 \rangle$ denotes the market return of venture W_1 when it chooses the first option σ_1 and, simultaneously, the venture W_2 selects its available possibility ζ_1. Similarly, it should be observed that the return on W_2 is expressed as $\langle \mathfrak{s}_1, \mathfrak{s}_5 \rangle$ with same set of possibilities σ_1 and ζ_1 opted for by the two ventures (players) W_1 and W_2, respectively. The subsequent entries in the payoffs can be interpreted in a similar pattern. The problem for the two players are as follows:

For player I

$$\begin{aligned}
\textbf{(PI-1)} \quad & \min\{\mathcal{F}\} \\
\text{s.t.} \quad & x_1\left[\alpha \ln\left(1 - \frac{25}{64}\right) + (1-\alpha)\ln\left(\frac{1}{8}\right)\right] \\
& \quad + x_2\left[\alpha \ln\left(1 - \frac{0}{64}\right) + (1-\alpha)\ln\left(\frac{3}{8}\right)\right] \\
& \quad + x_3\left[\alpha \ln\left(1 - \frac{0}{64}\right) + (1-\alpha)\ln\left(\frac{2}{8}\right)\right] \le \mathcal{F}, \\
& x_1\left[\alpha \ln\left(1 - \frac{16}{64}\right) + (1-\alpha)\ln\left(\frac{1}{8}\right)\right] \\
& \quad + x_2\left[\alpha \ln\left(1 - \frac{49}{64}\right) + (1-\alpha)\ln\left(\frac{1}{8}\right)\right] \\
& \quad + x_3\left[\alpha \ln\left(1 - \frac{1}{64}\right) + (1-\alpha)\ln\left(\frac{4}{8}\right)\right] \le \mathcal{F}, \\
& x_1\left[\alpha \ln\left(1 - \frac{1}{64}\right) + (1-\alpha)\ln\left(\frac{4}{8}\right)\right]
\end{aligned}$$

$$+ x_2\left[\alpha\ln\left(1-\frac{9}{64}\right)+(1-\alpha)\ln\left(\frac{1}{8}\right)\right]$$
$$+ x_3\left[\alpha\ln\left(1-\frac{4}{64}\right)+(1-\alpha)\ln\left(\frac{1}{8}\right)\right]\leq \digamma,$$
$$\digamma\leq 0,$$
$$x_1+x_2+x_3=1,\ x_1,\ x_2,\ x_3\geq 0.$$

For player II

$$\textbf{(PII-1)}\quad \max\{\wp\}$$
$$\text{s.t.}\quad y_1\left[\alpha\ln\left(1-\frac{25}{64}\right)+(1-\alpha)\ln\left(\frac{1}{8}\right)\right]$$
$$+ y_2\left[\alpha\ln\left(1-\frac{16}{64}\right)+(1-\alpha)\ln\left(\frac{1}{8}\right)\right]$$
$$+ y_3\left[\alpha\ln\left(1-\frac{1}{64}\right)+(1-\alpha)\ln\left(\frac{4}{8}\right)\right]\geq \wp,$$
$$y_1\left[\alpha\ln\left(1-\frac{0}{64}\right)+(1-\alpha)\ln\left(\frac{3}{8}\right)\right]$$
$$+ y_2\left[\alpha\ln\left(1-\frac{49}{64}\right)+(1-\alpha)\ln\left(\frac{1}{8}\right)\right]$$
$$+ y_3\left[\alpha\ln\left(1-\frac{9}{64}\right)+(1-\alpha)\ln\left(\frac{1}{8}\right)\right]\geq \wp,$$
$$y_1\left[\alpha\ln\left(1-\frac{0}{64}\right)+(1-\alpha)\ln\left(\frac{2}{8}\right)\right]$$
$$+ y_2\left[\alpha\ln\left(1-\frac{1}{64}\right)+(1-\alpha)\ln\left(\frac{4}{8}\right)\right]$$
$$+ y_3\left[\alpha\ln\left(1-\frac{4}{64}\right)+(1-\alpha)\ln\left(\frac{1}{8}\right)\right]\geq \wp,$$
$$\wp\leq 0,$$
$$y_1+y_2+y_3=1,\ y_1,\ y_2,\ y_3\geq 0.$$

Therefore, **(PI-1)** and **(PII-1)** can be resolved for different values of $\alpha\in[0,1]$, utilizing available tools for solving linear programming problems. In Table 6.1 the optimal solutions and the corresponding expected values of **(PI-1)** and **(PII-1)** are presented. Analogously, for exceptional cases, the non-linear programming models for the two players are derived as follows:

$$\textbf{(PI-2)}\quad \min\{\mathbb{U}\}$$
$$\text{s.t.}\quad \left[\left(1-\frac{25}{64}\right)^{\alpha}\left(\frac{1}{8}\right)^{(1-\alpha)}\right]^{x_1} * \left[\left(1-\frac{0}{64}\right)^{\alpha}\left(\frac{3}{8}\right)^{(1-\alpha)}\right]^{x_2}$$
$$* \left[\left(1-\frac{0}{64}\right)^{\alpha}\left(\frac{2}{8}\right)^{(1-\alpha)}\right]^{x_3}\leq \mathbb{U},$$
$$\left[\left(1-\frac{16}{64}\right)^{\alpha}\left(\frac{1}{8}\right)^{(1-\alpha)}\right]^{x_1} * \left[\left(1-\frac{49}{64}\right)^{\alpha}\left(\frac{1}{8}\right)^{(1-\alpha)}\right]^{x_2}$$

$$
\begin{aligned}
&* \left[\left(1-\frac{1}{64}\right)^{\alpha}\left(\frac{4}{8}\right)^{(1-\alpha)}\right]^{x_3} \leq \mathbb{U},\\
&\left[\left(1-\frac{1}{64}\right)^{\alpha}\left(\frac{4}{8}\right)^{(1-\alpha)}\right]^{x_1} * \left[\left(1-\frac{9}{64}\right)^{\alpha}\left(\frac{1}{8}\right)^{(1-\alpha)}\right]^{x_2}\\
&* \left[\left(1-\frac{4}{64}\right)^{\alpha}\left(\frac{1}{8}\right)^{(1-\alpha)}\right]^{x_3} \leq \mathbb{U},\\
&0 \leq \mathbb{U} \leq 1,\\
&x_1 + x_2 + x_3 = 1,\ x_1,\ x_2, x_3 \geq 0.
\end{aligned}
$$

and

$$
\begin{aligned}
\textbf{(PII-2)} \quad & \max\{\mathbb{V}\}\\
\text{s.t.} \quad & \left[\left(1-\frac{25}{64}\right)^{\alpha}\left(\frac{1}{8}\right)^{(1-\alpha)}\right]^{y_1} * \left[\left(1-\frac{16}{64}\right)^{\alpha}\left(\frac{1}{8}\right)^{(1-\alpha)}\right]^{y_2}\\
& * \left[\left(1-\frac{1}{64}\right)^{\alpha}\left(\frac{4}{8}\right)^{(1-\alpha)}\right]^{y_3} \geq \mathbb{V},\\
& \left[\left(1-\frac{0}{64}\right)^{\alpha}\left(\frac{3}{8}\right)^{(1-\alpha)}\right]^{y_1} * \left[\left(1-\frac{49}{64}\right)^{\alpha}\left(\frac{1}{8}\right)^{(1-\alpha)}\right]^{y_2}\\
& * \left[\left(1-\frac{9}{64}\right)^{\alpha}\left(\frac{1}{8}\right)^{(1-\alpha)}\right]^{y_3} \geq \mathbb{V},\\
& \left[\left(1-\frac{0}{64}\right)^{\alpha}\left(\frac{2}{8}\right)^{(1-\alpha)}\right]^{y_1} * \left[\left(1-\frac{1}{64}\right)^{\alpha}\left(\frac{4}{8}\right)^{(1-\alpha)}\right]^{y_2}\\
& * \left[\left(1-\frac{4}{64}\right)^{\alpha}\left(\frac{1}{8}\right)^{(1-\alpha)}\right]^{y_3} \geq \mathbb{V},\\
& 0 \leq \mathbb{V} \leq 1,\\
& y_1 + y_2 + y_3 = 1,\ y_1, y_2, y_3 \geq 0,
\end{aligned}
$$

respectively.

The non-linear programming models **(PI-2)** and **(PII-2)** are resolved using MATLAB (2021b) software. Their findings have been recorded in Table 6.2 for the ideal values and corresponding expected values.

The results described in Tables 6.1 and 6.2 demonstrate that as the quantity of α varies, distinct mixed strategies for Players I and II are generated, respectively. The values $\digamma^*$, $\wp^*$, $\mathbb{U}^*$ and $\mathbb{V}^*$ are monotonically non-decreasing with respect to α. Furthermore, maximin strategies x^* and minimax strategies y^* obtained via **(PI-1)** and **(PII-1)** are identical as attained by the models in **(PI-2)** and **(PII-2)**, respectively. In addition, $\mathbb{U}^* = e^{\digamma^*}$ and $\mathbb{V}^* = e^{\wp^*}$, with $\mathfrak{s}_{\phi_{a_{ij}}} \neq \mathfrak{s}_t$ and $\mathfrak{s}_{\psi_{a_{ij}}} \neq \mathfrak{s}_0$, $i, j = 1, 2, 3$.

TABLE 6.1
The Optimal Solutions and the Expected Value Attained for (PI-1) and (PII-1)

	Player I (W_1)		**Player II** (W_2)		
α	x^*	$\mathcal{F}^*$	y^*	$\wp^*$	$\text{Exp}_{\tilde{\mathcal{A}}}(x^*, y^*)$
0.10	(0.3713, 0.2271, 0.4015)	−1.4148	(0.4862, 0.1251, 0.3887)	−1.4148	$(\mathfrak{S}_{3.1687}, \mathfrak{S}_{1.6924})$
0.20	(0.3613, 0.2134, 0.4253)	−1.2759	(0.4908, 0.1090, 0.4001)	−1.2759	$(\mathfrak{S}_{3.0764}, \mathfrak{S}_{1.6891})$
0.30	(0.3497, 0.1984, 0.4519)	−1.1357	(0.4926, 0.0921, 0.4153)	−1.1357	$(\mathfrak{S}_{2.9754}, \mathfrak{S}_{1.6825})$
0.40	(0.3360, 0.1819, 0.4822)	−0.9938	(0.4904, 0.0741, 0.4355)	−0.9938	$(\mathfrak{S}_{2.7270}, \mathfrak{S}_{1.6717})$
0.50	(0.3195, 0.1634, 0.5171)	−0.8498	(0.4821, 0.0550, 0.4629)	−0.8498	$(\mathfrak{S}_{2.7354}, \mathfrak{S}_{1.6551})$
0.60	(0.2992, 0.1426, 0.5582)	−0.7032	(0.4649, 0.0348, 0.5003)	−0.7032	$(\mathfrak{S}_{2.5906}, \mathfrak{S}_{1.6268})$
0.70	(0.2735, 0.1186, 0.6080)	−0.5531	(0.4338, 0.0134, 0.5528)	−0.5531	$(\mathfrak{S}_{2.4261}, \mathfrak{S}_{1.5835})$

TABLE 6.2
The Optimal Solutions and the Expected Value of the Non-Linear Programming Problems (PI-2) and (PII-2)

	Player I (W_1)		**Player II** (W_2)		
α	x^*	$\mathbb{U}^*$	y^*	$\mathbb{V}^*$	$\text{Exp}_{\tilde{\mathcal{A}}}(x^*, y^*)$
0.10	(0.3713, 0.2271, 0.4015)	0.2430	(0.4862, 0.1251, 0.3887)	0.2430	$(\mathfrak{S}_{3.1687}, \mathfrak{S}_{1.6924})$
0.20	(0.3613, 0.2134, 0.4253)	0.2792	(0.4908, 0.1090, 0.4001)	0.2792	$(\mathfrak{S}_{3.0764}, \mathfrak{S}_{1.6891})$
0.30	(0.3497, 0.1984, 0.4519)	0.3212	(0.4926, 0.0921, 0.4153)	0.3212	$(\mathfrak{S}_{2.9754}, \mathfrak{S}_{1.6825})$
0.40	(0.3360, 0.1819, 0.4822)	0.3702	(0.4904, 0.0741, 0.4355)	0.3702	$(\mathfrak{S}_{2.7270}, \mathfrak{S}_{1.6717})$
0.50	(0.3195, 0.1634, 0.5171)	0.4275	(0.4821, 0.0550, 0.4629)	0.4275	$(\mathfrak{S}_{2.7354}, \mathfrak{S}_{1.6551})$
0.60	(0.2992, 0.1426, 0.5582)	0.4950	(0.4649, 0.0348, 0.5003)	0.4950	$(\mathfrak{S}_{2.5906}, \mathfrak{S}_{1.6268})$
0.70	(0.2735, 0.1186, 0.6080)	0.5751	(0.4338, 0.0134, 0.5528)	0.5751	$(\mathfrak{S}_{2.4261}, \mathfrak{S}_{1.5835})$

Example 6.4.2. Tourists Hotel Selection Problem
Suppose that two hotels, $\mathcal{H}_1$ and $\mathcal{H}_2$, are at a hill station. When the number of possible guests at one hotel rises, the number of potential guests at other hotels decreases. Owing to the very competing economic world, it is essential for hotel operators to comprehend how prospective guests select their hotel and which factors have been given preference. The two hostelries want to choose their course of action meticulously in order to achieve their objectives of inviting the most tourists. Although there appear to be many restrictions, the hostelry $\mathcal{H}_1$ privileged the following three key criteria:

(i) ambience and food quality (γ_1); (ii) level of price and value for money (γ_2); and (iii) cleanliness, comfort and service (γ_3).

On the contrary, hostelry $\mathcal{H}_2$ has three options based on priority: (i) location and noise level (ξ_1); (ii) bed quality and food quality (ξ_2); and (iii) cost range and value for money (ξ_3).

Due to their continuous struggles with scarce resources in hilly terrain, both hostelries are forced to select the best option available. The aforementioned difficult decision dilemma is therefore characterized as a matrix game. The hotels $\mathcal{H}_1$ and $\mathcal{H}_2$ are regarded as "player I" and "player II", respectively. The players' rewards rely on which player chooses the best criterion for its maximum number of allowable guests. Additionally, it is often unknown how prospective guests choose their hotel and which factors are given preference in their decision-making processes as it is primarily based on internet reviews of former guests. As a result, the rewards cannot be accurately represented by a single quantitative value for the yield. Consequently, a matrix game using linguistic Pythagorean fuzzy concepts may be able to accurately represent the game from the perspectives of both hostelries (players).

Considering the linguistic terms set $\mathbb{S} = \{\mathcal{s}_0$ = equal reward (ER), $\mathcal{s}_1$ = very very weak reward (VVWR), $\mathcal{s}_2$ = very weak reward (VWR), $\mathcal{s}_3$ = weak reward (WR), $\mathcal{s}_4$ = moderate reward (MR), $\mathcal{s}_5$ = strong reward (SR), $\mathcal{s}_6$ = very strong reward (VSR), $\mathcal{s}_7$ = very very strong reward (VVSR), $\mathcal{s}_8$ = excessive reward (ExR)$\}$, the payoff matrix for LPFVs, $\tilde{\mathcal{B}}$, displaying the specified combination of factors, shall be

$$\tilde{\mathcal{B}} = \begin{bmatrix} \langle \mathcal{s}_5, \mathcal{s}_1 \rangle & \langle \mathcal{s}_6, \mathcal{s}_1 \rangle & \langle \mathcal{s}_1, \mathcal{s}_6 \rangle \\ \langle \mathcal{s}_0, \mathcal{s}_4 \rangle & \langle \mathcal{s}_1, \mathcal{s}_7 \rangle & \langle \mathcal{s}_1, \mathcal{s}_3 \rangle \\ \langle \mathcal{s}_0, \mathcal{s}_3 \rangle & \langle \mathcal{s}_2, \mathcal{s}_5 \rangle & \langle \mathcal{s}_3, \mathcal{s}_2 \rangle \end{bmatrix}$$

It is significant here that the first element in the payoffs $\langle s_5, s_1 \rangle$ denotes the reward of hotel $\mathcal{H}_1$ when it chooses the first option γ_1 and simultaneously, the hostelry $\mathcal{H}_2$ selects his available criteria ξ_1. Similarly, it should be observed that the reward of $\mathcal{H}_2$ is expressed as $\langle s_1, s_5 \rangle$ with the same set of criterion γ_1 and ξ_1 opted by the two hotels (players) $\mathcal{H}_1$ and $\mathcal{H}_2$, respectively. A similar phenomenon can be used to analyze the payoffs of the subsequent entries.

The results of the problem are tabulated in Tables 6.3 and 6.4. It clearly shows that the criterion γ_3 is most preferred by hotel $\mathcal{H}_1$ (player I), i.e., cleanliness, comfort, and service ratherthan the ambience and food quality (γ_1), whereas the level of price

TABLE 6.3
The Optimal Solutions and the Expected Value Attained for Linear Programming Problems for Example 6.4.2

	Player I ($\mathcal{H}_1$)		Player II ($\mathcal{H}_2$)		
α	x^*	F^*	y^*	$\wp^*$	$\mathrm{Exp}_{\tilde{A}}(x^*, y^*)$
0.10	(0.3298, 0.0, 0.6702)	−0.9323	(0.0, 0.3966, 0.6034)	−0.9323	$(\mathcal{s}_{3.3255}, \mathcal{s}_{2.8999})$
0.20	(0.3199, 0.0, 0.6801)	−0.8495	(0.0, 0.3862, 0.6138)	−0.8495	$(\mathcal{s}_{3.2913}, \mathcal{s}_{2.8976})$
0.30	(0.3083, 0.0, 0.6917)	−0.7662	(0.0, 0.3740, 0.6260)	−0.7662	$(\mathcal{s}_{3.2525}, \mathcal{s}_{2.8930})$
0.40	(0.2946, 0.0, 0.7054)	−0.6822	(0.0, 0.3596, 0.6404)	−0.6822	$(\mathcal{s}_{3.2086}, \mathcal{s}_{2.8847})$
0.50	(0.2782, 0.0, 0.7218)	−0.5972	(0.0, 0.3423, 0.6577)	−0.5972	$(\mathcal{s}_{3.1588}, \mathcal{s}_{2.8707})$
0.60	(0.2582, 0.0, 0.7418)	−0.5110	(0.0, 0.3212, 0.6788)	−0.5110	$(\mathcal{s}_{3.1027}, \mathcal{s}_{2.8478})$
0.70	(0.2330, 0.0, 0.7670)	−0.4230	(0.0, 0.2947, 0.7053)	−0.4230	$(\mathcal{s}_{3.0397}, \mathcal{s}_{2.8101})$

TABLE 6.4
The Optimal Solutions and the Expected Value of the Non-Linear Programming Problems for Example 6.4.2

	Player I ($\mathcal{H}_1$)		Player II ($\mathcal{H}_2$)		
α	x^*	$\mathbb{U}^*$	y^*	$\mathbb{V}^*$	$\text{Exp}_{\tilde{\mathcal{A}}}(x^*, y^*)$
0.10	(0.3298, 0.0, 0.6702)	0.3937	(0.0, 0.3966, 0.6034)	0.3937	$(\mathcal{s}_{3.3255}, \mathcal{s}_{2.8999})$
0.20	(0.3199, 0.0, 0.6801)	0.4276	(0.0, 0.3862, 0.6138)	0.4276	$(\mathcal{s}_{3.2913}, \mathcal{s}_{2.8976})$
0.30	(0.3083, 0.0, 0.6917)	0.4648	(0.0, 0.3740, 0.6260)	0.4648	$(\mathcal{s}_{3.2525}, \mathcal{s}_{2.8930})$
0.40	(0.2946, 0.0, 0.7054)	0.5055	(0.0, 0.3596, 0.6404)	0.5055	$(\mathcal{s}_{3.2086}, \mathcal{s}_{2.8847})$
0.50	(0.2782, 0.0, 0.7218)	0.5504	(0.0, 0.3423, 0.6577)	0.5504	$(\mathcal{s}_{3.1588}, \mathcal{s}_{2.8707})$
0.60	(0.2582, 0.0, 0.7418)	0.5999	(0.0, 0.3212, 0.6788)	0.5999	$(\mathcal{s}_{3.1027}, \mathcal{s}_{2.8478})$
0.70	(0.2330, 0.0, 0.7670)	0.6551	(0.0, 0.2947, 0.7053)	0.6551	$(\mathcal{s}_{3.0397}, \mathcal{s}_{2.8101})$

(γ_2) is the least preferred for best reward. At the same time, $\mathcal{H}_2$ (player II) preferred criteria ξ_3, cost range, and value for money as compared to bed and food quality (ξ_2), whereas they gave least priority to location of the hotel (ξ_1) to maximize its reward.

6.5 COMPARATIVE STUDY

Verma and Aggarwal (2021) recently conducted a comprehensive study of matrix games incorporating linguistic intuitionistic fuzzy numbers as payoffs. They attained the solution by suggesting a novel aggregation operator based on the linguistic scale function. It is interesting to observe that our suggested method can even be used to resolve their problem. Considering, Example 6.4.1 in Section 6.4 of Verma and Aggarwal (2021), where LTS $\mathbb{S} = \{\mathcal{s}_0$ = very very low (VVL), $\mathcal{s}_1$ = very low (VL), $\mathcal{s}_2$ = moderately low (ML), $\mathcal{s}_3$ = slightly low (SL), $\mathcal{s}_4$ = average (Avg), $\mathcal{s}_5$ = slightly high (SH), $\mathcal{s}_6$ = moderately high (MH), $\mathcal{s}_7$ = very high (VH), $\mathcal{s}_8$ = very very high (VVH)$\}$. The two companies $\mathfrak{C}_1$ and $\mathfrak{C}_2$, considered as two players, are manufacturers of 3D printers. Then, **(LPFPI-4)** and **(LPFPII-4)**, can be solved for the two players respectively, except for the case when $\mathcal{s}_{\phi_{a_{ij}}} = \mathcal{s}_t$ and $\mathcal{s}_{\psi_{a_{ij}}} = \mathcal{s}_0$, i.e. $\frac{\phi_{a_{ij}}}{t} = 1$ and $\frac{\psi_{a_{ij}}}{t} = 0$. Particularly, when $\mathcal{s}_{\phi_{a_{ij}}} = \mathcal{s}_t$ and $\mathcal{s}_{\psi_{a_{ij}}} = \mathcal{s}_0$, we solve **(LPFPI-5)** and **(LPFPII-5)** for player I and player II, respectively. The results are recorded in Tables 6.5 and 6.6, respectively. The findings demonstrate that the proposed technique can be applied to Verma and Aggarwal (2021)'s example and the optimal strategies obtained by our proposed technique are quite similar to their strategies and have the same level of preference. According to the suggested approach, for $\lambda = 0.10$ for $\mathfrak{C}_1$ (player I) the mixed strategies obtained are (0.5593, 0.3059, 0.1347), which explains that player I preferred to play strategy x_1 over x_2 and x_3, while strategy x_3 is least favorable to achieve maximum market share. Similarly, for $\mathfrak{C}_2$ (player II), for $\lambda = 0.10$

TABLE 6.5
The Optimal Solutions and the Expected Value Attained for Linear Programming Problems (Example 6.4.1 in Verma and Aggarwal (2021))

	Player I ($\mathfrak{C}_1$)		Player II ($\mathfrak{C}_2$)		
α	x^*	$\digamma^*$	y^*	$\wp^*$	$Exp_{\tilde{\mathcal{A}}}(x^*, y^*)$
0.10	(0.5593, 0.3059, 0.1347)	−1.3774	(0.1373, 0.2817, 0.5810)	−1.3774	$(\mathfrak{s}_{5.4876}, \mathfrak{s}_{1.8585})$
0.20	(0.5345, 0.3058, 0.1597)	−1.2962	(0.1644, 0.2816, 0.5540)	−1.2962	$(\mathfrak{s}_{5.5437}, \mathfrak{s}_{1.8641})$
0.30	(0.5136, 0.3031, 0.1834)	−1.2168	(0.1897, 0.2791, 0.5312)	−1.2168	$(\mathfrak{s}_{5.5796}, \mathfrak{s}_{1.8712})$
0.40	(0.4961, 0.2977, 0.2062)	−1.1387	(0.2139, 0.2742, 0.5119)	−1.1387	$(\mathfrak{s}_{5.6009}, \mathfrak{s}_{1.8788})$
0.50	(0.4818, 0.2898, 0.2285)	−1.0616	(0.2372, 0.2669, 0.4960)	−1.0616	$(\mathfrak{s}_{5.6124}, \mathfrak{s}_{1.8840})$
0.60	(0.4704, 0.2791, 0.2505)	−0.9849	(0.2600, 0.2571, 0.4830)	−0.9849	$(\mathfrak{s}_{5.6159}, \mathfrak{s}_{1.8871})$

TABLE 6.6
The Optimal Solutions and the Expected Value for the Non-Linear Programming Problems (Example 6.4.1 in Verma and Aggarwal (2021))

	Player I ($\mathfrak{C}_1$)		Player II ($\mathfrak{C}_2$)		
α	x^*	$\mathbb{U}^*$	y^*	$\mathbb{V}^*$	$Exp_{\tilde{\mathcal{A}}}(x^*, y^*)$
0.10	(0.5593, 0.3059, 0.1347)	0.2522	(0.1373, 0.2817, 0.5810)	0.2522	$(\mathfrak{s}_{5.4876}, \mathfrak{s}_{1.8585})$
0.20	(0.5345, 0.3058, 0.1597)	0.2736	(0.1644, 0.2816, 0.5540)	0.2736	$(\mathfrak{s}_{5.5437}, \mathfrak{s}_{1.8641})$
0.30	(0.5136, 0.3031, 0.1834)	0.2962	(0.1897, 0.2791, 0.5312)	0.2962	$(\mathfrak{s}_{5.5796}, \mathfrak{s}_{1.8712})$
0.40	(0.4961, 0.2977, 0.2062)	0.3202	(0.2139, 0.2742, 0.5119)	0.3202	$(\mathfrak{s}_{5.6009}, \mathfrak{s}_{1.8788})$
0.50	(0.4818, 0.2898, 0.2285)	0.3459	(0.2372, 0.2669, 0.4960)	0.3459	$(\mathfrak{s}_{5.6124}, \mathfrak{s}_{1.8840})$
0.60	(0.4704, 0.2791, 0.2505)	0.3735	(0.2600, 0.2571, 0.4830)	0.3735	$(\mathfrak{s}_{5.6159}, \mathfrak{s}_{1.8871})$

the mixed strategies obtained are $(0.1373, 0.2817, 0.5810)$, these probabilities explicitly demonstrate that y_3 is most preferred strategy over y_2 and y_1, whereas y_1 is least preferred to maximize its market share. Thus, it is evident that due to LIFS being a specialized version of LPFS, our suggested technique can be utilized to find the best strategy in the LIFS matrix game.

6.6 CONCLUDING NOTE

Throughout the investigation, we explore the matrix games explicitly, in preference to a qualitative framework. Inspired by Garg (2018b), the matrix games incorporating uncertain payoffs characterized by linguistic Pythagorean fuzzy variables have been initiated. The expected value of the game is computed by implementing the linguistic Pythagorean fuzzy weighted average operator (LPFWA). Additionally, the non-linear bi-objective programming problems for each player are translated to a comparable pair of linear/non-linear programming problems using the order relationships between LPFVs to identify the maximin and minimax strategies for the two players. Conclusively, a real-life decision-making scenario is employed to validate the solution technique.

The significance of the proposed technique is that we might precisely define belongingness and non-belongingness of an entity by using LPFSs. Owing to the reduced probability of data loss or distortion during the resolution of these uncertain matrix games employing LPFSs, players can ultimately choose the optimum course of action. The proposed method could be used to analyze uncertain multiobjective matrix games involving LPFSs in the future. The suggested methodology can be extended to resolve matrix games involving interval-valued linguistic Pythagorean fuzzy numbers. With the proposed framework, even bi-matrix games incorporating LPFSs can be further explored.

REFERENCES

Aggarwal A, Dubey D, Chandra S, Mehra A (2012a) Application of Atanassov's I-fuzzy set theory to matrix games with fuzzy goals and fuzzy payoffs. Fuzzy Information and Engineering 4:401–414.

Aggarwal A, Mehra A, Chandra S (2012b) Application of linear programming with I-fuzzy sets to matrix games with I-fuzzy goals. Fuzzy Optimization and Decision Making 11:465–480.

Atanassov KT (1986) Intuitionistic fuzzy sets. Fuzzy Sets and Systems 20:87–96.

Atanassov KT (1995) Ideas for intuitionistic fuzzy equations, inequalities and optimization. Notes on Intuitionistic Fuzzy Sets 1:17–24.

Atanassov KT (1999) Intuitionistic Fuzzy Sets: Theory and Applications. Physica-Verlag HD.

Aubin J (1981) Cooperative fuzzy games. Mathematics of Operations Research 6(1):1–13.

Barron EN (2013) Game theory: An introduction. John Wiley & Sons.

Bector C, Chandra S (2005) Fuzzy mathematical programming and fuzzy matrix games. Springer-Verlag.

Bector C, Chandra S, Vidyottama V (2004a) Matrix games with fuzzy goals and fuzzy linear programming duality. Fuzzy Optimization and Decision Making 3:255–269.

Bector CR, Chandra S, Vijay V (2004b) Duality in linear programming with fuzzy parameters and matrix games with fuzzy pay-offs. Fuzzy Sets and Systems 146:253–269.

Bhaumik A, Roy SK (2021) Intuitionistic interval-valued hesitant fuzzy matrix games with a new aggregation operator for solving management problem. Granular Computing 6(2):359–375.

Bhaumik A, Roy SK, Li D (2017) Analysis of triangular intuitionistic fuzzy matrix games using robust ranking. Journal of Intelligent and Fuzzy Systems 33(1):327–336.

Bustince H, Kacprzyk J, Mohedano V (2000) Intuitionistic fuzzy generators application to intuitionistic fuzzy complementation. Fuzzy Sets and Systems 114(3):485–504.

Butnariu D (1978) Fuzzy games: A description of the concept. Fuzzy Sets and Systems 1(3):181–192.

Campos L (1989) Fuzzy linear programming models to solve fuzzy matrix games. Fuzzy Sets and Systems 32:275–289.

Chen TY (2019) Multiple criteria decision analysis under complex uncertainty: a Pearson-like correlation-based Pythagorean fuzzy compromise approach. International Journal of Intelligent Systems 34(1):114–151.

Collins WD, Hu C (2008) Studying interval valued matrix games with fuzzy logic. Soft Computing 12:147–155.

De SK, Biswas R, Roy AR (2001) An application of intuitionistic fuzzy sets in medical diagnosis. Fuzzy Sets and Systems 117:209–213.

Deng X, Wang J, Wei G (2019) Some 2-tuple linguistic Pythagorean Heronian mean operators and their application to multiple attribute decision-making. Journal of Experimental & Theoretical Artificial Intelligence 31(4):555–574.

Ding XF, Liu HC (2019) A new approach for emergency decision-making based on zero-sum game with Pythagorean fuzzy uncertain linguistic variables. Journal of Intelligent Systems 156:1–18.

Garg H (2016) A novel accuracy function under interval-valued Pythagorean fuzzy environment for solving multicriteria decision making problem. Journal of Intelligent Fuzzy Systems 31(1):529–540.

Garg H (2018a) Hesitant Pythagorean fuzzy sets and their aggregation operators in multiple attribute decision-making. International Journal of Uncertainty Quantification 8(3): 267–289.

Garg H (2018b) Linguistic Pythagorean fuzzy sets and its applications in multiattribute decision-making process. International Journal of Intelligent Systems 33(6): 1234–1263.

Garg H (2020) Neutrality operations-based Pythagorean fuzzy aggregation operators and its applications to multiple attribute group decision-making process. Journal of Ambient Intelligent Humaniz Computation 11(7):3021–3041.

Garg H, Kumar K (2019) A linear programming method based on an improved score function for interval-valued Pythagorean fuzzy numbers and its application to decision-making. IEEE Transactions on Fuzzy Sets 27:2302–2311.

Geng Y, Liu P, Teng F, Liu Z (2017) Pythagorean fuzzy uncertain linguistic TODIM method and their application to multiple criteria group decision making. Journal of Intelligent & Fuzzy Systems 33(6):3383–3395.

Han Y, Deng Y, Cao Z, Lin CT (2020) An interval-valued Pythagorean prioritized operator-based game theoretical framework with its applications in multicriteria group decision making. Neural Computing and Applications 32(12):7641–7659.

Herrera F, Martinez L (2000) A 2-tuple fuzzy linguistic represent model for computing with words. IEEE Transactions for Fuzzy Systems 8:746–752.

Herrera F, Martinez L (2001) A model based on linguistic 2-tuples for dealing with multigranular hierarchical linguistic contexts in multi-expert decision-making. IEEE Transactions on Systems, Man, and Cybernetics 31:227–234.

Hung WL, Yang MS (2008) On the J-divergence of intuitionistic fuzzy sets with its application to pattern recognition. Information Sciences 178(6):1641–1650.

Jana J, Roy SK (2022) Linguistic Pythagorean hesitant fuzzy matrix game and its application in multi-criteria decision making. Applied Intelligence: 1–22.

Kalpiński G, Tamošaitiene (2010) Game theory applications in construction engineering and management. Ukio Technologins ir Ekonominis Vystymas 16(2):348–363.

Khan I, Aggarwal A, Mehra A (2016) Solving I-fuzzy bi-matrix games with I-fuzzy goals by resolving indeterminacy. Journal of Uncertain Systems 10:204–222.

Khan I, Aggarwal A, Mehra A, Chandra S (2017) Solving matrix games with Atanassov's I-fuzzy goals via indeterminacy resolution approach. Journal of Information and Optimization Sciences 38:259–287.

Leng M, Parlar M (2005) Game theoretic applications in supply chain management: A review. INFOR 43(3):187–220.

Li DF (2010a) Mathematical-programming approach to matrix games with payoffs represented by Atanassov's interval-valued intuitionistic fuzzy sets. IEEE Transactions on Fuzzy Systems 18:1112–1128.

Li DF (2010b) TOPSIS-based nonlinear-programming methodology for multiattribute decision making with interval-valued intuitionistic fuzzy sets. IEEE Transactions on Fuzzy Systems 18:299–311.

Li DF, Chuntian C (2002) New similarity measures of intuitionistic fuzzy sets and application to pattern recognitions. Pattern Recognition Letters 23:221–225.

Li DF, Nan JX (2009) A nonlinear programming approach to matrix games with payoffs of Atanassov's intuitionistic fuzzy sets. International Journal of Uncertainty, Fuzziness and Knowledge-Based Systems 17:585–607.

Madani K (2010) Game theory and water resources. Journal of Hydrology 381(3-4):225–238.

Maeda T (2003) On characterization of equilibrium strategy of two-person zero-sum games with fuzzy payoffs. Fuzzy Sets and Systems 139:283–296.

Nan JX, Li DF, Zhang MJ (2010) A lexicographic method for matrix games with payoffs of triangular intuitionistic fuzzy numbers. International Journal of Computational Intelligence Systems 3:280–289.

Nan JX, Li DF, An JJ (2017) Solving bi-matrix games with intuitionistic fuzzy goals and intuitionistic fuzzy payoffs. Journal of Intelligent Fuzzy Systems 33:3723–3732.

Naqvi D, Aggarwal A, Sachdev G, Khan I (2021) Solving I-fuzzy two person zero-sum matrix games: Tanaka and Asai approach. Granular Computing 6(2):399–409.

Neumann V, Morgenstern O (1944) Theory of games and economic behavior. Princeton University Press.

Owen G (1982) Game theory. New York: Academic Press.

Peng X, Selvachandran G (2019) Pythagorean fuzzy set: state of the art and future directions. International Journal of Intelligent Systems 52:1873–1927.

Peng X, Yang Y (2016) Fundamental properties of interval-valued Pythagorean fuzzy aggregation operators. International Journal of Intelligent Systems 5:444–487.

Sakawa M, Nishizaki I (1994) Max-min solutions for multiobjective matrix games. Fuzzy Sets and Systems 61:265–275.

Seikh MR, Nayak PK, Pal M (2013) Matrix games in intuitionistic fuzzy environment. International Journal of Mathematics in Operational Research 5:693–708.

Seikh MR, Nayak PK, Pal M (2015) Solving bi-matrix games with pay-offs of triangular intuitionistic fuzzy numbers. European Journal of Pure and Applied Mathematics 8:153–171.

Singh A, Gupta A, Mehra A (2018) Matrix games with 2-tuple linguistic information. Annals of Operations Research 287:895–910.

Steuer R (1986) Multiple Criteria Optimization-theory, Computation and Application. New York, Wiley.

Szmidt E, Kacprzyk J (1996) Remarks on some applications of intuitionistic fuzzy sets in decision making. Notes on Intuitionistic Fuzzy Sets 2:22–31.

Tanaka H, Asai K (1984) Fuzzy linear programming problems with fuzzy numbers. Fuzzy sets and systems 13:1–10.

Verma R, Aggarwal A (2021) Matrix games with linguistic intuitionistic fuzzy payoffs: Basic results and solution methods. Artificial Intelligence Review 54:5127–5162.

Vijay V, Chandra S, Bector CR (2005) Matrix games with fuzzy goals and fuzzy payoffs. Omega 33(5):425–429.

Vijay V, Mehra A, Chandra S, Bector CR (2007) Fuzzy matrix games via a fuzzy relation approach. Fuzzy Optimization and Decision Making 6:299–314.

Wang J, Wei G, Gao H (2018) Approaches to multiple attribute decision making with interval-valued 2-tuple linguistic Pythagorean fuzzy information. Mathematics 6(10):201.
Xia M (2018) Interval-valued intuitionistic fuzzy matrix games based on archimedian t-conorm and t-norm. International Journal of General Systems 47:278–293.
Xing Y, Qiu D (2019) Solving triangular intuitionistic fuzzy matrix game by applying the accuracy function method. Symmetry 11(10):1258.
Xu C, Meng F, Zhang Q (2017) PN equilibrium strategy for matrix games with fuzzy payoffs. Journal of Intelligent and Fuzzy Systems 32:2195–2206.
Yager RR (2013) Pythagorean fuzzy subsets. In: 2013 joint IFSA world congress and NAFIPS annual meeting (IFSA/NAFIPS), IEEE: 57–61.
Yager RR (2014) Pythagorean membership grades in multicriteria decision making. IEEE Transactions on Fuzzy Systems 22:958–965.
Zadeh LA (1965) Fuzzy sets. Information and Control 8:338–353.
Zadeh LA (1975a) The concept of a linguistic variable and its application approximate reasoning-I. Information Sciences 8(3):199–240.
Zadeh LA (1975b) The concept of a linguistic variable and its application approximate reasoning-II. Information Sciences 8(4):301–357.
Zeng S, Chen J, Li X (2016) A hybrid method for Pythagorean fuzzy multiple-criteria decision making. International Journal of Information Technology & Decision Making 15(2): 403–422.
Zhang S, Zhang Y (2003) Introduction to game theory. Chinese Science Bulletin 48(9):841–846.
Zhang W, Ju Y, Liu X (2017) Interval-valued intuitionistic fuzzy programming technique for multicriteria group decision making based on shapley values and incomplete preference information. Soft Computing 21:5787–5804.
Zhao Y, Wang S, Cheng TE, Yang X, Huang Z (2010) Coordination of supply chains by option contracts: A cooperative game theory approach. European Journal of Operational Research 207(2):668–675.
Zhou X, Li W, Lin Z, Li S (2010) Two-person zero-sum matrix game based on intuitionistic fuzzy set. In: 2010 8th World Congress on Intelligent Control and Automation, IEEE: 2807–2810.

7 Cyclic Surgery Scheduling using Variations of Cohort Intelligence

Mandar S. Sapre[1], *Neil Dsouza*[1], *Ishaan R. Kale*[2], *Saksham Agarwal*[1] *and Abhishek Phadke*[1]
[1]Symbiosis Institute of Technology, Symbiosis International (Deemed University), Pune, MH, India
[2]Institute of Artificial Intelligence, MIT-WPU, Pune, MH, India

CONTENTS

7.1 INTRODUCTION

The problems from the engineering domain are complex in nature, and the traditional optimization techniques may not be suitable to solve such problems. So, the researchers proposed various heuristic and metaheuristic algorithms and employed them to solve complex problems. These techniques are generally inspired by nature. These methods are becoming increasingly popular because of their use of simple rules in the process of searching for optimal solutions to computational problems.

Healthcare scheduling is one of the real-world applications for providing an elite service to the patients with least cost. There are several types of problems considered in literature associated to the healthcare scheduling domain such as the patients' admission scheduling problem, nurse scheduling problem, operation room scheduling problem, and surgery scheduling problem. Specifically, this work is associated with reducing the waiting period of the patients in the recovery room after the surgery and ease of access to the medical services (Fei et al., 2010; Hall et al., 2012; Kulkarni et al., 2016). In order to achieve this objective, the hospital resources play a vital role in the

DOI: 10.1201/9781003329039-7

implementation of a hospital's benefit and service quality. It may also help to give the cost-effective service to the patients and to enhance the healthcare scheduling system (Gupta & Denton, 2008). There are several metaheuristic algorithms which were used to solve health care scheduling domain problems such as Particle Swarm Optimization (Cheung, 2009), Harmony Search (Awadallah et al., 2013), Tabu Search (Li et al., 2018), Simulated Annealing (Turhan & Bilgen, 2020), Evolutionary algorithm (Zhuo et al., 2015), Ant Colony Optimization (Ramli et al., 2019), Bee Colony (Rajeswari et al., 2017), etc.

The Cohort Intelligence (CI) algorithm is a socio-inspired metaheuristic that has proven to be suitable and efficient for solving unconstrained and constrained problems (Kulkarni et al., 2013). The algorithm is inspired by the self-supervised learning behavior of the candidates in a cohort. Every candidate aims for the same target and iteratively tries to improve their behavior by adopting peers' behavior. Patankar & Kulkarni (2018) proposed six variations of this CI algorithm: follow-best, follow-better, follow-worst, follow-itself, follow median, and alienate-and-random selection. These variations were validated by solving seven multimodal and three uni-modal unconstrained test functions. A multi CI algorithm developed by Shastri & Kulkarni (2018) focuses on intra-group and inter-group learning mechanisms amongst different cohorts. A new hybrid variant of Genetic Algorithm (GA), utilizing the sample space reduction technique of the CI algorithm, referred to as the Adaptive Range Genetic Algorithm (ARGA) was also developed (Iyer et al., 2019). The hybrid CI-SAPF-CBO overcomes the limitation of CI to avoid the setting of computational parameters (i.e., sampling space reduction factor r); whereas, the Self-Adaptive Penalty Function (SAPF) is the modified version of the Penalty Function approach which does not require tuning of penalty parameters (Kale & Kulkarni, 2021).

7.2 VARIATIONS OF CI

Patankar & Kulkarni (2018) developed six approaches for the following mechanism of CI, as explained below.

1. **Follow best rule:** Every candidate follows the best candidate in the cohort.
2. **Follow better rule:** Every candidate follows the subsequent best candidate and the best candidate follows itself.
3. **Follow worst rule:** Every candidate follows the worst candidate in the cohort.
4. **Follow itself rule:** Every candidate follows itself.
5. **Follow median rule:** Every candidate follows a candidate with the median probability.
6. **Alienation and random selection rule:** Every candidate alienates a randomly chosen candidate in the cohort, i.e., it does not follow a particular candidate at all. It randomly chooses another candidate to follow from the other candidates including itself.

The variations provide an insight into the varied applicability of the CI methodology. They can be applied to uni-modal as well as multimodal problem domains. No single optimization technique can solve all classes of problems and give desirable and efficient results. Choosing a suitable variation is the key to solving real-world problems to get the best possible results. In total, six variation of CI algorithm, viz. roulette wheel, follow best, follow better, alienate and random lection, follow worst and follow itself, were investigated for mesh smoothing of complex objects (Sapre et al., 2019). Four variations of Cohort intelligence are applied for optimization of AWJM process. The optimum values of function Kerf and Ra are obtained with corresponding process parameters (Gulia & Nargundkar, 2019). Kulkarni et al. (2016) further applied CI to solve combinatorial healthcare problems to develop a cyclic surgery schedule that minimized congestion in the recovery unit. It was also used to solve cross-border transportation problem. In this chapter, the variations of CI have been applied in the field of healthcare optimization to the problem of cyclic surgical scheduling in a hospital. The different parameters have been considered and the results have been compared with roulette wheel approach (Kulkarni et al., 2016). The flow chart of variations of CI is presented in Figure 7.1.

7.3 REAL-WORLD APPLICATION OF HEALTHCARE ASSIGNMENT PROBLEM

This variant of assignment problem has been applied in the field of healthcare for the surgical scheduling of patients using CI and is adopted from (Kulkarni et al., 2016). The mathematical formulation of the problem is discussed as follows:

Consider the case of a surgical unit. n surgeons operate on patients in the surgery unit, who are thereafter sent to the recovery unit. It is required to schedule them over a period of n days. The cyclic implies that the schedule is repeated every n days. It is assumed that the surgery unit is operated every day; that exactly one type of surgery must occur in each time period; and those patients do not stay more than n days in the recovery unit. The variables associated are the number of doctors, nurses, beds, equipment costs, etc. It is required to minimize the maximum number of patients on any given day in the recovery unit so as to minimize the variable costs associated with the patients.

Nomenclature:

C	A $n \times n$ circular matrix $\left(C_{ij}\right)$ where $ij = 1, \dots, n$
N	Set of integers $[1, 2, \dots, n]$
Π	A permutation of set N
C^{π}	An $n \times n$ matrix obtained by shifting each element of row i of matrix C for $i = 1, \dots, n$ by $(\pi(i) - 1)$ positions to the right in a circular manner
I_k	The sum of the kth column of matrix C_{π}
Z	The maximum column sum of matrix C_{π}
x_{ij}	A binary variable equal to 1 if $\pi(i) = j$; and 0, otherwise

The variant of the assignment problem uses a circular matrix for generating the different permutations. Circular matrices which perform shifts by rotating the matrix elements give rise to a number of permutations. A circulant matrix is a special kind

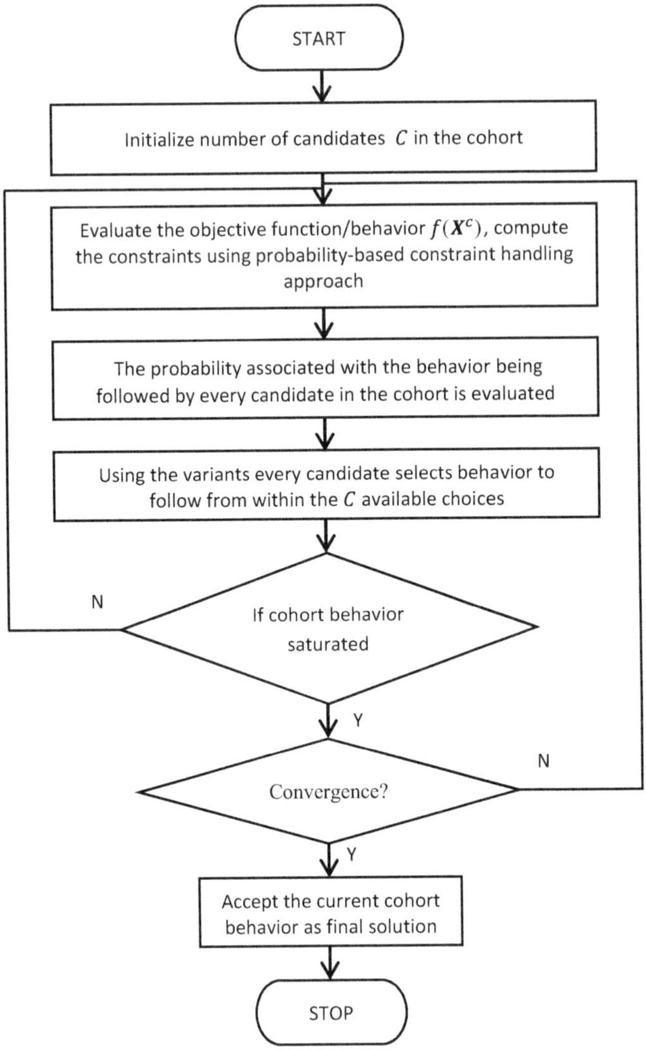

FIGURE 7.1 Flow chart of variations of CI

of Toeplitz matrix where each row vector is rotated one element to the right relative to the preceding row vector. An $n \times n$ circulant matrix can be defined as:

$$C = \begin{bmatrix} c_0 & c_{n-1} & \cdots & c_2 & c_1 \\ c_1 & c_0 & c_{n-1} & \cdots & c_2 \\ \vdots & c_1 & c_0 & \ddots & \vdots \\ c_{n-2} & \cdots & \ddots & \ddots & c_{n-1} \\ c_{n-1} & c_{n-2} & \cdots & c_1 & c_0 \end{bmatrix}$$

For a given $n \times n$ circular matrix $C = C_{ij}$ aimed to minimize the maximum column sum of matrix C.

$$Z = max_{k=1}^{n} \sum_{i=1}^{n} C_{i,k} - \pi(i) + 1$$

where $\pi = (\pi(1), \pi(2), \ldots, \pi(n))$ is a permutation of the set $N = \{1, 2, \ldots, n\}$.

This kind of a problem is referred to as a Cyclic Bottleneck Assignment Problem (CBAP). Cyclic refers to the row circularity of the circular matrix, bottleneck refers to the objective function to be optimized and assignment refers to a variant of the assignment problem and its similarities to the classical assignment problem. Every row of matrix C is rotated by a permutation π to obtain the rotated matrix C^{π}. The objective of this study is to find a permutation that minimizes the maximum column sum of the rotated matrix C^{π}.

The decision variables for this problem can be defined by the following decision variables:

$$x_{i,j} = \begin{cases} 1 & if^{j=\pi(i)}_{otherwise} \\ 0 & \end{cases}$$

The model is expressed mathematically by Kulkarni et al. (2016) as follows:

Minimize objective function Z

Subject to the constraints:

$$\text{Constraint 1}: \sum_{i=1}^{n} x_{i,j} = 1, \ \forall 1 \le j \le n$$

$$\text{Constraint 2}: \sum_{j=1}^{n} x_{i,j} = 1, \ \forall 1 \le i \le n$$

$$\text{Constraint 3}: I_k = \sum_{i=1}^{n} \sum_{j=1}^{n} C_{i,k-j+1} x_{i,j} = \sum_{i=1}^{n} \sum_{j=1}^{n} C_{i,k-j+1} x_{i,j} + \sum_{i=1}^{n} \sum_{j=k+1}^{n} C_{i,k-j+1+n} x_{i,j}, \ \forall 1 \le k \le n$$

$$\text{Constraint 4}: Z \ge I_k, \quad \forall 1 \le k \le n$$

$$x_{i,j} \in \{0, 1\}, \quad \forall 1 \le i \le n, 1 \le j \le n$$

The objective function minimizes the maximum column sum of the rotated matrix C^{π}. Constraint 1 ensures that for each j there exists an i such that $j = \pi(i)$. Constraint 2 ensures that for each i there exists a j such that $i = \pi(j)$. Constraints 3 computes the sum of the kth column of the rotated matrix C^{π}. Constraint 4 sets the value of the objective function equal to the maximum column sum of the rotated matrix C^{π}.

Patankar & Kulkarni (2018) have developed six variations of CI. The aim of this paper is to implement these variations individually with the main CI code and to compare the results of the variations of CI with the results obtained using CI by Kulkarni & Shabir (2016). Also, the function evaluations of all the variations have been compared. It is intended to investigate which variation is better for solving this problem.

7.4 RESULTS AND DISCUSSION

The algorithm is coded in MATLAB® R2013a on Windows Platform with Intel Core i7 with 8GB RAM. It is validated by solving 16 problems of sizes ranging from 5×5 upto 50×50 with 10 instances for each case. Every problem in these test cases is solved 30 times. In the context of the CI algorithm the elements of the rearrangement/permutation vector where $\pi = (\pi(1), \pi(2), \ldots, \pi(n))$ are considered the qualities that candidates in the cohort select and are associated with.

The procedure began with the initialization of the number of cohort candidates, and the permutation of every candidate. The circular matrices are imported into the code. The interval reduction factor is chosen to be 0.99 based on the analysis carried out in Kale & Kulkarni (2018). In this chapter, for the very first-time variations of CI are investigated solving the cyclic surgery unit scheduling problem aimed to be minimize the congestion of patients in the recovery unit. However, Roulette wheel selection, the method used by Kulkarni et al. (2016), is used as a reference and the results are compared with it. In the present paper, the follow best rule, follow better rule, follow itself rule and, alienation and random selection rule are compared with the reference method. The follow worst rule and follow median rule didn't give satisfactory results and hence the infusible solutions obtained from these approaches are discarded. The objective function values and function evaluations are noted. These results are presented in Table 7.1 individually for each size of the matrix.

TABLE 7.1
Comparison of Results of Variations of CI Algorithm with CI Algorithm

Matrix Dimension	Z- Values: CI (Kulkarni et al., 2016)	Z- Values: Follow Better	Z- Values: Follow Best	Z- Values: Alienation and Random Selection	Z- Values: Follow Itself	FE (Kulkarni et al., 2016)	FE Current Work
5×5	828	812	812	812	812	1620	1750
6×6	940	899	899	899	899	3169	2100
7×7	1157	1062	1062	1062	1062	5070	2450
8×8	1309	1196	1196	1196	1196	6988	2800
9×9	1621	1442	1442	1443	1440	9136	3150
10×10	1766	1645	1645	1645	1644	10750	3500
11×11	1736	1603	1602	1602	1600	12190	3850
12×12	1984	1721	1717	1717	1716	13601	4200
13×13	2172	1975	1982	1975	1972	15477	4550
15×15	2331	2198	2192	2190	2180	17525	5250
20×20	3258	3108	3110	3110	3108	26085	7000
25×25	3961	3825	3828	3822	3820	35435	8750
30×30	4725	4658	4662	4660	4650	43715	10500
35×35	5428	5130	5125	5128	5122	51145	12250
40×40	6202	6113	6114	6110	6108	61610	14000
45×45	6927	6840	6839	6841	6832	71835	15750
50×50	7756	7590	7588	7600	7588	81207	17500

The function values obtained from the follow better, follow best, alienation and random section and follow itself are better compared to those from CI. The results obtained by follow worst and follow median are not feasible, hence they are discarded in this work. In the CI algorithm, the roulette wheel approach was used, where the candidates are randomly followed to improve their individual behavior. Due to this the convergence rate significantly decreased, which results in an increase in the function evaluations. Whereas in the variations of CI presented in Table 7.1, the candidate follows particular instances, which helps in decreasing the number of function evaluations. The table shows that the results of follow itself and alienation are better than the other two variations.

The sample convergence plots for all the 16 problem sizes using the follow best approach are presented in 7.2–7.10. Figure 7.11 shows the variation in standard deviation with the matrix dimension.

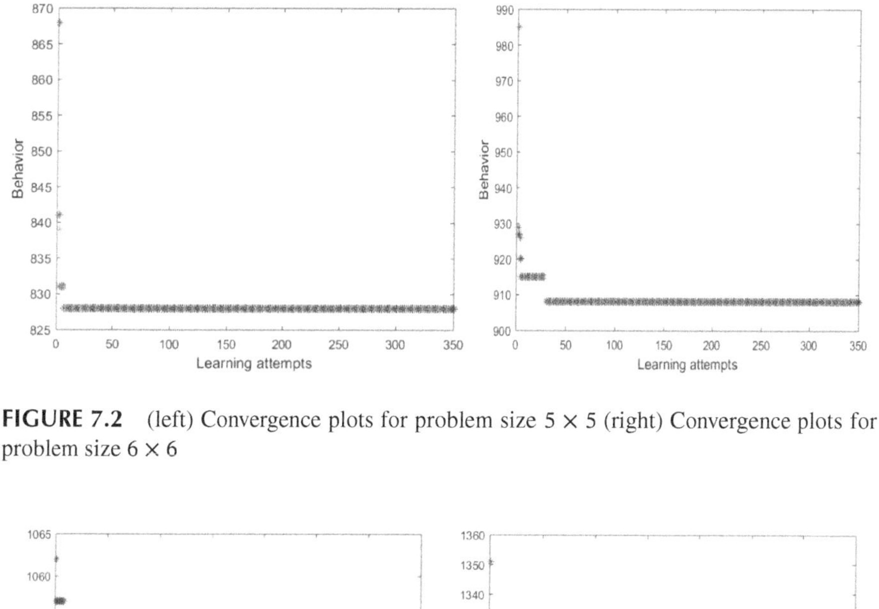

FIGURE 7.2 (left) Convergence plots for problem size 5 × 5 (right) Convergence plots for problem size 6 × 6

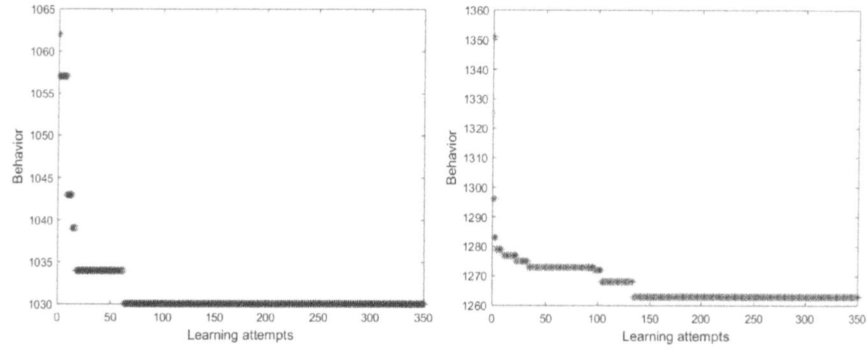

FIGURE 7.3 (left) Convergence plots for problem size 7 × 7 (right) Convergence plots for problem size 8 × 8

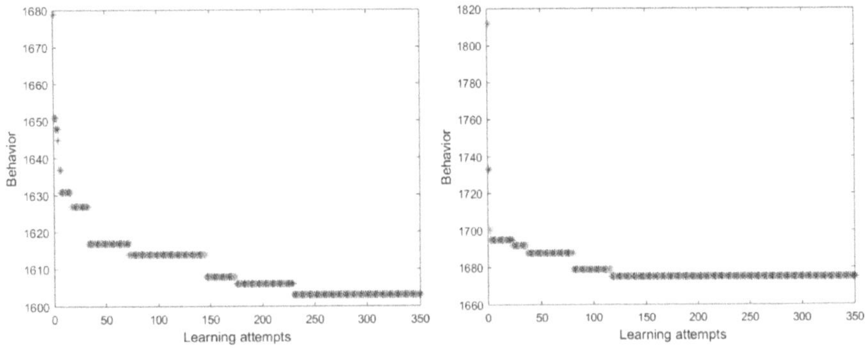

FIGURE 7.4 (left) Convergence plots for problem size 9 × 9 (right) Convergence plots for problem size 10 × 10

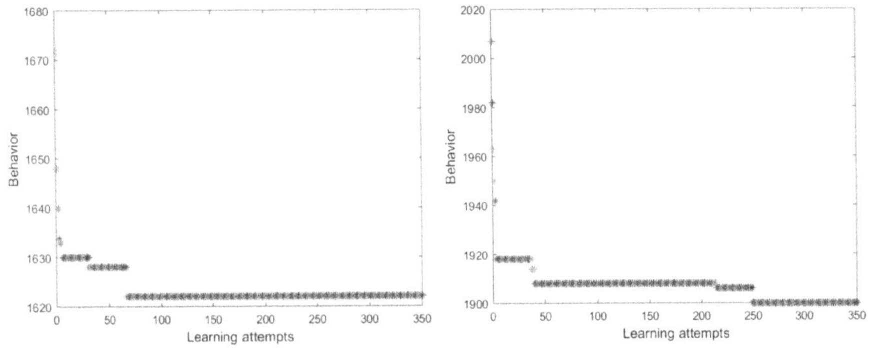

FIGURE 7.5 (left) Convergence plots for problem size ×1111 (right) Convergence plots for problem size 12 × 12

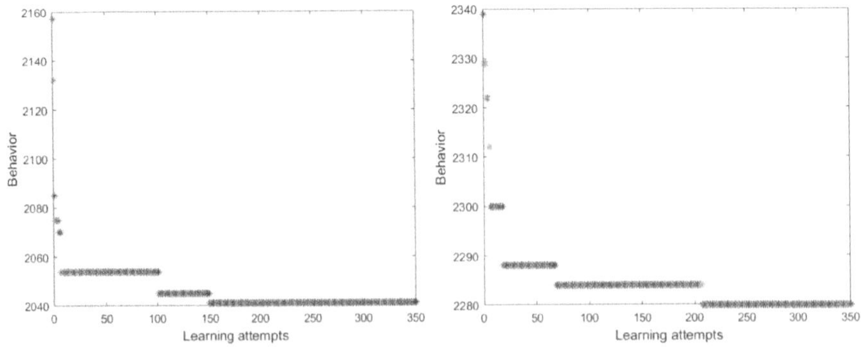

FIGURE 7.6 (left) Convergence plots for problem size 13 × 13 (right) Convergence plots for problem size 15 × 15

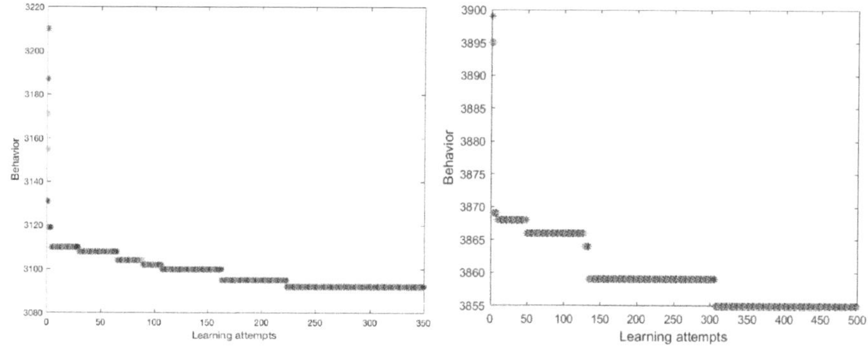

FIGURE 7.7 (left) Convergence plots for problem size 20 × 20 (right) Convergence plots for problem size 25 × 25

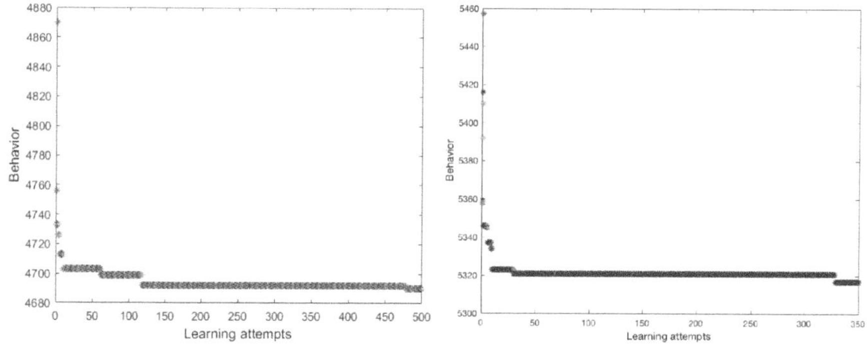

FIGURE 7.8 (left) Convergence plots for problem size 30 × 30 (right) Convergence plots for problem size 35 × 35

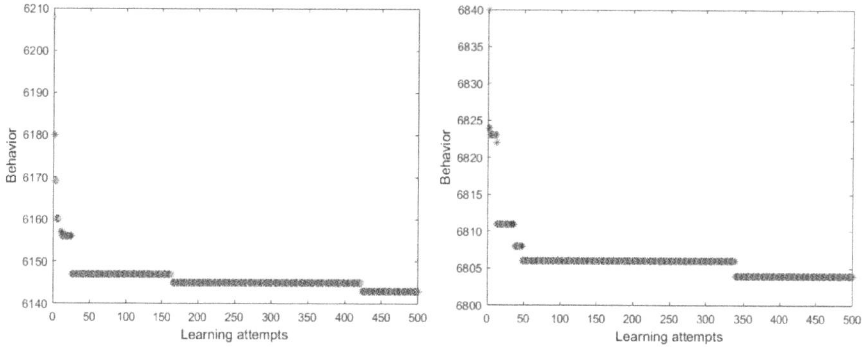

FIGURE 7.9 (left) Convergence plots for problem size 40 × 40 (right) Convergence plots for problem size 45 × 45

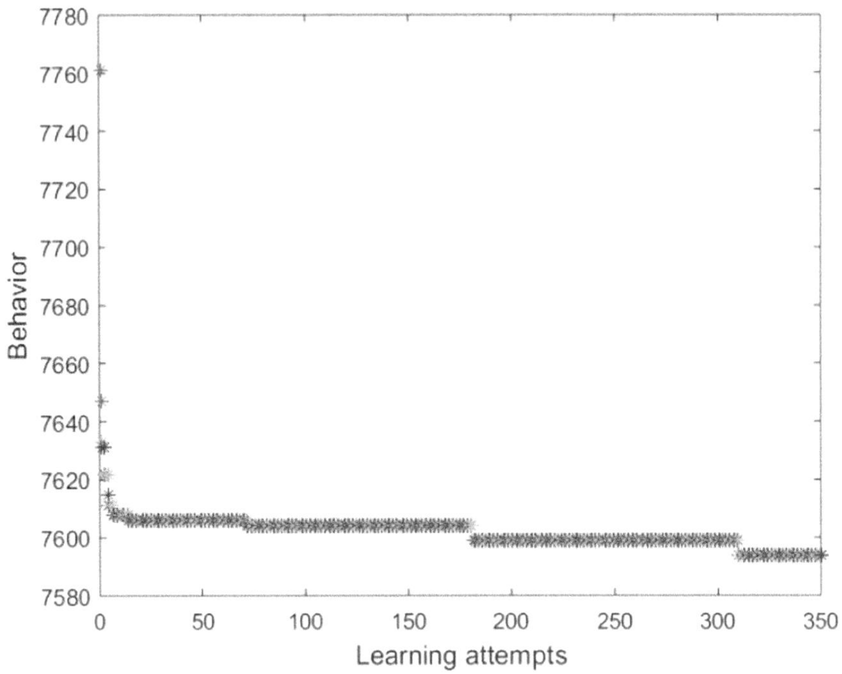

FIGURE 7.10 Convergence plots for problem size 50 × 50

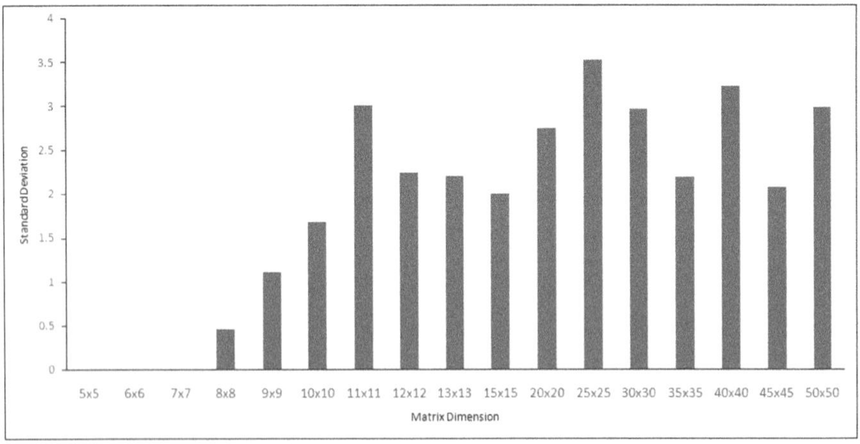

FIGURE 7.11 Graph for standard deviation vs problem size

7.5 CONCLUSION

The six variations of CI have been investigated to solve CBAP, the new variant of the assignment problem. It is found that the follow itself and alienation variation gave the best results and also showed improvements compared to the roulette wheel approach. Some limitations are also observed. The follow worst rule and follow median rule didn't give satisfactory results and this is in accordance with the free lunch theorem. From the results, it is observed that four out of six variations are successfully applicable to solve the problem and the maximum percentage improvement in the objective function is 13.51%. In future, several metaheuristic algorithms can be applied for the solution of this problem. In addition to the assumptions in the current formulation, more constraints could be incorporated to make the problem more generic.

REFERENCES

Awadallah, M.A., Khader, A.T., Al-Betar, M.A. and Bolaji, A.L.A., 2013, April. Hybrid harmony search for nurse rostering problems. In 2013 IEEE Symposium on Computational Intelligence in Scheduling (CISched) (pp. 60–67).

Cheung, G., 2009. A Discrete Stereotyped Particle Swarm Optimization Algorithm for Quadratic Assignment Problems. State University of New York at Binghamton.

Fei H, Meskens N, Chu C., 2010. A planning and scheduling problem for an operating theatre using an open scheduling strategy. Computers and Industrial Engineering 58(2), pp. 221–30.

Gulia, V. and Nargundkar, A., 2019. Optimization of process parameters of abrasive water jet machining using variations of cohort intelligence (CI). In Applications of Artificial Intelligence Techniques in Engineering (pp. 467–474).

Gupta D, Denton B., 2008. Appointment scheduling in health care: Challenges and opportunities. IIE transactions. 40(9), pp. 800–819.

Hall RW, et al. Handbook of Healthcare System Scheduling. Springer, 2012.

Iyer, V.H., Mahesh, S., Malpani, R., Sapre, M. and Kulkarni, A.J., 2019. Adaptive range genetic algorithm: a hybrid optimization approach and its application in the design and economic optimization of shell-and-tube heat exchanger. *Engineering Applications of Artificial Intelligence*, 85, pp. 444–461.

Kale, I.R. and Kulkarni, A.J., 2018. Cohort intelligence algorithm for discrete and mixed variable engineering problems. *International Journal of Parallel, Emergent and Distributed Systems*, 33(6), pp. 627–662.

Kale, I.R. and Kulkarni, A.J., 2021. Cohort intelligence with self-adaptive penalty function approach hybridized with colliding bodies optimization algorithm for discrete and mixed variable constrained problems. Complex & Intelligent Systems, 7(3), pp. 1565–1596.

Kulkarni, A.J., Baki, M.F. and Chaouch, B.A., 2016. Application of the Cohort-Intelligence Optimization method to three selected combinatorial optimization problems. European Journal of Operational Research, 250(2), pp. 427–447.

Kulkarni, A.J., Durugkar, I.P. and Kumar, M., 2013, October. Cohort Intelligence: a self-supervised learning behavior. In 2013 IEEE International Conference on Sstems, Mn, and Cbernetics pp. 1396–1400.

Kulkarni, A.J. and Shabir, H., 2016. Solving 0–1 Knapsack Problem using Cohort Intelligence Algorithm. International Journal of Machine Learning and Cybernetics, 7(3), pp. 427–441.

Li, X., Zhu, L., Baki, F. and Chaouch, A.B., 2018. Tabu search and iterated local search for the cyclic bottleneck assignment problem. Computers & Operations Research, 96, pp. 120–130.

Patankar, N.S. and Kulkarni, A.J., 2018. Variations of Cohort Intelligence. Soft Computing, 22(6), pp. 1731–1747.

Rajeswari, M., Amudhavel, J., Pothula, S. and Dhavachelvan, P., 2017. Directed bee colony optimization algorithm to solve the nurse rostering problem. Computational Intelligence and Neuroscience, 2017.

Ramli, R., Abd Rahman, R. and Rohim, N., 2019. A hybrid ant colony optimization algorithm for solving a highly constrained nurse rostering problem. Journal of Information and Communication Technology, 18(3), pp. 305–326.

Sapre, M.S., Kulkarni, A.J., Chettiar, L., Deshpande, I. and Piprikar, B., 2021. Mesh smoothing of complex geometry using Variations of Cohort Intelligence Algorithm. Evolutionary Intelligence, 14(2), pp. 227–242.

Shastri, A.S. and Kulkarni, A.J., 2018. Multi-Cohort Intelligence Algorithm: an intra-and inter-group learning behaviour-based socio-inspired optimisation methodology. International Journal of Parallel, Emergent and Distributed Systems, 33(6), pp. 675–715.

Turhan, A.M. and Bilgen, B., 2020. A hybrid fix-and-optimize and simulated annealing approaches for nurse rostering problem. Computers & Industrial Engineering, 145, p. 106531.

Zhuo, X., Huang, H., Cai, Z. and Hu, H., 2015, May. An hybrid evolutionary algorithm with scout bee global search strategy for chinese nurse rostering problems. In 2015 IEEE Congress on Evolutionary Computation (CEC) (pp. 769–775). IEEE.

8 Cone Method for Uncertain Multiobjective Optimization Problems with Minmax Robustness

Ashutosh Upadhayay, Debdas Ghosh, Jauny and Nand Kishor
Department of Mathematical Sciences, Indian Institute of Technology (Banaras Hindu University), Varanasi, Uttar Pradesh, India

CONTENTS

8.1 INTRODUCTION

Although conventional optimization methods have been around for a long time, it is still hard to connect theory with practice, which can be seen in the fact that they are not being applied to many real-world problems. Two primary reasons for real-world optimization problems may fail to lead to practical solutions. Firstly, most real-world problems are multiobjective in nature, and secondly, there is no prior knowledge

DOI: 10.1201/9781003329039-8

of the input data. Both issues are extensively examined in multiobjective and robust optimization (RO) [1, 2]. Nevertheless, only a little research has been carried out on combining these two aspects, and there is still scope for further advancement.

Most real-life optimization problems may have uncertain parameter values because of prediction and estimation errors or not having enough information while making decisions. As a result, solving such uncertain optimization problems (UOPs) is crucial for decision-makers. Soyster [3] first presented robust linear programs with uncertain coefficients in 1973. The idea is to look for a feasible solution for all the scenarios from the uncertainty set, based on the assumption that the problem coefficients belong to the set of scenarios. In the 1990s, RO regained popularity (see, e.g., [4–6]). Ben-Tal et al. [1] presented numerous significant results in robust linear optimization, multistage optimization, and conic optimization in 2009. The RO approach has proven to be an effective tool for solving UOPs.

The multiobjective optimization problem arises in various applications, including transportation, financial services, and communication. Uncertain data impact a multiobjective optimization problem (MOP) in the same way as a single-objective optimization problem (SOP). Multiobjective optimization is primarily concerned with finding the Pareto optimal solutions. A point is termed as *Pareto optimal* if there are no alternative points whose objective function values are lesser or equal, with a decrease in at least one objective [2, 7]. It is, therefore, crucial to find robust efficient (RE) solutions that are less affected by a slight change in variables when solving MOPs with data uncertainty.

Deb and Gupta [8] introduced two robust multiobjective optimization methods in 2006. First, all objective functions are replaced by their mean functions, and a robust solution is defined as the efficient solution to the transformed deterministic optimization problem. In the second, the original objective functions are optimized by adding constraints to the predefined limit. Recent work by Kuroiwa and Lee [9] described three types of RE solutions that differ from those proposed by Deb and Gupta [8] for uncertain MOPs. Additionally, they introduced necessary optimality theorems for RE solutions and provided scalarization methods for RE solution of MOPs. Ehrgott et al. [2] gave the extension of minmax robustness for MOPs, inspired by the work done by Ben-Tal et al. [1]. They introduced robust Pareto efficiency and discussed a way to find RE solutions to uncertain MOPs. Recent research in [10] provides a sufficient and necessary condition for robust efficiency, which Ehrgott et al. [2] studied for multiobjective optimization for uncertain data. In addition, these results were applied to the field of oncology [10]. Inspired by the works cited in this paper, we study the MOP with objective-wise uncertainty. We show that the cone method [11–13] can be used to capture the RE solution of uncertain MOPs.

It has been reported in [2] that the implementation of weighted sum and epsilon-constraint may not provide all the RE solutions of an uncertain MOP. The primary reason that supports the recommendation of the cone method for uncertain MOPs is its appropriacy for deterministic non-convex optimization problems. If the robust counterpart of an uncertain MOP is non-convex, then the cone method may be a better alternative than other existing scalarization methods.

The framework of the chapter is structured as follows. In Section 8.2, we have described the deterministic and uncertain MOPs. The cone method for deterministic

MOPs and method of minmax robustness for single-objective UOPs is also shown. In Section 8.3, we show the robust counterpart of an uncertain MOP and discuss the concept of robust efficiency for uncertain MOPs. Section 8.4 describes the formulation of parametric SOP for a UOP using the cone method. Interpretation of robust counterpart of an uncertain MOP using the idea of objective-wise worst case is also shown. Finally, we conclude with a few future directions in Section 8.5.

8.2 PREREQUISITE

8.2.1 (DETERMINISTIC) MOP AND CONE METHOD

Mathematically, a deterministic MOP can be expressed as follows:

$$\left.\begin{array}{lll} \text{minimize} & F(x) = (f_1(x), f_2(x), \dots, f_m(x))^\top, & m \geq 2, \\ \text{subject to} & l_p(x) \leq 0, & p = 1, 2, \dots, n_1, \\ & u_q(x) = 0, & q = 1, 2, \dots, n_2, \end{array}\right\} \quad (8.1)$$

where $x = (x_1, x_2, \dots, x_n)^\top \in \mathbb{R}^n$, $F : \mathbb{R}^n \to \mathbb{R}^m$ and each f_s, $s = 1, 2, \dots, m$, l_p and u_q are real-valued functions.

We set $\mathcal{X} = \{x \in \mathbb{R}^n : l_p(x) \leq 0,\ u_q(x) = 0,\ p = 1, 2, \dots, n_1,\ q = 1, 2, \dots, n_2\}$ as the feasible set in the decision space $\mathbb{R}^n$, and $\mathcal{Y} = F(\mathcal{X})$ is the feasible region in the objective space $\mathbb{R}^m$.

In order to gain a better understanding of MOPs' concept of optimality, it is necessary to understand a dominance structure in $\mathbb{R}^m$. For a given $y = (y_1, y_2, \dots, y_m)^\top$ in $\mathbb{R}^m$, we set the following notions:

$\mathbb{R}^m_{\geqq} = \{y \in \mathbb{R}^m : y \geqq 0\}$, the nonnegative orthant of $\mathbb{R}^m$.
$\mathbb{R}^m_{\geq} = \{y \in \mathbb{R}^m : y \geq 0\}$ where $y \geq 0$ denotes $y \geqq 0$ but $y \neq 0$.
$\mathbb{R}^m_{>} = \{y \in \mathbb{R}^m : y > 0\}$ represents the interior of $\mathbb{R}^m_{\geqq}$, where $y > 0$ means $y_s > 0$ for all $s = 1, 2, \dots, m$.

Definition 8.2.1.1. *A feasible point $\hat{x} \in \mathcal{X}$ is called Pareto optimal or efficient if there does not exist any other $x \in \mathcal{X}$ such that $F(\hat{x}) \geq F(x)$.*

Definition 8.2.1.2. *A feasible point $\hat{x} \in \mathcal{X}$ is called weakly Pareto optimal if there does not exist any other $x \in \mathcal{X}$ such that $F(\hat{x}) > F(x)$.*

Definition 8.2.1.3. *A feasible point $\hat{x} \in \mathcal{X}$ is called strictly Pareto optimal if does not exist $x \in \mathcal{X}$ such that $F(\hat{x}) \geqq F(x)$.*

We note that

$$\hat{x} \text{ is (weakly) efficient} \iff \text{there does not exist any } x \in \mathcal{X} \text{ such that } F(\hat{x}) - \mathbb{R}^m_{(>)\geq} \supset F(x) \text{ with } x \neq \hat{x}. \quad (8.2)$$

Later, we will introduce the robust efficiency based on this viewpoint for efficiency.

To obtain a (weakly) efficient point of the problem (8.1), the cone method [12–14] solves the following SOP:

$$IC(\hat{\beta}) \left\{ \begin{array}{lll} \text{minimize} & t & \\ \text{subject to} & t\hat{\beta} \geqq F(x) & \\ & l_p(x) \leq 0, & p = 1, 2, \ldots, n_1, \\ & u_q(x) = 0, & q = 1, 2, \ldots, n_2, \\ & t \geq 0. & \end{array} \right\} \tag{8.3}$$

To obtain a discrete approximation of the whole Pareto set (see [11–13]) of the MOP (8.1), we need to solve (8.3) for each

$$\hat{\beta} = \left(\cos\theta_1,\ \cos\theta_2 \sin\theta_1,\ \cos\theta_3 \sin\theta_2 \sin\theta_i,\ \ldots,\ \cos\theta_{r-1} \prod_{i=1}^{m-2} \sin\theta_1,\ \prod_{i-1}^{m-1} \sin\theta_i \right) \in \mathbb{S}_{\geqq}^{m-1}, \tag{8.4}$$

where $\theta_i \in [0, \frac{\pi}{2}]$, $i = 1, 2, \ldots, (m-1)$ and $\mathbb{S}_{\geqq}^{m-1} = S^{m-1} \cap \mathbb{R}_{\geqq}^{m}$ (where S^{m-1} is the unit sphere in $\mathbb{R}^m$). In order to obtain a discrete approximation of the whole Pareto set, we follow the suggestion in [12, 13] regarding the number of grid points for θ_i's.

8.2.2 UOPs and Minmax Robustness

The current part presents some notions from RO. Given an uncertainty set $\mathcal{M} \subseteq \mathbb{R}^r$, also called the set of scenarios, we denote the UOP by $\mathcal{P}(\mathcal{M})$, a family of optimization problems, i.e., $\mathcal{P}(\mathcal{M}) = \{\mathcal{P}(\zeta) : \zeta \in \mathcal{M}\}$, where

$$\mathcal{P}(\zeta) \left\{ \begin{array}{lll} \text{minimize} & f(x, \zeta) & \\ \text{subject to} & l_p(x) \leq 0, & p = 1, 2, \ldots, n_1, \\ & u_q(x) = 0, & q = 1, 2, \ldots, n_2, \end{array} \right\} \tag{8.5}$$

where $f : \mathbb{R}^n \times \mathcal{M} \to \mathbb{R}$, and l and u are the same as in (8.1). We refer to ζ as a scenario and $\mathcal{P}(\zeta)$ an instance of $\mathcal{P}(\mathcal{M})$. Note that the problem (8.5) contains objective-wise uncertainties. The feasible set is assumed to be deterministic.

$\mathcal{P}(\mathcal{M})$ indicates an UOP which is a family of different optimization problems, and they differ on the basis of their scenario $\zeta \in \mathcal{M}$. To find the optimal solution of UOP $\mathcal{P}(\mathcal{M})$, which we usually refer to as *robust optimal solution*, we must define an appropriate robust counterpart to $\mathcal{P}(\mathcal{M})$, which has to be a deterministic optimization problem corresponding to $\mathcal{P}(\mathcal{M})$. In this chapter, we adopt the famous *minmax robustness* technique which was first proposed by Soyster [3] to obtain a robust counter part of an UOP. One may also refer to [1] for more details on minmax robustness. By using this strategy, we attempt to minimize the objective function over all the feasible solutions for the worst-case scenario. Toward this goal, we convert the

UOP $\mathcal{P}(\mathcal{M})$ into its *robust counterpart*, which leads to a deterministic optimization problem. It can be formulated as

$$\left.\begin{array}{lll} \text{minimize} & \sup\limits_{\zeta\in\mathcal{M}} f(x,\zeta) & \\ \text{subject to} & l_p(x) \leq 0, & p = 1,2,\dots,n_1, \\ & u_q(x) = 0, & q = 1,2,\dots,n_2. \end{array}\right\}$$

We refer to the solution of the above problem as *robust optimal* to (8.5). Note that a solution is called robust if it is feasible for all the scenarios.

The novelty of this work is to solve the uncertain MOP. In this chapter, we only investigate the problems with the following properties:

- there is uncertainty in the objectives,
- the feasible set is assumed to be deterministic, and
- the objective feasible region lies in the non-negative orthant.

8.3 ROBUST COUNTERPARTS FOR UNCERTAIN MOPS

The current section aims to extend the idea of minmax robustness to multiobjective optimization. We consider the problem (8.1). We assume that f in (8.1) may depend upon scenarios ζ which are uncertain or unknown. Using the same idea as given in subsection 8.2.2 for uncertain SOP, we can extend the idea of minmax for a multiobjective case. For a given set of uncertainty $\mathcal{M} \subseteq \mathbb{R}^r$, an uncertain multiobjective optimization $\mathcal{P}(\mathcal{M})$ is given by the family $\{\mathcal{P}(\zeta) : \zeta \in \mathcal{M}\}$ of MOPs:

$$\mathcal{P}(\zeta)$$
$$\left\{\begin{array}{lll} \text{minimize} & F(x,\zeta) = (f_1(x,\zeta), f_2(x,\zeta), \dots, f_m(x,\zeta))^{\mathsf{T}}, & m \geq 2, \\ \text{subject to} & l_p(x) \leq 0, & p = 1,2,\dots,n_1, \\ & u_q(x) = 0, & q = 1,2,\dots,n_2, \end{array}\right\} \tag{8.6}$$

where $F : \mathbb{R}^n \times \mathcal{M} \to \mathbb{R}^m$ and $x \in \mathbb{R}^n$. Thus, ζ is a scenario and $\mathcal{P}(\zeta)$ is an instance of $\mathcal{P}(\mathcal{M})$.

We can easily note that if $|\mathcal{M}| = 1$, the family $\mathcal{P}(\mathcal{M})$ reduces to a deterministic MOP and if $m = 1$, then $\mathcal{P}(\mathcal{M})$ can be regarded as an uncertain SOP.

As in single-objective uncertain optimization, the same question arises with an uncertain MOP, $\mathcal{P}(\mathcal{M})$, which is how one can evaluate a feasible solution $\hat{x} \in \mathcal{X}$. It is difficult to evaluate the solutions for uncertain MOPs by simply considering the worst case across all scenarios since each scenario leads to a vector of objectives.

Hereafter, we denote the set of objective values at x by

$$F_{\mathcal{M}}(x) = \{F(x,\zeta) : \zeta \in \mathcal{M}\} \subseteq \mathbb{R}^m.$$

In light of this notion, the concept of efficiency similar to (8.2) has been extended in [2]. Recall that for a deterministic MOP, a solution is said to be $\hat{x} \in \mathcal{X}$ (weakly) efficient if there is no alternative $x \in \mathcal{X}$ with $x \neq \hat{x}$ such that $F(x) \in F(\hat{x}) - \mathbb{R}^m_{(>)\geq}$. This concept can be extended in the following way:

We call a feasible solution $\hat{x} \in \mathcal{X}$ to be RE if there does not exist any $x \in \mathcal{X}$ such that $F_{\mathcal{M}}(\hat{x}) - \mathbb{R}^m_{\geq} \supset F_{\mathcal{M}}(x)$ with $x \neq \hat{x}$. Accordingly, we take into account all possible objective values for a solution $\hat{x}$ under all possible scenarios, namely the set $F_{\mathcal{M}}(\hat{x})$. Formally, robust efficiency is defined as follows.

Definition 8.3.4. *(See [2]). For a given uncertain MOP* (8.6), *we denote a feasible solution $\hat{x} \in \mathcal{X}$ as a*

- robust weakly efficient (RWE) *if there does not exist $x \in \mathcal{X} \setminus \{\hat{x}\}$ such that $F_{\mathcal{M}}(x) \subseteq F_{\mathcal{M}}(\hat{x}) - \mathbb{R}^m_{>}$, and*
- RE *if there does not exist $x \in \mathcal{X} \setminus \{\hat{x}\}$ such that $F_{\mathcal{M}}(x) \subseteq F_{\mathcal{M}}(\hat{x}) - \mathbb{R}^m_{\geq}$.*

The following lemma specifies a relation between RE solution and RWE solutions.

Lemma 8.3.1. *(See [2]). Let $\mathcal{P}(\mathcal{M})$ be an uncertain MOP. Then, every RE solution is a RWE.*

We provide a geometrical example to illustrate Definition 8.3.4.

Example 8.3.1. *Consider an uncertain MOP with two objectives $f_1(x,\zeta)$ and $f_2(x,\zeta)$, $x \in \mathbb{R}^n$ and $\zeta \in \mathcal{M}$. The left picture in Figure 8.1 shows an uncertain MOP whose feasible set is $\mathcal{X} = \{x_1, x_2, x_3\}$, and the three sets $F_{\mathcal{M}}(x_1)$, $F_{\mathcal{M}}(x_2)$, and $F_{\mathcal{M}}(x_3)$. In the right figure, we see that neither $F_{\mathcal{M}}(x_1) - \mathbb{R}^2_{\geq}$ nor $F_{\mathcal{M}}(x_2) - \mathbb{R}^2_{\geqq}$ contains any other set $F_{\mathcal{M}}(x_i)$, and thus x_1 and x_2 are both RE. In contrast, $F_{\mathcal{M}}(x_3) - \mathbb{R}^2_{\geqq}$ contains $F_{\mathcal{M}}(x_1)$ and $F_{\mathcal{M}}(x_2)$, and thus x_3 is not an RE.*

The robust counterpart of the uncertain MOP (8.6) to capture the (weakly) efficient solution may be defined as follows:

$$\left.\begin{array}{lll} \text{minimize} & \sup\limits_{\zeta \in \mathcal{M}} F(x,\zeta) = (f_1(x,\zeta), f_2(x,\zeta), \dots, f_m(x,\zeta))^\top, & m \geq 2, \\ \text{subject to} & l_p(x) \leq 0, & p = 1,2,\dots,n_1, \\ & u_q(x) = 0, & q = 1,2,\dots,n_2, \end{array}\right\} \tag{8.7}$$

where $\sup\limits_{\zeta \in \mathcal{M}} F(x,\zeta)$ is defined as the set of (weakly) efficient solution of the MOP

$$\left.\begin{array}{lll} \text{maximize} & F(x,\zeta) = (f_1(x,\zeta), f_2(x,\zeta), \dots, f_m(x,\zeta))^\top, & m \geq 2, \\ \text{subject to} & \zeta \in \mathcal{M}. & \end{array}\right\}$$

Following the definition of RE solutions for uncertain MOPs, in the next section, we examine the cone method in order to find the RE solution for the uncertain MOP (8.6).

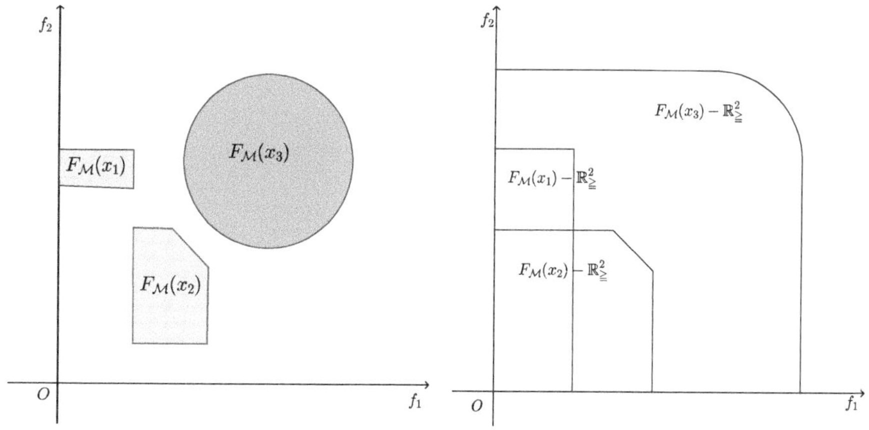

FIGURE 8.1 Illustration of RE solutions

8.4 CONE METHOD FOR UNCERTAIN MOPS

We apply the cone method to reduce an uncertain MOP (8.7) to a set of parametric single-objective UOPs. Based on the formulation in (8.3), the *robust form* of the cone method for the problem (8.7) is given by

$$IC_{\mathcal{M}}(\hat{\beta})\left\{\begin{array}{lll}\text{minimize} & t & \\ \text{subject to} & t\hat{\beta} \geqq \sup F(x,\zeta) & \text{for all } \zeta \in \mathcal{M} \\ & l_p(x) \leq 0, & p = 1,2,\ldots,n_1, \\ & u_q(x) = 0, & q = 1,2,\ldots,n_2, \\ & t \geq 0. & \end{array}\right\} \tag{8.8}$$

Theorem: 8.4.1. *Let $\hat{\beta} \in \mathbb{S}^{m-1}_{\geqq}$. If (x^*, t^*) is a RE solution of problem $IC_{\mathcal{M}}(\hat{\beta})$, then x^* is an RE solution of the problem* (8.7).

Proof. For a given $\hat{\beta} \in \mathbb{S}^{m-1}_{\geqq}$, let (x^*, t^*) be an RE solution of the problem $IC_{\mathcal{M}}(\hat{\beta})$. Then, clearly x^* is feasible to the problem (8.7) and $t^*\hat{\beta} \geqq \sup F(x^*, \zeta)$. Suppose that x^* is not efficient for the problem (8.7). Then, there exists $\bar{x}$, which is feasible for the problem (8.7), and for all $\zeta \in \mathcal{M}$, we have

$$\sup F(\bar{x}, \zeta) \leqq \sup F(x^*, \zeta) \leqq t^*\hat{\beta}.$$

Then, for a given $\hat{\beta} \in \mathbb{S}^{m-1}_{\geqq}$, there exists $\bar{t} > 0$ such that $\sup F(\bar{x}, \zeta) = \bar{t}\hat{\beta}$. Thus, we have

$$\bar{t}\hat{\beta} = \sup F(\bar{x}, \zeta) \leqq \sup F(x^*, \zeta) \leqq t^*\hat{\beta}.$$

So, we get $\bar{t} \leq t^*$. Therefore, (x^*, t^*) is not an efficient solution of $IC_{\mathcal{M}}(\hat{\beta})$, which is a contradiction. Hence, the result follows. □

The following example illustrates the cone method.

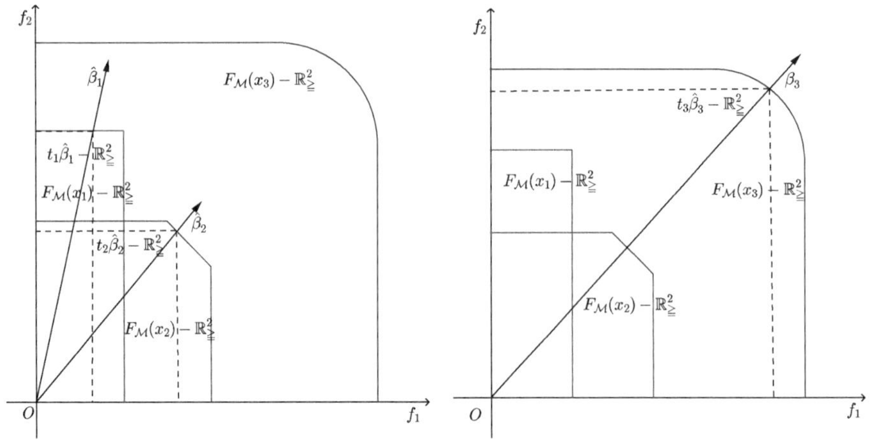

FIGURE 8.2 Illustration of cone method

Example 8.4.2. *Consider the problem in Example 8.3.1. We aim to minimize the supremum of the bi-objective function over the sets* $F_{\mathcal{M}}(x_i) - \mathbb{R}^2_{\geqq}$ *for* $i = 1, 2, 3$. *Toward this aim, we consider the directions* $\hat{\beta}_i$ *for* $i \in \{1, 2, 3\}$ *(see Figure 8.2) to obtain the cone* $t_i\hat{\beta}_i - \mathbb{R}^2_{\geqq}$ *for a given positive real number* $t_i > 0$ *for* $i = \{1, 2, 3\}$. *From the left picture in Figure 8.2, we get* x_1 *and* x_2 *as optimal solutions because none of* $t_1\hat{\beta}_1 - \mathbb{R}^2_{\geqq}$ *and* $t_2\hat{\beta}_2 - \mathbb{R}^2_{\geqq}$ *contains any other set* $F_{\mathcal{M}}(x_i) - \mathbb{R}^2_{\geqq}$. *Thus,* x_1 *and* x_2 *are RE. From the right picture, we obtain that* x_3 *is not a RE because the set* $t_3\hat{\beta}_3 - \mathbb{R}^2_{\geqq}$ *contains both* $F_{\mathcal{M}}(x_1) - \mathbb{R}^2_{\geqq}$ *and* $F_{\mathcal{M}}(x_2) - \mathbb{R}^2_{\geqq}$.

In Algorithm 8.4.1, a sequential procedure is presented for obtaining the RE solutions to a bi-objective optimization problem. A parametric uncertain SOP is first constructed from the uncertain MOP (see (8.10)). In solving a a bi-objective optimization problem, it is required to run a "for" loops (corresponding to θ, and when we have to solve an MOP with m criteria, the number of "for" loops will be $m - 1$ (see [12]). As the criterion feasible region lies in the non-negative quadrant, thus for each $0 \leq \theta \leq \pi/2$, we get a direction $\hat{\beta} \in \mathbb{S}^1_{\geqq}$. It is important to mention here that Algorithm 1 does not require the convexity of uncertain objectives or constraints.

Rather than interpreting the supremum via its robust counterpart to the uncertain MOP $\mathcal{P}(\mathcal{M})$ as a MOP, this could be interpreted as a point rather than a set [2]. This entails formulating a new problem by the objective-wise worst case, which is given by

$$\mathcal{OWC}_{\mathcal{P}(\mathcal{M})} \left\{\begin{array}{lll} \text{minimize} & F^{owc}_{\mathcal{M}}(x) & m \geq 2, \\ \text{subject to} & l_p(x) \leq 0, & p = 1, 2, \ldots, n_1, \\ & u_q(x) = 0, & q = 1, 2, \ldots, n_2, \end{array}\right\} \tag{8.9}$$

Algorithm 8.4.1 Cone method to generate robust optimal solutions of (8.6) with two objectives

Aim: To generate a discrete approximation of the robust optimal solution of the problem (8.6)

Provide n, and the functions l_i and u_j's for the deterministic feasible set $\mathcal{X} = \{x \in \mathbb{R}^n : l_p(x) \leq 0, u_q(x) = 0, p = 1, 2, \dots, n_1, q = 1, 2, \dots, n_2\}$

Provide $F(x, \zeta) = (f_1(x, \zeta), f_2(x, \zeta))^\top$, where each of f_1, and f_2 is an uncertain objective function with $x \in \mathbb{R}^n$ and $\zeta \in \mathcal{M}$

Choose m, the number of grid points for θ

for $\theta = 0 : \frac{\pi}{2m} : \frac{\pi}{2}$ **do**

  (By the equation (8.3)) Set $\hat{\beta} \leftarrow (\cos\theta, \sin\theta)$

  Solve $IC_{\mathcal{M}}(\hat{\beta})$

  return x^* as a robust optimal solution in the direction of $\hat{\beta}$

  Update set $\mathcal{S} \leftarrow \mathcal{S} \bigcup \{F(x^*, \zeta)\}$ for all $\zeta \in \mathcal{M}$

end for

where

$$F_{\mathcal{M}}^{owc}(x) = \begin{pmatrix} \sup\limits_{\zeta \in \mathcal{M}} f_1(x, \zeta) \\ \sup\limits_{\zeta \in \mathcal{M}} f_2(x, \zeta) \\ \vdots \\ \sup\limits_{\zeta \in \mathcal{M}} f_m(x, \zeta) \end{pmatrix}$$

$\mathcal{OWC}_{\mathcal{P}(\mathcal{M})}$ is a (deterministic) multiobjective minimization of the objective-wise supremum. This interpretation of robustness has been proposed by Kuroiwa and Lee [9] for MOPs. Calculation of $F_{\mathcal{M}}^{owc}(x)$ for a given x is obviously simpler than solving the MOP, $\max\{F(x, \zeta) : \zeta \in \mathcal{M}\}$. It only requires solving m deterministic single-objective optimization problems. Therefore, $\mathcal{OWC}_{\mathcal{P}(\mathcal{M})}$ is a deterministic MOP, which can be solved using any deterministic multiobjective optimization technique. To put it differently, the $\mathcal{OWC}$-method implies that instead of considering set dominance, as required in Definition 8.3.4, we should consider standard dominance of points (the objective-wise worst case scenario), see Figure 8.3. By using this method, we can reduce the uncertain MOP to a deterministic one.

In Figure 8.3, $F_{\mathcal{M}}^{owc}(x_1), F_{\mathcal{M}}^{owc}(x_2)$ and $F_{\mathcal{M}}^{owc}(x_3)$ are the vectors obtained by applying the objective-wise supremum to the uncertain MOP with two objectives $f_1(x, \zeta)$ and $f_2(x, \zeta)$. From Figure 8.3, we observe that $F_{\mathcal{M}}^{owc}(x_1) - \mathbb{R}^2_{\geq}$ and $F_{\mathcal{M}}^{owc}(x_2) - \mathbb{R}^2_{\geq}$ does not contain any other set $F_{\mathcal{M}}^{owc}(x_i)$ and thus x_1 and x_2 are robust efficient. On the other hand, $F_{\mathcal{M}}^{owc}(x_3) - \mathbb{R}^2_{\geq}$ contains $F_{\mathcal{M}}^{owc}(x_1)$ and $F_{\mathcal{M}}^{owc}(x_2)$. Thus, x_3 is not a robust efficient solution.

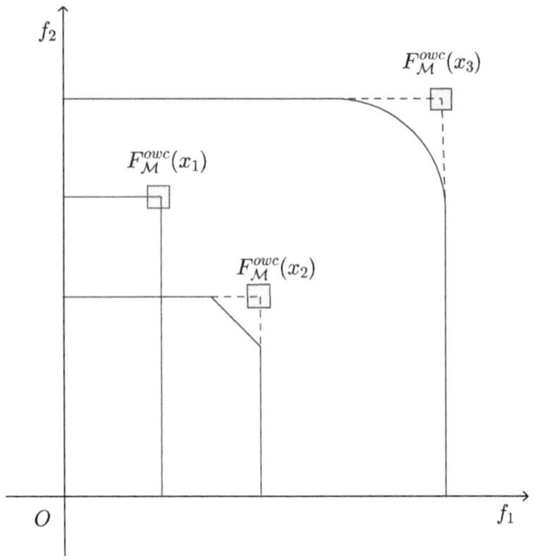

FIGURE 8.3 Objective-wise worst case scenario

Based on the formulation in (8.3), one can obtain the robust version of the cone method for the $\mathcal{OWC}_{\mathcal{P}(\mathcal{M})}$ problem (8.7) by

$$IC^{owc}_{\mathcal{M}}(\hat{\beta})\begin{cases} \text{minimize} & t & \\ \text{subject to} & t\hat{\beta} \geqq F^{owc}_{\mathcal{M}}(x) & \text{for all } \zeta \in \mathcal{M} \\ & l_p(x) \leq 0, & p = 1, 2, \ldots, n_1, \\ & u_q(x) = 0, & q = 1, 2, \ldots, n_2, \\ & t \geq 0. & \end{cases} \tag{8.10}$$

8.5 CONCLUSION

There has been much research on multiobjective optimization and robust optimization, but their combination has rarely been explored. In this work, we have considered MOPs with objective-wise uncertainty. Finally, we have implemented the cone method to find the RE solutions of uncertain MOPs using minmax robustness.

To find the RE solutions, we have first found the robust counterpart of the given uncertain MOP. Then, the cone method is applied on the MOP of robust counterpart to find the set of deterministic parametric single-objective optimization problems. For each parameter $\hat{\beta}$, the solution of the resultant single-objective problems is an RE solution. We have shown in Theorem 8.4.1 that if (x^*, t^*) is a solution of a parametric single objective optimization problem for some $\hat{\beta} \in \mathbb{S}^{m-1}_{\geqq}$, then x^* is an RE solution of the robust counterpart MOP, which is obtained by applying minmax robustness to an uncertain MOP. A step-wise procedure is shown in Algorithm 8.4.1 to obtain a robust optimal solution to the problem (8.6). As an alternative approach to solve the uncertain MOP (8.6), we have also discussed an $\mathcal{OWC}$-method, which showed that instead of considering the set dominance, as required in Definition 8.3.4, we should consider standard dominance of points (the objective-wise worst case scenario) (refer to Figure 8.3.

8.5.1 Pros and Cons of Minmax Robustness

- **Pros:** When an attempt is made to solve an uncertain optimization problem by minmax robustness, we get an optimal value c (say) of the objective function. Then, for any given uncertainty, the optimal value of that objective function never exceeds c, i.e., the optimal value c is strictly valid for any chosen uncertainty. Hence, when we work with sensitive uncertain optimization problems like the case of airplanes, and nuclear plants [15] where a failurc is not tolerable, we can use minmax robustness.
- **Cons:** To solve the deterministic problem obtained by applying the minmax approach, we need some priorly available methods. As the function $\max f(x,\zeta), \zeta \in \mathcal{M}$ is inherently nonsmooth, we cannot apply numerical methods, such as interior-point methods, Newton methods, etc.. This is a significant drawback of the minmax approach.

8.5.2 Future Directions

As future research, we would like to focus on the following directions:

- The current study motivates us to analyze several conventional algorithms in greater depth for finding RE solutions. Various algorithms may serve as valuable tools for capturing RE solutions.
- In addition to minmax robustness, one can also implement various new robustnesses such as adjustable robustness, recovery robustness, or light robustness for uncertain MOPs.
- Moreover, future studies can explore the theoretical properties as well as the experimental performance of the cone method for robust MOPs.

ACKNOWLEDGEMENTS

The authors give sincere thanks to the reviewers for their assistance in improving the article. D. Ghosh acknowledges the financial support of the research project MATRICS (MAT/2021/000696) by SERB, India.

REFERENCES

1. Ben-Tal, A., El Ghaoui, L., Nemirovski, A.: Robust Optimization. Princeton University Press (2009).
2. Ehrgott, M., Ide, J., Schöbel, A.: Minmax robustness for multi-objective optimization problems. Eur. J. Oper. Res. 239(1), 17–31 (2014).
3. Soyster, A.L.: Convex programming with set-inclusive constraint and application to inexact linear programming. Oper. Res. 21(5), 1154–1157 (1973).

4. Ben-Tal, A., Nemirovski, A.: Robust convex optimization. Math. Oper. Res. 23(4), 769–805 (1998).
5. Ben-Tal, A., Nemirovski, A.: Robust solutions of uncertain linear programs. Oper. Res. Lett. 25(1), 1–13 (1999).
6. Bertsimas, D., Sim, M.: The price of robustness. Oper. Res. 52(1), 35–53 (2004).
7. Miettinen, K.: Nonlinear Multiobjective Optimization. Kluwer Academic, Boston (1999).
8. Deb, K., Gupta, H.: Introducing robustness in multiobjective optimization. Evol. Comput. 14(4), 463–494 (2006).
9. Kuroiwa, D., Lee, G.: On robust multiobjective optimization. Vietnam J. Math. 40, 305–317 (2012).
10. Bokrantz, R., Fredriksson, A.: Necessary and sufficient conditions for Pareto efficiency in robust multiobjective optimization. Eur. J. Oper. Res. 262, 682–692 (2017).
11. Ghosh, D., Chakraborty, D.: A direction based classical method to obtain complete Pareto set of multi-criteria optimization. Opsearch. 52(2), 340–366 (2015).
12. Ghosh, D., Chakraborty, D.: A new Pareto set generating method for multi-criteria optimization problems. Oper. Res. Lett. 42(8), 514–521 (2014).
13. Upadhayay, A., Ghosh, D., Ansari, Q. H., Jauny: Augmented Lagrangian cone method for multiobjective optimization problems with an application to an optimal control problem. Optim Eng., 1–33 (2022) https://doi.org/10.1007/s11081-022-09747-y.
14. Pascoletti, A., Serafini, P.: Scalarizing vector optimization problems. J. Optim. Theory Appl. 42(4), 499–524 (1984).
15. García, J., Peña, A. Robust optimization: concepts and applications. Nature-Inspired Methods for Stochastic, Robust and Dynamic Optimization, 7, (2018).
16. Bertsekas, D.P.: Convex Optimization Algorithms. Belmont, Massachusetts: Athena Scientific (2015).
17. Birgin, E.G., Martinez, J.M.: Complexity and performance of an augmented Lagrangian algorithm. Optim. Methods Softw. 35(5):885–920 (2020).
18. Birgin, E.G., Martinez, J.M.: Practical augmented Lagrangian methods for constrained optimization. SIAM Philadelphia (2014).

9 Solving Multi-Index Transportation Problem with Axial Constraints Having Impaired Flow

Archana Khurana[1] *and Veena Adlakha*[2]
[1]Williams E. Kirwan Hall, Department of Mathematics, University of Maryland, College Park, Maryland, United States
[2]Department of Management and International Business, Merrick School of Business, University of Baltimore, Maryland, United States

CONTENTS

9.1 INTRODUCTION

The classical transportation problem is a mathematical model with a special structure. In the standard classical transportation problem, a commodity is to be transported from each of m sources to each of n destinations. The total sum of the available amounts at the sources is equal to the sum of the demands at the destinations. The goal is to determine the amounts of a commodity to be transported over all routes (i, j) such that the total transportation cost is minimized. The transportation problem considered in the classical transportation problem is generally a two-dimensional linear transportation problem.

The solid transportation problem arises when we need to transport heterogeneous commodities of products. A three-dimensional transportation problem with planar constraints was first introduced by Haley (1962), who also gave a solution procedure. A solid transportation problem equivalent to Schell's "Three Axial Sums" problem was formulated in another paper by Haley (1963). The solution method presented

DOI: 10.1201/9781003329039-9

by Haley is an extension of the MODI method. Later, Corban (1964) studied a multi-dimensional transportation problem. Appa (1973) studied variants of the conventional transportation problem. Another type of variant of the standard transportation problem was studied by Bridgen (1974) by considering mixed-type constraints. Klingman and Russel (1974) investigated the effect of an extra linear constraint on the transportation problem. Khanna et al. (1981) discussed the controlled transportation problem and later in 1983 they gave an algorithm to solve the transportation problem with mixed constraints. Bandopadhyaya and Puri (1988) discussed the impaired flow multi-index transportation problem with axial constraints. Thirwani et al. (1997) gave an algorithm for solving a fixed-charge bi-criterion transportation problem with restricted flow.

Arora and Khurana (2001) studied the paradox in an indefinite quadratic transportation problem.

Ahuja and Arora (2001) studied the solid fixed charge bi-criterion transportation problem. Later Khurana and Arora (2004) gave a solution method fo ar three-dimensional fixed charge quadratic transportation problem. The sum of a linear and linear fractional transportation problem with restricted and enhanced flow was discussed by Khurana and Arora (2006). Khurana et al. (2006) studied a bi-criterion quadratic transportation problem with restricted flow. Khurana and Arora (2011a) studied a three-dimensional linear transshipment problem and transshipment problem with mixed constraints. Khurana and Arora (2011b) also studied a fixed charge bi-criterion indefinite quadratic transportation problem with enhanced flow. Khurana and Adlakha (2015) discussed a multi-index transportation problem. Javed (2018) discussed a method for solving the transportation problem. Khurana et al. (2018) discussed multi-index constrained transportation with bounds on availabilities, requirements and commodities. The optimal value range problem for the interval (immune) transportation problem was discussed by Ambrosio et al. (2020). Recently, Agarwal and Sharma (2022) discussed the more-for-less paradox in time minimization transportation problem with mixed constraints.

The "Three Axial Sums" problem deals with the transport of various commodities from a set of m warehouses to n different markets, whose total demands are specified. Thus, the multi-index transportation problem with axial constraints can be analytically written as:

$$\textbf{(P1)}\quad \text{Minimize } \sum_I \sum_J \sum_K c_{ijk} x_{ijk}$$

subject to

$$\sum_J \sum_K x_{ikj} = a_i, \qquad i \in I$$

$$\sum_I \sum_K x_{ikj} = b_j, \qquad j \in J$$

$$\sum_I \sum_J x_{ikj} = e_k, \qquad k \in K$$

$$x_{ijk} \geq 0$$

where a_i is the total availability at the ith warehouse, b_j is the total demand at the jth market, e_k is the total availability of the kth commodity, c_{ijk} is the per unit cost of transporting the kth product from the ith warehouse to the jth market, and x_{ijk} is the amount of kth product transported on the above route. $I = \{1, 2, \dots \dots ., m\}$ is the index set of m warehouses, $J = \{1, 2, \dots \dots, n\}$ is the index set of n markets and $K = \{1, 2, \dots \dots ., p\}$ is the index set of p different products.

Obviously for any feasible solution of (**P1**) $\sum_{I} a_i = \sum_{J} b_j = \sum_{K} e_k = \mathrm{N}$ (say).

Sometimes, when one wishes to keep reserve stocks at the sources for emergencies, thereby restricting the total transportation flow to a known specified level, it results in a transportation problem with impaired flow. Situations, like financial problems, may sometimes compel one to curtail the flow. For example, a manufacturer may hold the goods, viz. medicines, food grains, and other items, at warehouses for emergencies. In this situation, some of the warehouses are forced to close or are made to operate below their original operational level, while some continue to maintain their original supply behavior. The net supply of each product and the demand at each destination are also consequently affected, thereby restricting the total flow to a known specified level say, F (< N). This flow constraint breaks the transportation structure of the problem.

In the present paper we develop a method to find an optimal basic feasible solution of a three-dimensional transportation problem with axial constraints having restricted flow. The problem is transformed into a related transportation problem, by adding a row, a column, and a commodity, which is equivalent to a standard axial sum problem and can be solved by transforming the problem into an equivalent multi-index transportation problem (Haley, 1962). Also, we show that the optimal basic feasible solution obtained by our method is betterthan the existing Bandopadhyaya and Puri (1988) algorithm.

9.2 THEORETICAL DEVELOPMENT

9.2.1 Impaired Flow

The three-dimensional transportation problem when the impaired flow is exactly known, is defined by restricting the total flow to a known specified level, F. Thus, the flow constraint $\sum_{I} \sum_{J} \sum_{K} x_{ijk} = \mathrm{F}$ is also introduced into the system and generates the following transportation problem

$$(\mathbf{P2}) \quad \text{Minimize} \quad \sum_{I} \sum_{J} \sum_{K} c_{ijk} x_{ijk}$$

subject to

$$\sum_{J} \sum_{K} x_{ikj} \le a_i, \qquad i \in I$$

$$\sum_{I} \sum_{K} x_{ikj} \le b_j, \qquad j \in J$$

$$\sum_{I} \sum_{J} x_{ikj} \le e_k, \qquad k \in K$$

$$\sum_I \sum_J \sum_K x_{ijk} = F \qquad (F < N)$$

$$x_{ijk} \geq 0$$

$$\text{where } F < \min\left\{\sum_i a_i, \sum_j b_j, \sum_k e_k\right\}$$

Note: In the case of a balanced capacitated transportation problem, $\sum_I a_i = \sum_J b_j = \sum_K e_k$.

To solve the above problem (**P2**), a related solid problem is formulated, with a dummy supply point, a dummy destination, and an extra commodity. The related three-dimensional transportation problem is given as follows:

$$\text{(P3)} \quad \text{Minimize } \sum_{I'} \sum_{J'} \sum_{K'} c'_{ijk} y_{ijk}$$

subject to

$$\sum_{J'} \sum_{K'} y_{ikj} = a'_i, \qquad i \in I' = I \cup \{m+1\}$$

$$\sum_{I'} \sum_{K'} y_{ikj} = b'_j, \qquad j \in J' = J \cup \{n+1\}$$

$$\sum_{I'} \sum_{J'} y_{ikj} = e'_k, \qquad k \in K' = K \cup \{p+1\}$$

$$y_{ijk} \geq 0$$

where

$$a'_i = a_i, \quad i \in I, \quad a'_{m+1} = \sum_J b_j - F$$

$$b'_j = b_j, \quad j \in J, \quad b'_{n+1} = \sum_I a_i - F$$

$$e'_k = e_k, \quad k \in K, \quad e'_{p+1} = \sum_I a_i + \sum_J b_j - \sum_K e_k - F$$

The above transformations will transform problem **P2** into a balanced transportation problem. Also,

$$\left.\begin{array}{ll} c'_{ijk} = c_{ijk} & \forall i \in I, j \in J, k \in K \\ c'_{ijp+1} = M & \forall i \in I, j \in J \\ c'_{in+1k} = c'_{m+1jk} = 0 & \forall i \in I, j \in J, k \in K \\ c'_{in+1p+1} = c'_{m+1jp+1} = 0 & \forall i \in I', j \in J' \\ c_{m+1n+1k} = M & \forall k \in K' \end{array}\right\} \tag{9.1}$$

where **M** is a sufficiently large positive integer, much larger than each of c'_{ijk}.

Problem (**P3**) is a "Three Axial Sum" problem which can be reformulated as a transformed multi-index transportation problem by using the following definitions (Haley, 1963):

$$\begin{aligned} c''_{ijk} &= c'_{ijk}\,(i \le m+1, j \le n+1, k \le p+1), c''_{m+2n+2p+2} = 0 \\ c''_{ijp+2} &= 0(i \le m+1, j \le n+1), c''_{in+2p+2} = M(i \le m+1), \\ c''_{in+2k} &= 0(i \le m+1, k \le p+1), c''_{m+2jp+2} = M(j \le n+1), \\ c''_{m+2jk} &= 0(j \le n+1, k \le p+1), c''_{m+2n+2k} = M(k \le p+1) \end{aligned}$$

Let $R = \underset{i,j,k}{Max}\left(a'_i, b'_j, e'_k\right)$

$$\begin{aligned} A_{jk} &= R\,(j \le n+1; k \le p+1), & B_{ki} &= R\,(k \le p+1; i \le m+1), \\ E_{ij} &= R\,(i \le m+1; j \le n+1), & A_{n+2k} &= (m+1)R - e'_k\,(k \le p+1), \\ A_{jp+2} &= (m+1)R - b'_j\,(j \le n+1), & B_{p+2i} &= (n+1)R - a'_i\,(i \le m+1), \\ B_{km+2} &= (n+1)R - e'_k\,(k \le p+1), & E_{m+2j} &= (p+1)R - b'_j\,(j \le n+1), \\ E_{in+2} &= (p+1)R - a'_i\,(i \le m+1), & A_{n+2p+2} &= B_{p+2m+2} = E_{m+2n+2} = R \end{aligned}$$

The multi-index transportation problem formed by above transformations can be solved easily by any regular method (Haley, 1962).

Definition: M-feasible solution of (P3)

A feasible solution $\{y_{ijk}\}_{i\in I', j\in J', k\in K'}$ of (**P3**) is called an M-feasible solution, if $y_{ijk} = 0$ whenever $c'_{ijk} = M, i \in I', j \in J', k \in K'$.

Theorem 9.2.1.1. Any M-feasible solution (**P3**) gives a feasible solution of (**P2**) and vice versa.

Proof. Let $\{y_{ijk}\}_{i\in I', j\in J', k\in K'}$ be an M-feasible solution of (**P3**). Then $\{x_{ijk} = y_{ijk}\}_{i\in I, j\in J, k\in K}$ is a feasible solution of (**P2**), because $x_{ijk} \ge 0$ and

$$\begin{aligned} \sum_J \sum_K x_{ijk} &= \sum_J \sum_K y_{ijk} \\ &= \sum_J \left(\sum_{K'} y_{ijk} - y_{ijp+1} \right) \\ &= \sum_J \sum_{K'} y_{ijk} - \sum_J y_{ijp+1} \\ &= \sum_{J'} \sum_{K'} y_{ijk} - \sum_{K'} y_{in+1k} - \sum_J y_{ijp+1} \\ &= a'_i - \sum_{K'} y_{in+1k} - \sum_J y_{ijp+1} \end{aligned}$$

$$= a'_i - \sum_{K'} y_{in+1k}$$

as $c_{ijp+1} = M, y_{ijp+1} = 0,$ by above definition,

$$\leq a'_i = a_i, \; for\, i \in I$$

Thus, $\sum_J \sum_K x_{ijk} \leq a_i, i \in I$ □

Similarly, $\sum_I \sum_K x_{ijk} \leq b_j, j \in J$ and $\sum_I \sum_J x_{ijk} \leq e_k, k \in K$. Also,

$$\begin{aligned}
\sum_I \sum_J \sum_K x_{ijk} &= \sum_I \sum_J \sum_K y_{ijk} \\
&= \sum_I \sum_J \left(\sum_{K'} y_{ijk} - y_{ijp+1} \right) \\
&= \sum_I \sum_J \sum_{K'} y_{ijk} - \sum_I \sum_J y_{ijp+1} \\
&= \sum_{K'} \sum_I \left(\sum_{J'} y_{ijk} - y_{in+1k} \right) - \sum_I \left(\sum_{J'} y_{ijp+1} - y_{in+1p+1} \right) \\
&= \sum_{K'} \sum_I \sum_{J'} y_{ijk} - \sum_{K'} \sum_I y_{in+1k} - \sum_I \sum_{J'} y_{ijp+1} + \sum_I y_{in+1p+1} \\
&= \sum_{K'} \sum_{J'} \left(\sum_{I'} y_{ijk} - y_{m+1jk} \right) - \sum_{K'} \left(\sum_{I'} y_{in+1k} - y_{m+1n+1k} \right) \\
&\quad - \sum_{J'} \left(\sum_{I'} y_{ijp+1} - y_{m+1jp+1} \right) + \sum_I y_{in+1p+1} \\
&= \sum_{I'} \sum_{J'} \sum_{K'} y_{ijk} - \sum_{K'} \sum_{J'} y_{m+1jk} - \sum_{K'} \sum_{I'} y_{in+1k} + \sum_{K'} y_{m+1n+1k} \\
&\quad - \sum_{J'} \sum_{I'} y_{ijp+1} + \sum_{J'} y_{m+1jp+1} + \sum_I y_{in+1p+1} \\
&= \sum_{I'} a'_i - a'_{m+1} - b'_{n+1} - \sum_{J'} \sum_{I'} y_{ijp+1} + \sum_{J'} y_{m+1jp+1} + \sum_I y_{in+1p+1} \\
&\qquad (as\; y_{m+1n+1k} = 0\; for\; k \in K'\; by\; M - feasibility) \\
&= \sum_I a_i - b'_{n+1} - \sum_I \sum_J y_{ijp+1} \\
&= \sum_I a_i - \sum_I a_i + F \\
&\quad (as\; y_{ijp+1} = 0\; for\; i \in I, j \in J\; by\; M \\
&\qquad - feasibility\; and\; definition\; of\; b'_{n+1}) \\
&= F
\end{aligned}$$

Thus, to every M-feasible solution of (**P3**), there corresponds a feasible solution of (**P2**).

Conversely, given if $\{x_{ijk}\}_{i\in I, j\in J, k\in K}$ is a feasible solution of (**P2**), then $\{y_{ijk}\}_{i\in I', j\in J', k\in K'}$ is a feasible solution of (**P3**), where

$$\begin{aligned} y_{ijk} &= x_{ijk}, \quad i\in I, j\in J, k\in K \\ &= \bar{x}_{ijk}, \quad (i,j,k)\in(I',J',K')-(I,J,K) \end{aligned}$$

where $\{\bar{x}_{ijk}\}_{i\in I', j\in J', k\in K'}$ is an M-feasible solution of the "Three Axial Sums" problem as follows:

(**P4**) Minimize $\sum_{I'}\sum_{J'}\sum_{K'} c'_{ijk}\bar{x}_{ijk}$

subject to

$$\sum_{J'}\sum_{K'}\bar{x}_{ijk} = \bar{a}_i, \quad i\in I'$$

$$\sum_{I'}\sum_{K'}\bar{x}_{ijk} = \bar{b}_j, \quad j\in J'$$

$$\sum_{I'}\sum_{J'}\bar{x}_{ijk} = \bar{e}_k, \quad k\in K'$$

$$\bar{x}_{ijk} \geq 0$$

$$\text{where } \bar{c}_{ijk} = \begin{cases} c_{ijk}, & (i,j,k)\in(I,J,K) \\ c'_{ijk}, & (i,j,k)\in(I',J',K')-(I,J,K) \end{cases}$$

where M is a sufficiently large positive integer as defined earlier and

$$\bar{a}_i = a_i - \sum_{J}\sum_{K} x_{ijk} \quad i\in I, \quad \bar{a}_{m+1} = a'_{m+1}$$

$$\bar{b}_j = b_j - \sum_{I}\sum_{K} x_{ijk} \quad j\in J, \quad \bar{b}_{n+1} = b'_{n+1}$$

$$\bar{e}_k = e_k - \sum_{I}\sum_{J} x_{ijk} \quad k\in K, \quad \bar{e}_{p+1} = e'_{p+1}$$

Let $i\in I$. Then

$$\begin{aligned} \sum_{J'}\sum_{K'} y_{ijk} &= \sum_{J'}\left(\sum_{K} y_{ijk} + y_{ijp+1}\right) \\ &= \sum_{J}\sum_{K} y_{ijk} + \sum_{K} y_{in+1k} + \sum_{J'} y_{ijp+1} \\ &= \sum_{J}\sum_{K} x_{ijk} + \sum_{K} \bar{x}_{in+1k} + \sum_{J'} \bar{x}_{ijp+1} \\ &= \sum_{J}\sum_{K} x_{ijk} + \sum_{J'}\sum_{K'} \bar{x}_{ijk} \\ &= \sum_{J}\sum_{K} x_{ijk} + \bar{a}_i \\ &= \sum_{J}\sum_{K} x_{ijk} + \left(a_i - \sum_{J}\sum_{K} x_{ijk}\right) = a_i = a'_i \end{aligned}$$

Also, $\sum_{J'}\sum_{K'} y_{m+1jk} = \sum_{J'}\sum_{K'} \bar{x}_{m+1k} = \bar{a}_{m+1} = a'_{m+1}$

Hence, $\sum_{J'}\sum_{K'} y_{ijk} = a'_i, i \in I'$.

Similarly, $\sum_{I'}\sum_{K'} y_{ijk} = b'_j, j \in J'$ and $\sum_{I'}\sum_{J'} y_{ijk} = e'_k, k \in K'$.

Obviously, $y_{ijk} \geq 0, i \in I', j \in J', k \in K'$. Thus, $\{y_{ijk}\}_{i\in I', j\in J', k\in K'}$ is a feasible solution to (**P3**).

Remark 9.2.1.1. It is obvious that a feasible solution of (**P3**), which is not an M-feasible solution can never correspond to a feasible solution of (**P2**). Henceforth, for following discussions, only M-feasible solutions of (**P3**) will be considered.

Theorem 9.2.1.2. The objective function value at an M-feasible solution of (**P3**) is equal to that of (**P2**) at its corresponding feasible solution.

Proof. Let $\{y_{ijk}\}_{I'\times J'\times K'}$ and $\{x_{ijk}\}_{I\times J\times K}$ be corresponding feasible solutions of problems (**P3**) and (**P2**) respectively. Then

$Z =$ objective function value of (**P3**) at $\{y_{ijk}\}_{I'\times J'\times K'}$

$$= \sum_{I'}\sum_{J'}\sum_{K'} c'_{ijk}y_{ijk}$$

$$= \sum_{I}\sum_{J}\sum_{K} c'_{ijk}y_{ijk} + \sum_{I}\sum_{K} c'_{in+1k}y_{in+1k} + \sum_{J}\sum_{K} c'_{m+1jk}y_{m+1jk}$$

$$+ \sum_{I}\sum_{J} c'_{ijp+1}y_{ijp+1} + \sum_{J} c'_{m+1jp+1}y_{m+1jp+1} + \sum_{I} c'_{in+1p+1}y_{in+1p+1}$$

$$+ \sum_{K} c'_{m+1n+1k}y_{m+1n+1k} + c'_{m+1n+1p+1}y_{m+1n+1p+1}$$

$$= \sum_{I}\sum_{J}\sum_{K} c'_{ijk}y_{ijk} + \sum_{I}\sum_{J} c'_{ijp+1}y_{ijp+1} + c'_{m+1n+1p+1}y_{m+1n+1p+1} \quad (from\ (1))$$

$$= \sum_{I}\sum_{J}\sum_{K} c_{ijk}y_{ijk} \quad \text{(by M-feasibility and as } c'_{ijk} = c_{ijk}\ \forall i \in I, j \in J, k \in K$$

$=$ objective function value of (**P2**) at $\{x_{ijk}\}_{I\times J\times K}$.

□

Theorem 9.2.1.3. There exists one-to-one correspondence between optimal feasible solutions to (**P2**) and optimal feasible solutions to (**P3**).

Proof. Let $\{x^0_{ijk}\}_{I\times J\times K}$ be optimal solution to (**P2**) yielding objective function value and $\{y^0_{ijk}\}_{I'\times J'\times K'}$ be the corresponding feasible solution to (**P3**). The value yielded by $\{y^0_{ijk}\}$ is z^0 (refer to Theorem 9.2.1.2). If possible, let $\{y^0_{ijk}\}$ be not an optimal solution to (**P3**). Therefore, there exists a feasible solution $\{y'_{ijk}\}$, say to (**P3**) with the value $z' < z^0$. Let $\{x'_{ijk}\}$ be the corresponding feasible solution to (**P2**). Then by theorem 9.2.1.2, $\sum_{I}\sum_{J}\sum_{K} c_{ijk}x'_{ijk} = z'$, a contradiction to the assumption that

$\{x^0_{ijk}\}_{I\times J\times K}$ is an optimal solution to (**P2**). Similarly, an optimal feasible solution to (**P3**) will give an optimal solution to (**P2**). □

Remark 9.2.1.2. Theorems 9.2.1.1, 9.2.1.2, 9.2.1.3 prove that the optimal solution of (**P3**) gives an optimal solution of (**P2**).

Remark 9.2.1.3. As (**P3**) is a standard "Three Axial Sums" problem, from Remark 9.2.1.2, it follows that there exists an integer optimal basic feasible solution of (**P2**).

9.3 NUMERICAL EXAMPLES

Example 9.3.1. Consider the multi-index transportation problem with axial constraints having restricted flow with data as per Bandopadhyaya and Puri (1988).

$$\text{Minimize } \sum_I \sum_J \sum_K c_{ijk} x_{ijk}$$

subject to

$$\sum_J \sum_K x_{ikj} \le a_i, \quad i \in I = \{1,2,3,4\}$$

$$\sum_I \sum_K x_{ikj} \le b_j, \quad j \in J = \{1,2,3,4\}$$

$$\sum_I \sum_J x_{ikj} \le e_k, \quad k \in K = \{1,2,3\}$$

$$\sum_I \sum_J \sum_K x_{ijk} = 60 \qquad (\text{F} = 60 < 66 = \text{N})$$

$$x_{ijk} \ge 0$$

where i, j, and k represent warehouses, markets, and different products, respectively. Note that a_i is the total availability at the ith warehouse, b_j is the total demand at the jth market, and e_k is the total availability of the kth commodity. These quantities are given as $a_1 = 24$, $a_2 = 14$, $a_3 = 18$, $a_4 = 10$, $b_1 = 17$, $b_2 = 19$, $b_3 = 21$, $b_4 = 9$, $e_1 = 17$, $e_2 = 31$, $e_3 = 18$.

The cost matrix $[c_{ijk}]$ for the three-dimensional transportation problem with axial constraints having restricted flow is given in Figure 9.1.

To solve the above problem, a related solid problem is formulated, with a dummy supply point, a dummy destination, and an extra commodity. A related three-dimensional transportation problem is given as follows:

$$\text{Minimize } \sum_{I'} \sum_{J'} \sum_{K'} c'_{ijk} y_{ijk}$$

subject to

$$\sum_{J'} \sum_{K'} y_{ikj} = a'_i, \quad i \in I' = I \cup \{5\}$$

	j=1			j=2			j=3			j=4			a_i
i=1	15	18	7	9	14	11	10	4	12	8	10	13	24
i=2	19	20	17	15	21	14	20	24	18	16	19	23	14
i=3	14	11	8	24	20	13	17	16	15	11	13	8	18
i=4	24	28	31	14	19	21	18	21	14	9	10	20	10
b_j	17			19			21			9			

FIGURE 9.1 Cost matrix for the three-dimensional transportation problem

$$\sum_{I'}\sum_{K'} y_{ikj} = b'_j, \quad j \in J' = J \cup \{5\}$$

$$\sum_{I'}\sum_{J'} y_{ikj} = e'_k, \quad k \in K' = K \cup \{4\}$$

$$y_{ijk} \geq 0$$

where

$a_1 = 24,\ a_2 = 14,\ a_3 = 18,\ a_4 = 10,\ b_1 = 17,\ b_2 = 19,\ b_3 = 21,\ b_4 = 9,\ e_1 = 17,$
$e_2 = 31,\ e_3 = 18,\ a_5 = 66 - 60 = 6,\ b_5 = 66 - 60 = 6,\ e_4 = 66 - 60 = 6$

The cost for the related transportation problem is given in the three-dimensional matrix $\left[c'_{ijk}\right]$ in Figure 9.2.

The related transportation problem is a "Three Axial Sum" problem which can be reformulated as a multi-index problem by applying the Haley method (Haley, 1962). Thus, to solve the related three-dimensional transportation problem we solve an equivalent multi-index transformed transportation problem whose cost matrix is given in Figure 9.3.

By solving the transformed transportation problem, the optimal solution of the related transportation problem can be found. From the optimal solution of the related transportation problem the optimal solution of the three-dimensional transportation problem with axial constraints having impaired flow is found and given as $x_{121} = 3$, $x_{132} = 21$, $x_{221} = 8$, $x_{313} = 17$, $x_{323} = 1$, $x_{421} = 1$, $x_{441} = 5$, $x_{442} = 4$ with the objective function value for the given problem as $z = 479$ which is better than the cost, 527, obtained by Bandopadhyaya and Puri (1988). This shows that the solution obtained by them is not optimal.

	j=1	j=2	j=3	j=4	j=5	a_i
i=1	15, 18, 7, M	9, 14, 11, M	10, 4, 12, M	8, 10, 13, M	0, 0, 0, 0	24
i=2	19, 20, 17, M	15, 21, 14, M	20, 24, 18, M	16, 19, 23, M	0, 0, 0, 0	14
i=3	14, 11, 8, M	24, 20, 13, M	17, 16, 15, M	11, 13, 8, M	0, 0, 0, 0	18
i=4	24, 28, 31, M	14, 19, 21, M	18, 21, 14, M	9, 10, 20, M	0, 0, 0, 0	10
i=5	0, 0, 0, 0	0, 0, 0, 0	0, 0, 0, 0	0, 0, 0, 0	M, M, M, M	6
b_j	17	19	21	9	6	

FIGURE 9.2 The cost matrix for the related transportation problem (axial sum problem)

							Bki↓
E11→	15, 18, 7, M, 0; 31	9, 14, 11, M, 0; 31	10, 4, 12, M, 0; 31	8, 10, 13, M, 0; 31	0, 0, 0, 0, 0; 31	0, 0, 0, 0, M; 100	31, 31, 31, 31, 131
	19, 20, 17, M, 0; 31	15, 21, 14, M, 0; 31	20, 24, 18, M, 0; 31	16, 19, 23, M, 0; 31	0, 0, 0, 0, 0; 31	0, 0, 0, 0, M; 110	31, 31, 31, 31, 141
	14, 11, 8, M, 0; 31	24, 20, 13, M, 0; 31	17, 16, 15, M, 0; 31	11, 13, 8, M, 0; 31	0, 0, 0, 0, 0; 31	0, 0, 0, 0, M; 106	31, 31, 31, 31, 137
	24, 28, 31, M, 0; 31	14, 19, 21, M, 0; 31	18, 21, 14, M, 0; 31	9, 10, 20, M, 0; 31	0, 0, 0, 0, 0; 31	0, 0, 0, 0, M; 114	31, 31, 31, 31, 145
	0, 0, 0, 0, 0	0, 0, 0, 0, 0	0, 0, 0, 0, 0	0, 0, 0, 0, 0	M, M, M, M, 0	0, 0, 0, 0, M; 118	31, 31, 31, 31, 149
	0, 0, 0, 0, M; 107	0, 0, 0, 0, M; 105	0, 0, 0, 0, M; 103	0, 0, 0, 0, M; 115	0, 0, 0, 0, M; 118	M, M, M, M, 0; 31	138, 124, 137, 149, 31
Ajk→	31, 31, 31, 31, 138	31, 31, 31, 31, 136	31, 31, 31, 31, 134	31, 31, 31, 31, 146	31, 31, 31, 31, 149	138, 124, 137, 149, 31	

FIGURE 9.3 The cost matrix for the transformed transportation problem

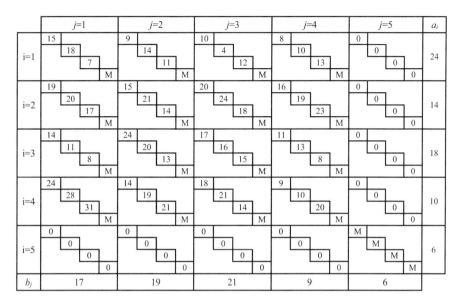

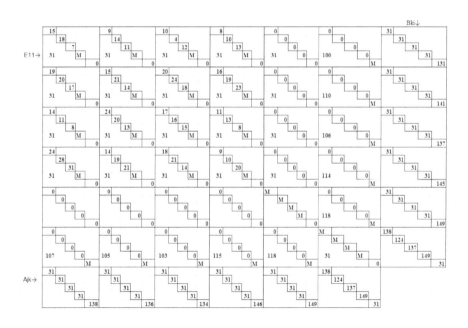

	j=1	j=2	j=3	j=4	j=5
i=1	15, 18, 7, 15, 12	9, 14, 11, 6, 8	10, 4, 12, 4, 13	8, 10, 13, 10, 7	3, 11, 5, 8, 10
i=2	19, 20, 17, 12, 15	15, 21, 14, 10, 13	20, 24, 18, 11, 16	16, 19, 23, 12, 17	11, 12, 17, 8, 13
i=3	14, 11, 8, 10, 12	24, 20, 13, 11, 14	17, 16, 15, 12, 13	11, 13, 7, 13, 10	8, 11, 10, 16, 11
i=4	24, 28, 31, 15, 22	14, 19, 21, 20, 17	18, 21, 14, 12, 14	9, 10, 20, 15, 12	12, 11, 11, 12, 14
i=5	10, 8, 13, 15, 15	12, 6, 11, 19, 14	8, 15, 12, 10, 16	12, 4, 20, 12, 13	11, 14, 15, 17, 16

FIGURE 9.4 Cost matrix for Problems 3 and 4

9.4 ADDITIONAL COMPUTATIONAL RESULTS

We solved various test problems on GAMS (General Algebraic Modeling System) on a PC with an Intel Pentium Processor 1.70 GHz having 8 GB RAM, and it took less than 5 seconds to solve the problems. For test problem 1 and 2 of order $4 \times 4 \times 3$, the cost matrix is the same as for Example 9.3.1. Figure 9.4 provides the cost matrix $[c_{ijk}]$ for the test Problem 3 of order $5 \times 5 \times 4$ and test Problem 4 of order $5 \times 5 \times 5$, where k corresponds to diagonal elements in each cell. Ignore the $k = 5$ to get the cost matrix for test Problem 3.

Table 9.1 provides the computational results of various test problems by using GAMS software

9.5 CONCLUSION

In the present paper we studied a multi-index transportation problem with axial constraints having restricted flow and provided its solution method. We find out the optimal basic feasible solution of the given problem by solving the equivalent transformed transportation problem by adding two additional rows, two additional columns, and two additional commodities, making it equivalent to a standard axial sum problem. The solution method is useful for transporting heterogenous commodities. It also helps to deal with production allocation problems when the total flow in the market needs to be restricted due to budgetary, political, and emergency situations, making it critical for a manager to reevaluate allocations. The solution method is very simple from a computational point of view and is easy to understand. The

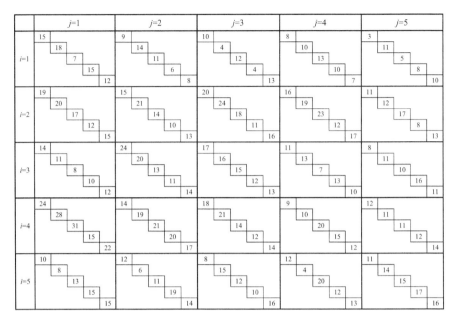

TABLE 9.1
Results of Flow Constrained Three-Dimensional Transportation Problems

Test Problems No.	Type of Problem	Order of Original Problem	Order of Transformed Problem (as of Haley, 1963)	Original Availabilities and Demand			Impaired Flow (IF) Value	Optimal Objective Function Value for IF	Exe. Time (in sec.) of Transformed Problem for IF and EF
				Availability	Demand	Commodity			
4.1	Impaired Flow	$4 \times 4 \times 3$	$6 \times 6 \times 5$	$a_1 = 10$, $a_2 = 16$, $a_3 = 15$, $a_4 = 11$	$b_1 = 16$, $b_2 = 11$, $b_3 = 13$, $b_4 = 12$	$e_1 = 12$, $e_2 = 15$, $e_3 = 25$	45	385	0.033
4.2 (un-balanced)	Impaired Flow	$4 \times 4 \times 3$	$6 \times 6 \times 5$	$a_1 = 15$, $a_2 = 22$, $a_3 = 17$, $a_4 = 10$	$b_1 = 21$, $b_2 = 16$, $b_3 = 18$, $b_4 = 15$	$e_1 = 20$, $e_2 = 19$, $e_3 = 16$	50	409	0.026
4.3	Impaired Flow	$5 \times 5 \times 4$	$7 \times 7 \times 6$	$a_1 = 10$, $a_2 = 15$, $a_3 = 21$, $a_4 = 12$, $a_5 = 16$	$b_1 = 22$, $b_2 = 14$, $b_3 = 11$, $b_4 = 17$, $b_5 = 10$	$e_1 = 15$, $e_2 = 25$, $e_3 = 20$, $e_4 = 14$	70	492	0.026
4.4	Impaired Flow	$5 \times 5 \times 5$	$7 \times 7 \times 7$	$a_1 = 10$, $a_2 = 15$, $a_3 = 21$, $a_4 = 12$, $a_5 = 16$	$b_1 = 22$, $b_2 = 14$, $b_3 = 11$, $b_4 = 17$, $b_5 = 10$	$e_1 = 15$, $e_2 = 22$, $e_3 = 13$, $e_4 = 14$, $e_5 = 10$	65	456	0.011
4.5 (un-balanced)	Impaired Flow	$5 \times 5 \times 5$	$7 \times 7 \times 7$	$a_1 = 10$, $a_2 = 15$, $a_3 = 12$, $a_4 = 12$, $a_5 = 16$	$b_1 = 17$, $b_2 = 13$, $b_3 = 11$, $b_4 = 17$, $b_5 = 12$	$e_1 = 15$, $e_2 = 25$, $e_3 = 11$, $e_4 = 14$, $e_5 = 10$	60	405	0.005

execution time to solve the transformed problem is less than 5 seconds for various data problems. Also, we found out that the optimal cost obtained by our procedure for impaired flow multi-index transportation problem is 479 which is better than the cost of 527 obtained by Bandopadhyaya and Puri (1988). Thus the optimal basic feasible solution obtained by existing approaches is not optimal and our procedure provided a better solution than other contemporary algorithms/techniques.

REFERENCES

Agarwal S, Sharma S (2022) More-for-less paradox in time minimization transportation problem with mixed constraints, Am. J. Math. Manag. Sci., published online 27 Jan 2022.

Ahuja A, Arora SR (2001) Multi-index fixed charge bicriterion transportation problem. Indian J. Pure Appl. Math. 32(5): 739–746.

Ambrosio CD, Gentill M, Cerulli R (2020) The optimal value range problem for the Interval (immune) Transportation Problem. Omega. 95(102059): 1–9.

Appa GM (1973) The transportation problem and its variants. Opns. Res. Quart. 24: 79–99.

Arora SR, Khurana A (2001) A paradox in an indefinite quadratic transportation problem. Int. J. Manag. Sci. 7(2): 13–30.

Bandopadhyaya L, Puri MC (1988) Impaired flow multi-index transportation problem with axial constraints. J. Austral. Math. Soc. B 29: 296–309.

Bridgen MEB (1974) A variant of the transportation problem in which the constraints are of mixed type. Opns. Res. Quart. 25(3): 437–446.

Corban A (1964) Multi-dimensional transportation problem. Rev. Roumaine Math. Pures Appl. 9(8): 721–735.

Haley KB (1962) The solid transportation problem. Opns. Res. 10: 448–463.

Haley KB (1963) The multi-index problem. Opns. Res. Quart. 11(3): 368–379.

Javed S (2018) A method for solving the transportation problem. Int. J. Stat. Manag., 21(5): 817–837.

Khanna S, Bakshi HC, Puri MC (1981) On controlling total flow in transportation problems. NK Jiaswal (Ed) *Scientific Management of Transport Systems*. North-Holland Publishing Company Amsterdam: 293–303.

Khanna S, Puri MC (1983) Solving transportation problem with mixed constraints and a specified transportation flow. Opsearch 20(1): 16–24.

Khurana A, Adlakha V (2015) Multi index bi-criterion transportation problem. Opsearch 52(4): 733–745.

Khurana A, Adlakha V, Lev B (2018) Multi-index constrained transportation problem with bounds on availabilities, requirements and commodities. Oper. Res. Perspectives, 319–333.

Khurana A, Arora SR (2004) Three-dimensional fixed charge bi-criterion Indefinite quadratic transportation problem. Yugosl. J. Oper. Res. 14: 83–97.

Khurana A, Arora SR (2006) The sum of a linear and linear fractional transportation problem with restricted and enhanced flow. J. Interdiscip. Math. 9(2): 373–383.

Khurana A, Arora SR (2011a) Solving transshipment problems with mixed constraints. Int. J. Manag. Sci. Eng. Manag. 6(4): 292–297.

Khurana A, Arora SR (2011b) An algorithm for solving three-dimensional transshipment problem. Int. J. Math. Oper. Res. 4(2): 97–113.

Khurana A, Arora SR (2011c) Fixed charge bi-criterion indefinite quadratic transportation problem with enhanced flow. Revista Investigacion Operacional 32: 133–145.

Khurana A, Thirwani D, Arora SR (2009) An algorithm for solving fixed charge bi-criterion indefinite quadratic transportation problem with restricted flow. Int. J. Optim.: Theory Methods Appl. 1(4): 367–380.

Klingman D, Russel R (1974) The transportation problem with mixed constraints. Opns. Res. Quart. 3: 447–455.

Mishra S, Dass C (1981) Solid transportation problem with lower and upper bounds on rim conditions-a note. New Zealand Oper. Res. 9(2): 137–140.

Thirwani D, Arora SR, Khanna S (1997) An algorithm for solving fixed-charge bi-criterion transportation problem with restricted flow. Optimization 40(2): 193–206.

10 STAR Heuristic Method: A Novel Approach and Its Comparative Analysis with CI Algorithm to Solve CBAP in Healthcare

Sharayu Dosalwar[1], *Tanishq Varshney*[1], *Ambika Patidar*[1], *Rishab Koul*[1], *Anand J. Kulkarni*[2], *Madhura Phatak*[3] and *Bhavana Tiple*[3]
[1]UG Student, School of CET, MITWPU, Pune, India
[2] Professor & Associate Director, Institute of Artificial Intelligence, MITWPU, Pune, India
[3]Professor, School of CET, MITWPU, Pune, India

CONTENTS

DOI: 10.1201/9781003329039-10

10.1 INTRODUCTION

In this paper, we have developed the STAR Heuristic method to solve the Cyclic Bottleneck Assignment Problem. In Cohort Intelligence the process involves cyclic shifting of the elements of the matrix [1], but in the STAR Heuristic method the approach followed is not cyclic and the shifting is done on the basis of the minimum intermediate column sum. In hospitals, there is a problem of CBAP in healthcare which has been solved using Cohort Intelligence (CI) but the STAR Heuristic method solves it more efficiently in terms of maximum column sum and time consumption. In healthcare systems, a bottleneck analysis was used to show processes and identify obstacles to the adoption of standard precautions [2]. Personal protective equipment (PPE) procurement, storage/inventory, in-hospital distribution, in-department distribution, usage/monitoring, and recycling are the typical sequential tasks of standard precaution implementation. A new variant of real-world Combinatorial Problems is the Cyclic Bottleneck Assignment Problem. One application of this problem is in healthcare to reduce congestion in recovery units [9].

10.1.1 Cyclic Bottleneck Assignment Problem (CBAP)

Suppose we seek to schedule 'n' surgeons/doctors over a planning horizon of 'n' days. The recovery time for each operated patient in the recovery room varies according to the type of surgery. When building cyclic surgery schedules, one important objective is to minimize congestion in the recovery room. That is, we want to minimize the maximum number of patients in the recovery unit on any given day of the planning horizon so that the costs associated with important resources such as nurses, space, beds, and equipment are also minimized. This basic problem has already been solved using cohort intelligence. Our objective is to develop a method to provide a better solution than the existing solutions in terms of minimizing the maximum column sum, time consumption, and find exceptions to the problem.

10.1.2 STAR Heuristic Method

The STAR Heuristic method is a novel approach developed to solve the Cyclic Bottleneck Assignment Problem (CBAP). In this method, by interchanging the values in rows we tend to decrease the maximum column sum as much as possible. The input matrix will be the work day schedule of the doctors in the hospitals set by the admin. There are major two steps in this algorithm. The first step is to make the optimized matrix by calculating the intermediate column sum of each row. The second step is to reduce the optimized matrix further to get the final solution.

Advantages of STAR

- Non-cyclic novel approach
- Less time-consuming
- Outperforms CI algorithm in minimizing maximum column sum

10.1.3 Cohort Intelligence (CI)

Anand J. Kulkarni developed a novel optimization technique known as Cohort Intelligence (CI) in 2013 based on Artificial intelligence (AI) [1]. It tries to simulate the behavior seen in self-organizing systems where candidates in a cohort interact and compete with one another to attain common goals. By monitoring the conduct of the other candidates in the cohort, each applicant seeks to enhance his or her own behavior. Each candidate in the cohort exhibits specific conduct that may lead to an improvement in their own behavior. When a candidate tries to emulate a specific behavior marked by certain characteristics, it frequently adopts those characteristics in ways that benefit its own purpose. Candidates in the cohort learn from one another in this way, which helps the entire group improve its behavior over time. If, after a significant number of learning attempts, the individual behavior of all candidates does not improve significantly, making it difficult to discriminate between them, the cohort's behavior is said to have attained saturation (convergence). In other words, the differences in the candidates' individual behaviors become insignificant. As a result, the final answer is accepted as any candidate's present behavior.

10.1.4 Comparison with CI

- Minimizing the maximum column sum
- It takes less time and is faster as the matrix size increases.

We undertook to solve the Cyclic Bottleneck Problem using our own STAR heuristic method and compared the outcomes with previously known algorithms such as CI. We are attempting to make the most of every resource available in the hospitals as part of this endeavor. The problem comes while trying to keep the recovery unit as free of congestion as possible. Our primary focus was on limiting the number of patients to the smallest number possible in order to reduce the need for beds, nurses, and other variable costs. We evaluated our approach on a range of 5×5 to 50×50 matrices, each with 10 test data, and compared the results to previously published CI results. According to the computational results, our provided solution performs admirably in the majority of our problem circumstances.

10.2 LITERATURE REVIEW

Cohort intelligence is defined as a self-supervised learning behavior in which other members of the cohort better themselves by monitoring the actions of others in the cohort [1]. We may leverage this intelligence, as well as the fundamentals of this approach, to build a model that improves over time on real-time data and can solve the cyclic bottleneck problem for a large number of datasets [1]. In healthcare systems, a bottleneck analysis can be used to illustrate workflow and highlight constraints to the implementation of standard precautions [2]. To address the standard precaution bottlenecks, institutional modifications such as targeted provider training, adjusting providers' workloads, and budget allocation are suggested [2]. The research was conducted in China, but the methodology could be utilized to enhance health services in other countries and could be developed further [2].

A novel bottleneck detection approach based on maximum operation capacity has been proposed by the researchers and numerically validated in complex production assembly lines [4]. Extending the strategies for detecting bottlenecks in assembly networks with variable operating capacity is a difficult mathematical task [4]. Xiangyong Li et al investigated the cyclic bottleneck assignment problem (CBAP) in this research [6]. As a result, the authors provided two algorithms that can be utilized to efficiently tackle the problem: the Tabu Search approach is used in one algorithm, while the iterated local search scheme is used in the other [6]. Their algorithms outperform the current methods. However, they have not taken into account any exceptions, such as a patient's emergency admission to a hospital, which we may work on [6].

Xiangyong Li et al. investigated the dynamic lot-sizing problem with product returns and remanufacturing (DLRR) in this research work [5]. As a result, they provide the Tabu Search method, which is an efficient solution to this problem. The earlier research work utilized a similar approach to solving the CBAP [6]. This research work added to our understanding of the algorithm [5]. The performance of the eight distinct dominance relations employed in this research is evaluated to the strong dominance relation that has been proposed and when this technique is limited to a small number of iterations, however, it does not perform well [7]. There is a pressing need for an algorithm that can retrieve the appropriate number of nondominated solutions (NDS) in the given number of iterations, such as merging it with an artificial neural network (ANN) to avoid overfitting [7].

The impact of other features such as direct waiting time (patients' waiting time at the clinic) and the patient-to-provider ratio [8]. The performance of the proposed variants of the opposition-based self-adaptive cohort intelligence (OSACI) algorithm can be investigated on other datasets and problem domains [8]. Furthermore, it is also interesting to examine the impact of employing other OBL types, such as Center-based Sampling, and Quasi-Reflection opposition-based learning (OBL), for initialization and cohort update on the convergence speed, especially with highly-dimensional datasets which could be the subject of future research [8]. CI algorithm lacks consideration of multi-criteria and multi-objectives to investigate discrete and mixed variable nonlinear constrained optimization problems [9]. Therefore the optimization problem could be resolved effectively using a generic probability-based constraint handling technique [9].

10.3 DATASET

The dataset for this problem is in matrix format. A wide range of 5×5 to 50x50 matrices with 10 test cases each has been used as input data to the source code [1]. The rows in the matrix represent the number of patients to be operated on or treated while the column represents the 7 days of the week. Collectively the matrix has the weekly schedule for the doctor and the number of patients to be operated on/treated on that particular day. We have tested our algorithm on 170 different matrices of sizes, like 5×5, 6×6, 7×7, 10×10, 25×25, 30×30, etc.

10.4 STAR HEURISTIC ALGORITHM

The STAR Heuristic method is a novel approach which has been developed to solve the Cyclic Bottleneck Problem by minimizing the congestion in the recovery unit of hospitals. Let us consider a Matrix C of size $r \times c$ where r is the total no of rows and c is the total no of columns and we are considering the index to start from 1, i.e., the element of the matrix in the first row and the first column will be considered as $C[1,1]$ as compared to $C[0,0]$, which is usually done

Part 1: Generating the Optimized Matrix

Step 1: Write the first row as $C[1,1]$, $C[1,2]$, ..., $C[1,c]$

Step 2: Put the lowest value in the second row below the highest value in the first row, the second-lowest below the second largest, and so on…until you put the largest value below the smallest.

Step 3: Calculate the intermediate column sums untll this stage, let them be $\sum_{x=1}^{i} C_{x,1} \sum_{x=1}^{i} C_{x,2}, \dots, \sum_{x=1}^{i} C_{x,c}$, where i is the row, until the intermediate column sums are calculated $(1 \leq i \leq r)$.

Step 4: Now put the next row's lowest value below the column with the highest intermediate column sum, the second-lowest below the second-largest intermediate column sum, and so on ... until you put the largest value below the smallest intermediate column sum

Step 5: Repeat Steps 3 & 4 until all the rows are placed and now the obtained matrix will be called the optimized matrix, that is $C_{\text{optimized}}$

Part 2: Reduce the Optimized Matrix

Step 1: Calculate every column's sum of the matrix and put them in an array, and let that array of column sums be Z, that is $\sum_{x=1}^{r} C_{x,1} \sum_{x=1}^{r} C_{x,2}, \dots, \sum_{x=1}^{r} C_{x,c} = Z$

Step 2:

Take a note of the column sum that is max out of Z, and let it be called Z_{max}

Find the first column from left to right in the matrix C with the largest column sum, i.e., Z_{max}, and let this column be called c_{max} and the corresponding column will be called $j(1 \leq j \leq c)$. Then start iterating on each element $C[i,j]$ of c_{max} from top to bottom, where i will be starting from 1 and ending on r. For each $C[i,j]$ try to swap it with an element that is in the same row (i) and is smaller than $C[i,j]$, giving the priority from left to right in that row (i). Also, both the column sums of the swapped positions (that is the sum of column in which $C[i,j]$ is present and the sum of the column in which that element is present that is swapped by $C[i,j]$ after the swap) should be less than the max column sum Z_{max} obtained earlier. Then only swap these elements permanently and immediately move to Step 3 before going to other elements in the column c_{max}.

If all the elements in the column c_{max} are finished and no swap was possible terminate the algorithm.

Step 3:

Do Step 1 & Step 2 until Step 2 asks you to terminate the algorithm.

The resulting matrix will be the reduced matrix, that is, $C_{reduced}$

Solving an Example with STAR Algorithm:

Consider the matrix

$$\begin{bmatrix} 6 & 4 & 2 \\ 8 & 8 & 8 \\ 7 & 7 & 0 \end{bmatrix}$$

Part 1: Making the Optimized Matrix

Step 1: Write the first row

$$\begin{bmatrix} 6 & 4 & 2 \end{bmatrix}$$

Step 2: Put the lowest value in the second row below the highest value in the first row, the second lowest below the second largest, and so on…until you put the largest value below the smallest.

$$\begin{bmatrix} 6 & 4 & 2 \\ 8 & 8 & 8 \end{bmatrix}$$

Step 3: Calculate the intermediate column sums until this stage.

$$\begin{bmatrix} 6 & 4 & 2 \\ 8 & 8 & 8 \end{bmatrix}$$
$$\begin{bmatrix} 14 & 12 & 10 \end{bmatrix}$$

Step 4: Now put the next row's lowest value below the column with the highest Intermediate column sum, the second lowest below the second largest intermediate column sum, and so on ..., until you put the largest value below the smallest intermediate column sum.

$$\begin{bmatrix} 6 & 4 & 2 \\ 8 & 8 & 8 \\ 0 & 7 & 7 \end{bmatrix}$$

Step 5: Repeat Step 3 & 4 until all the rows are placed and now the obtained matrix will be called the optimized matrix.

$$\begin{bmatrix} 6 & 4 & 2 \\ 8 & 8 & 8 \\ 0 & 7 & 7 \end{bmatrix}$$

Part 2: Reduce the Optimized Matrix

Step 1: Calculate the column sum of the matrix

Step 1 Iteration 1:

$$\begin{bmatrix} 6 & 4 & 2 \\ 8 & 8 & 8 \\ 0 & 7 & 7 \end{bmatrix}$$
$$\begin{bmatrix} 14 & 19 & 17 \end{bmatrix}$$

Step 2:

Take a note of the max column sum
Find the first column from left to right with the largest column sum and then start iterating on each element of that column from top to bottom for each element. Try to swap with the element that is in the same row and is smaller than the element. Also, the column sum of the swapped position after the swap should be less than the max column sum obtained earlier. Then only swap these elements permanently and immediately move to Step 3 before going to other elements in the column.

if all the elements in the column are finished and no swap was possible terminate the algorithm.

Step 3:

Repeat Step 1 & Step 2 until Step 2 asks you to terminate the algorithm.

The resulting matrix will be the reduced matrix.

$$\begin{bmatrix} 6 & 4 & 2 \\ 8 & 8 & 8 \\ 0 & 7 & 7 \end{bmatrix}$$
$$\begin{bmatrix} 14 & 19 & 17 \end{bmatrix} \quad -\text{Column Sum}$$

10.5 FLOWCHART

The flowchart diagram gives a brief summary of the STAR Heuristic method. From Figure 10.1 it is clearly understandable that the STAR Hueristic algorithm is divided

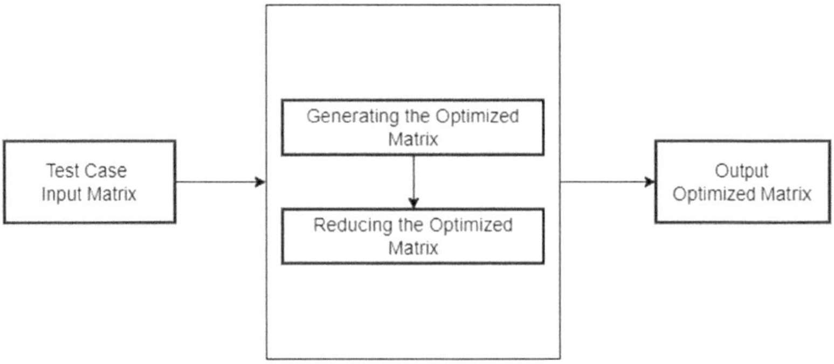

FIGURE 10.1 Block diagram of STAR Heuristic method

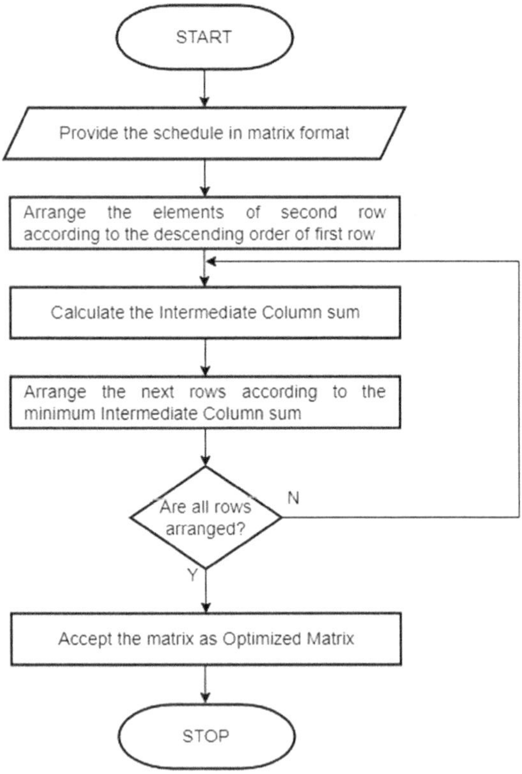

FIGURE 10.2 Generating the optimized matrix

into two parts, i.e., generating the optimized matrix, and then reducing the optimized matrix to get the desired output.

Figure 10.2 explains how to get the optimized matrix after giving the original weekly schedule of doctors in the matrix format as an input. The output Optimized matrix of the Figure 10.2 will be given as input to the second part of the algorithm.

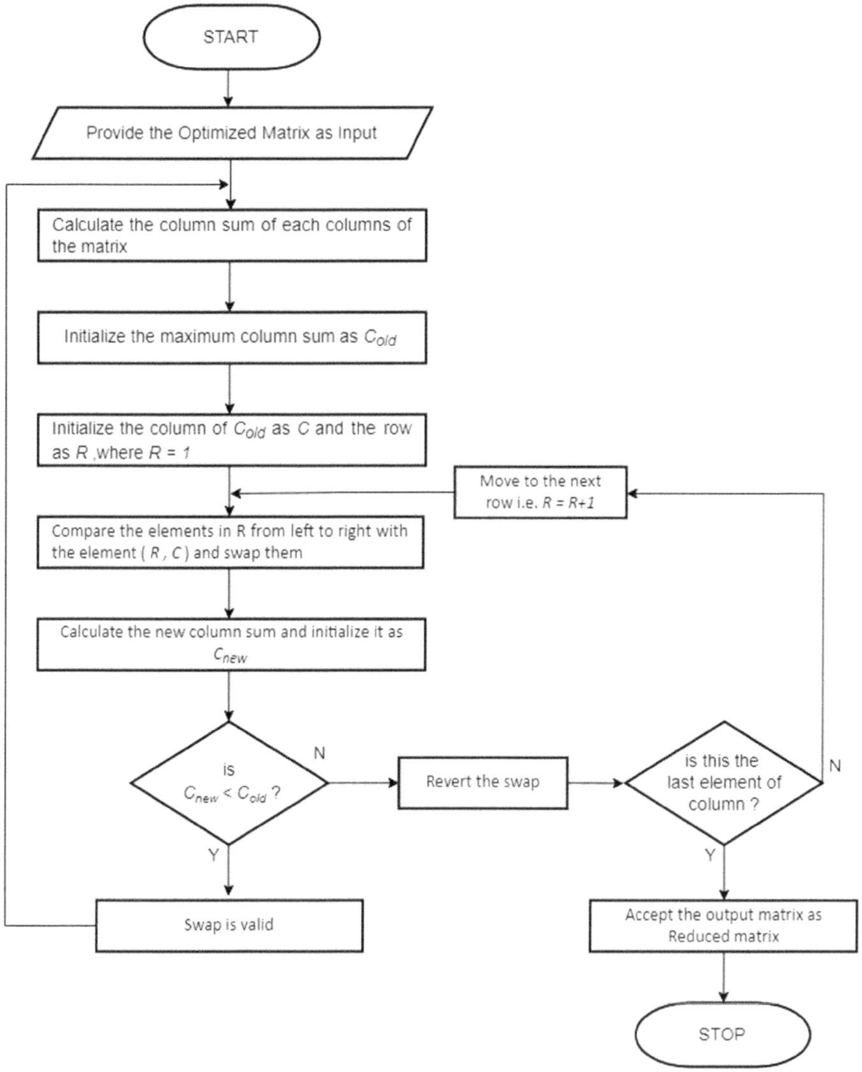

FIGURE 10.3 Reducing the optimized matrix

In Figure 10.3 to proceed further the output of Figure 10.2, i.e., the optimized matrix will be given as an input. The final matrix received will be the reduced optimized matrix which terminates the novel STAR Heuristic method.

10.6 IMPLEMENTATION RESULTS

There will be limited number of iterations, as according to the algorithm, it will automatically get terminated after a certain number of iterations based on the matrix provided. Figure 10.4 shows the graph of Max column vs No. of iterations.

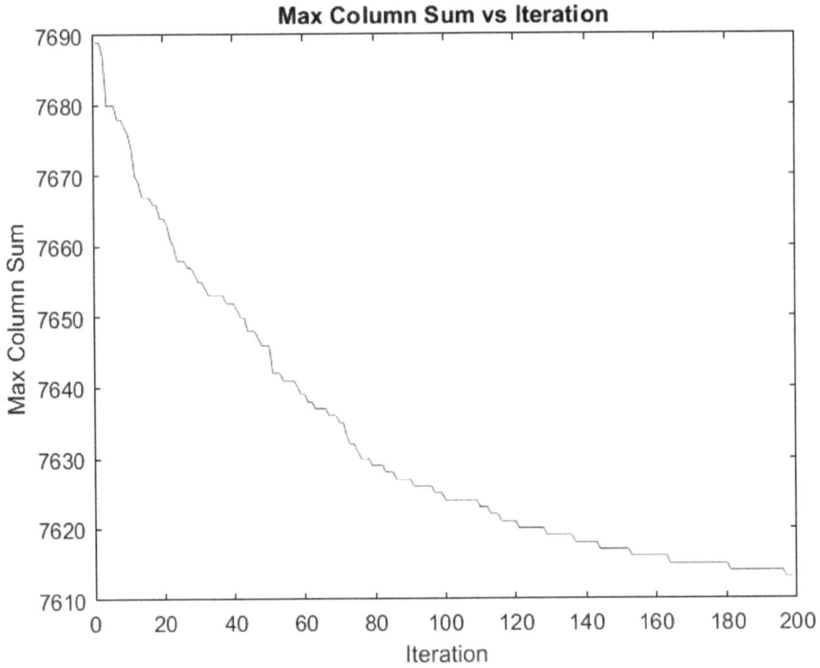

FIGURE 10.4 Max column sum vs no. of iterations for a 50x50 matrix

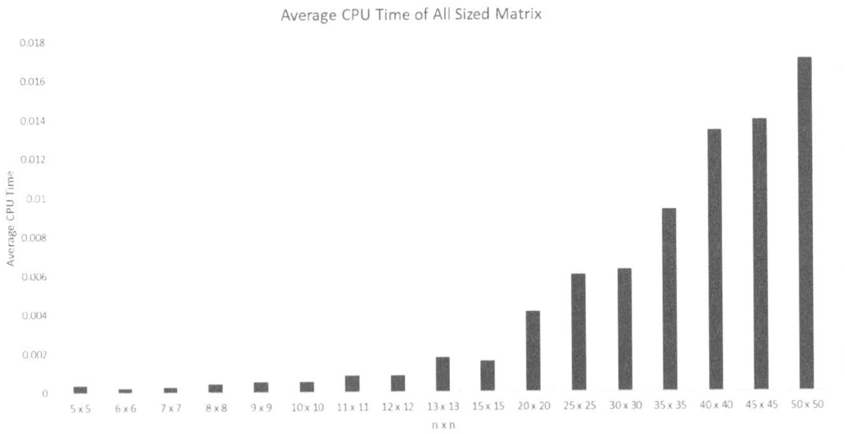

FIGURE 10.5 Average CPU time of all sized matrix after running Algorithm

It is clearly visible from Table 10.1 that the average CPU time increases with the increase in the size of the matrix. The STAR Heuristic method gives 92.53% lower results than the CI algorithm when provided with the same input test cases.

Figure 10.5 shows the graph of the Max column vs No. of iterations. The rise in the average CPU time is exponential as the size of the matrix increases.

TABLE 10.1
Comparison of STAR Heuristic and CI Algorithms on Different Parameters

Parameters	Cohort Intelligence		STAR Heuristic	
Percentage of lower results (max column sum) than the other algorithm	7.64%		92.53%	
Average CPU time				
	5 × 5	0.0381	5 × 5	0.0004
	6 × 6	0.0825	6 × 6	0.0002
	7 × 7	0.1416	7 × 7	0.0003
	8 × 8	0.2060	8 × 8	0.0004
	9 × 9	0.3011	9 × 9	0.0005
	10 × 10	0.3598	10 × 10	0.0005
	11 × 11	0.4434	11 × 11	0.0008
	12 × 12	0.5370	12 × 12	0.0008
	13 × 13	0.6633	13 × 13	0.0018
	15 × 15	0.8315	15 × 15	0.0016
	20 × 20	1.6600	20 × 20	0.0041
	25 × 25	3.1387	25 × 25	0.0060
	30 × 30	5.0054	30 × 30	0.0063
	35 × 35	7.8427	35 × 35	0.0093
	40 × 40	11.9725	40 × 40	0.0134
	45 × 45	17.0199	45 × 45	0.0139
	50 × 50	23.9809	50 × 50	0.0170
Cost	MATLAB® Annual License 66000 INR		MATLAB Annual License 66000 INR	

10.7 RESULT DISCUSSION

In the implementation results from Table 10.1, we can see that for a particular test case as the number of iterations increases the max column sum decreases, which is the expected result. The average CPU time is calculated by the tick-tock method. In Figure 10.4, the maximum column sum is shown to be minimized after a predetermined number of iterations, and the algorithm is terminated when the minimized value is reached. In contrast, the CI algorithm can be run for as many iterations as necessary, but after a predetermined number of iterations, it will reach saturation and stop further minimizing the column sum. From Figure 10.5, we can conclude that as the size of the matrix increases the average CPU time also increases, which is similar to CI and other algorithms used to solve this problem. Also, while comparing CI with STAR for this particular problem, we can see that the STAR algorithm performs better as the average CPU time is less for all of the test cases and also there is alarger number of lower results for max column sum for STAR. Some real-world applications of the STAR Heuristic method are within the healthcare industry and supply chain management.

10.8 BOUNDARY CASE

In this problem, boundary cases would be emergency case patients. If an emergency case patient arrives on a particular day, it will be difficult to adjust for that patient when all the doctors are occupied. Boundary cases can be solved in two ways:

1. Every hospital has an emergency ward so the patients requiring urgent needs like heart attack patients, accidents, etc. can be assisted with proper care.
2. Patients who do not require immediate care but need it the same day can be sent to a doctor with the fewest patients.

10.9 CONCLUSIONS AND FUTURE SCOPE

In conclusion, we have developed a novel STAR Heuristic algorithm that is capable of solving the CBAP and is able to provide an efficient way to organize the weekly schedule for the doctors in multispecialty hospitals by swapping the rows and columns of the matrix on the basis of intermediate column sum which we intend to reduce to its lowest value possible. In addition, this will help hospitals to maximize the use of resources such as beds, paramedical staff, and other necessary machines and wards. The STAR Heuristic method gives a more efficient and less time-consuming solution compared to other previously presented solutions. In this paper, we have compared our STAR heuristic method to Cohort Intelligence. The percentage of lower results (max column sum) of the STAR Heuristic algorithm is higher than the CI algorithm which provides an efficient solution for minimizing the maximum column sum. We have calculated the average CPU time of our algorithm by using the tick-tock method on our algorithm. In comparison with the CI algorithm, the STAR algorithm is much faster and is more efficient in the case of both small and large size matrices. The time taken by CI and STAR is increasing as the size of the matrix increases but the rate of increase for CI is much higher than that for the STAR heuristic approach. The 50×50 matrix takes 23.98 secs while given as an input to CI whereas the STAR heuristic method takes 0.01 sec to analyze the matrix and provide a more efficient solution to solve CBAP.

To tackle more real-world situations, we must consider each and every possibility, and take count of all the resources and facilities of the hospital. For this purpose, the method might be modified to address both limited and combinatorial problems. With some more specialized emergency scenarios, the time issue might be considered. Some of the problems that might be solved in the future include a 3D matrix solution considering more parameters like the capacity of doctors, time, etc. Anpther possible factor that can be taken into consideration is the number of empty rooms that can be converted to emergency wards in the case of any emergencies. The application of the STAR to the problems with multiple criteria [10] is underway.

REFERENCES

1. Anand J. Kulkarni, Ishan P. Durugkar, Mrinal Kumar, "Cohort Intelligence: A Self Supervised Learning Behavior", 2013 IEEE International Conference on Systems, Man, and Cybernetics, 2013.
2. Chunqing Lin, Li Li, Liang Chen, Yunjiao Pan, Jihui Guan, "Using bottleneck analysis to examine the implementation of standard precautions in hospitals", American Journal of Infection Control Volume 48, Issue 7, 2020.
3. Dongping Zhao, Xitian Tian, and Junhao Geng, "A Bottleneck Detection Algorithm for Complex Product Assembly Line Based on Maximum Operation Capacity", High-Performance Computing Strategies for Complex Engineering Optimization Problems, Volume 2014.
4. Helena Ramalhinho Louren, Olivier C. Martin, Thomas Stützle, "Iterated Local Search: Framework and Applications", International Series in Operations Research & Management Science Book Series, Volume 272, 2018.
5. Xiangyong Li, Fazle Baki Peng Tian Ben A. Chaouch, "A robust block-chain based tabu search algorithm for the dynamic lot sizing problem with product returns and remanufacturing", Omega Volume 42, 2014.
6. Xiangyong Li, Lanjian Zhu, Fazle Baki, A.B. Chaouch, "Tabu search and iterated local search for the cyclic bottleneck assignment problem", Computers & Operations Research Volume 96, 2018.
7. Mohammed Aladeemy, Linda Adwan, Amy Booth, Mohammad T. Khasawneh, Srikanth Poranki, "New feature selection methods based on opposition-based learning and self-adaptive cohort intelligence for predicting patient no-shows", Applied Soft Computing Volume 86, 2020.
8. Ishaan R. Kale, Anand J. Kulkarni, "Cohort intelligence algorithm for discrete and mixed variable engineering problems", International Journal of Parallel, Emergent and Distributed Systems Volume 33, 2018.
9. Anand J. Kulkarni, M.F. Baki, Ben, A. Chaouch, "Application of the cohort-intelligence optimization method to three selected combinatorial optimization problems", European Journal of Operational Research Volume 250, Issue 2, 2016.
10. Anand J. Kulkarni, "Multiple Criteria Decision Making – Techniques, Analysis and Applications", Studies in Systems, Decision and Control, Springer, 407, 2022.

11 Development and Optimization of Quadratic Programming Problems with Intuitionistic Fuzzy Parameters

Manisha Malik and S. K. Gupta
Department of Mathematics, Indian Institute of Technology Roorkee, India

CONTENTS

11.1 INTRODUCTION

The existing conventional optimization algorithms were designed to solve various problems having well-defined/crisp values of the involved parameters. However, to formulate a realistic problem as a mathematical optimization model, the decision-maker often encounters the situation of hesitation and vagueness in deciding the values of the input parameters. Consequently, fuzzy sets and logic are being greatly applied to analyze, formulate, and to design a solution algorithm for the practical problems so as to successfully incorporate the human reasoning. The notion of fuzzy theory is invoked to handle the uncertainty in data of the real-world problems. To enrich the optimization theory under uncertainty, the notion of fuzzy sets was further

DOI: 10.1201/9781003329039-11

extended to the intuitionistic fuzzy (IF) sets which associates a rejection degree along with the acceptance degree for each element of the set. On the other hand, quadratic programming problems (QPPs) are a special kind of optimization problem having numerous applications in various fields of engineering modeling, game theory, logistics, portfolio, regression, facility allocation, water resources management, irrigation water allocation, etc. Therefore, the construction and development of a solution algorithm for the intuitionistic fuzzy quadratic programming problems (IF-QPPs) are of the utmost importance.

In recent years, fuzzy theory has been tremendously used to model the systems which are ill-defined, complex, poorly understood, and uncertain. Fuzzy logic is a mathematical representation of this ambiguity. The fuzzy sets were first introduced by Zadeh (1965) and further examined by Dubois and Prade (1980). Later on, several authors devised different techniques to solve the various optimization problems under a fuzzy environment. Dorn (1960) introduced the concept of duality for QPPs. Further, Liu (2009) studied the fuzzy QPPs (FQPP) and using the duality theory, reduced the FQPP into two deterministic subprograms so as to obtain the bounds for the objective function value. Ammar (2008) explored the fuzzy random multi-objective QPPs, and thereby, solved a portfolio optimization problem. The sensitivity analysis on FQPPs was investigated by Kheirfam and Verdegay (2012). Thereafter, Silva et al. (2013) developed a solution algorithm to optimize the FQPPs having fuzzy data only in cost function by transforming the FQPP into a parametric multi-objective QPP. Zhou (2014) employed the ranking function and duality theory to handle the FQPPs. Later on, Mirmohseni and Nasseri (2017) gave a numerical method for solving the FQPPs where all the constraint coefficients were expressed using triangular fuzzy numbers. By using the simplex algorithm and α-cut technique, Kumar and Jeyalakshmi (2017) proposed a solution methodology for FQPPs with interval numbers as parameters.

Recently, Khalifa et al. (2021) suggested a technique to solve the neutrosophic quadratic fractional programming problem using the score function and Taylor's series approach. In this context, Khalifa (2020) transformed a fully FQPP with parameters given by *LR*-type fuzzy numbers into a fully fuzzy linear programming problem using the Taylor's series. Next, Maheswari (2019) presented the Kuhn-Tucker conditions to solve the FQPPs. Cruz et al. (2011) devised a new parametric algorithm for solving the FQPPs. Silva et al. (2007) developed a two-phase method to optimize a class of FQPPs with fuzziness in constraints only. Thereafter, Mahajan and Gupta (2019) proposed two different approaches to convert the fully FQPP into a crisp QPP to obtain the fuzzy optimal of the FQPP.

Although, the literature is enriched with several distinct approaches to solve the crisp QPPs as well as the FQPPs, the problem of solving a QPP under IF conditions still needs to be investigated. Moreover, QPPs have a significant number of remarkable practical applicabilities such as a portfolio optimization problem (Ammar, 2008), a sustainable garbage disposal model for environmental management (Mahajan and Gupta, 2020), a product mix and production planning problem in the tea industry (Mahajan and Gupta, 2021), a waste management problem (Mahajan et al., 2021), a water quality management model (Huang, 1996), a multi-product newsvendor problem (Abdel-Malek and Areeratchakul, 2007), and several others. These real-world optimization problems can be handled and solved with more flexibility using the IF environment. Consequently, the main focus of the present study is to develop a computationally efficient solution algorithm for convex IF-QPPs. Motivated by the IF approach given by Angelov (1997) to solve the optimization models, the proposed methodology uses α and β cuts for the objective function and the constraints to reduce the original IF-QPP to two subproblems giving the lower bound and upper bound for the objective value. The lower bound problem can be easily tackled, however, to handle the upper bound problem, we have developed two distinct approaches. Firstly, the duality theory is employed to obtain the optimal upper bound, and secondly, a direct, simplified and computationally easy to apply equivalent model has been proposed for the upper bound problem. This direct approach has significantly reduced the number of variables and constraints that are involved while solving the upper bound problem via a dual formulation. This reduction leads to a much simpler methodology along with curtailment of computation time to solve the upper bound problem. The rest of this chapter is as follows: Section 2 presents the basic preliminaries. The mathematical formulation of an IF-QPP model and a detailed description of the proposed solution algorithm is provided in Section 3. A numerical example illustrating the developed technique is given in Section 4. Finally, concluding remarks with future scope points are summed up.

11.2 PRELIMINARIES

Here, we present the basic definitions related to intuitionistic fuzzy (IF) theory and fundamental arithmetic operations on IF numbers.

Definition 11.2.1. (Mahajan and Gupta, 2019) A fuzzy set $\tilde{A}$ in a non-empty universal set X is a set of ordered pairs: $\{(x, \mu_{\tilde{A}}(x)) : x \in X\}$ where $\mu_{\tilde{A}}(x) \in [0,1]$ denotes the degree of membership of the element $x \in X$ being in $\tilde{A}$, and $\mu_{\tilde{A}} : X \to [0,1]$ is called the membership function.

Definition 11.2.2. (Singh and Yadav, 2015) An IF set $\tilde{A}^I$ in X is a set of the form $\tilde{A}^I = \{(x, \mu_{\tilde{A}^I}(x), \nu_{\tilde{A}^I}(x)) : x \in X\}$, where $\mu_{\tilde{A}^I}(x) : X \to [0,1]$ and $\nu_{\tilde{A}^I}(x) : X \to [0,1]$ represent the acceptance and rejection degrees of the element $x \in X$ being in $\tilde{A}^I$, respectively, provided $0 \leq \mu_{\tilde{A}^I}(x) + \nu_{\tilde{A}^I}(x) \leq 1,\ \forall\, x \in X$. The value of $\pi_{\tilde{A}^I}(x) = 1 - \mu_{\tilde{A}^I}(x) - \nu_{\tilde{A}^I}(x)$, is called the degree of indeterminacy of the element $x \in X$.

Definition 11.2.3. (Malik et al., 2021) An IF set $\tilde{A}^I = \{(x, \mu_{\tilde{A}^I}(x), \nu_{\tilde{A}^I}(x)) : x \in X\}$ in X

- is normal if there exists $x_0,\ x_1 \in X$ such that $\mu_{\tilde{A}^I}(x_0) = 1$ and $\nu_{\tilde{A}^I}(x_1) = 1$.
- is convex if $\forall\ x_1, x_2 \in X,\ 0 \leq \lambda \leq 1$,

$$\mu_{\tilde{A}^I}(\lambda x_1 + (1-\lambda)x_2) \geq \min\{\mu_{\tilde{A}^I}(x_1), \mu_{\tilde{A}^I}(x_2)\} \text{ and}$$
$$\nu_{\tilde{A}^I}(\lambda x_1 + (1-\lambda)x_2) \leq \max\{\nu_{\tilde{A}^I}(x_1), \nu_{\tilde{A}^I}(x_2)\}.$$

Definition 11.2.4. (Singh and Yadav, 2015) An IF set $\tilde{A}^I = \{(x, \mu_{\tilde{A}^I}(x), \nu_{\tilde{A}^I}(x)) : x \in \mathbb{R}\}$ of the real number $\mathbb{R}$ is called an intuitionistic fuzzy number (IFN) if

(i) $\tilde{A}^I$ is a normal IF set,
(ii) $\tilde{A}^I$ is a convex IF set, and
(iii) $\mu_{\tilde{A}^I}$ and $\nu_{\tilde{A}^I}$ are piecewise continuous functions from $\mathbb{R}$ to the closed interval $[0,1]$ such that $0 \leq \mu_{\tilde{A}^I}(x) + \nu_{\tilde{A}^I}(x) \leq 1,\ \forall\ x \in \mathbb{R}$.

Definition 11.2.5. (Singh and Yadav, 2015) An IF number $\tilde{A}^I$ is said to be a triangular intuitionistic fuzzy number (TIFN) if its membership and non-membership functions are, respectively, defined as

$$\mu_{\tilde{A}^I}(x) = \begin{cases} \frac{x-\gamma_1}{\gamma_2-\gamma_1}, & \text{if } \gamma_1 < x \leq \gamma_2, \\ \frac{\gamma_3-x}{\gamma_3-\gamma_2}, & \text{if } \gamma_2 < x \leq \gamma_3, \\ 0, & \text{otherwise} \end{cases} \quad \text{and} \quad \nu_{\tilde{A}^I}(x) = \begin{cases} \frac{\gamma_2-x}{\gamma_2-\gamma_1'}, & \text{if } \gamma_1' < x \leq \gamma_2, \\ \frac{x-\gamma_2}{\gamma_3'-\gamma_2}, & \text{if } \gamma_2 < x \leq \gamma_3', \\ 1, & \text{otherwise} \end{cases}$$

where $\gamma_1' \leq \gamma_1 \leq \gamma_2 \leq \gamma_3 \leq \gamma_3'$. This TIFN $\tilde{A}^I$ is denoted by $(\gamma_1, \gamma_2, \gamma_3; \gamma_1', \gamma_2, \gamma_3')$ and its graphical representation is given by Figure 11.1.

Definition 11.2.6. (Malik et al., 2021) An TIFN $\tilde{A}^I = (\gamma_1, \gamma_2, \gamma_3; \gamma_1', \gamma_2, \gamma_3')$ is said to be

- non-negative ($\succeq 0$) if and only if $\gamma_1' \geq 0$.
- positive ($\succ 0$) if and only if $\gamma_1' > 0$.
- non-positive ($\preceq 0$) if and only if $\gamma_3' \leq 0$.
- negative ($\prec 0$) if and only if $\gamma_3' < 0$.
- equal to a TIFN $\tilde{B}^I = (\delta_1, \delta_2, \delta_3; \delta_1', \delta_2, \delta_3')$ if and only if $\gamma_1 = \delta_1$, $\gamma_2 = \delta_2$, $\gamma_3 = \delta_3, \gamma_1' = \delta_1'$ and $\gamma_3' = \delta_3'$.

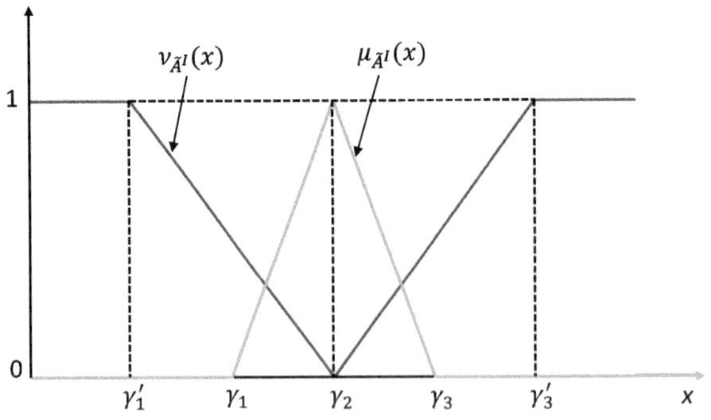

FIGURE 11.1 Triangular intuitionistic fuzzy number

Definition 11.2.7. (Singh and Yadav, 2015) Let IF($\mathbb{R}$) denote the set of all IF numbers defined on $\mathbb{R}$. Then, the accuracy function H : IF($\mathbb{R}$) $\rightarrow \mathbb{R}$ is a rule which maps each IF number onto a real number. The comparison on crisp real numbers is used to rank or compare IF numbers.

For a TIFN, $\tilde{A}^I = (\gamma_1, \gamma_2, \gamma_3; \gamma_1', \gamma_2, \gamma_3')$, the accuracy function is given by

$$H(\tilde{A}^I) = \frac{\gamma_1 + \gamma_3 + 4\gamma_2 + \gamma_1' + \gamma_3'}{8}.$$

Based upon this accuracy function, the ordering on IF numbers is defined as follows:

(i) $\tilde{A}^I \succeq \tilde{B}^I$ if and only if $H(\tilde{A}^I) \geq H(\tilde{B}^I)$,
(ii) $\tilde{A}^I \preceq \tilde{B}^I$ if and only if $H(\tilde{A}^I) \leq H(\tilde{B}^I)$.

Definition 11.2.8. (Malik et al., 2021) Let $\tilde{A}^I$ be an IF set. Then, the α-cut for the IF set $\tilde{A}^I$ is denoted by A_α and is given by $A_\alpha = \{x \in X : \mu_{\tilde{A}^I}(x) \geq \alpha\}$ and its β-cut is denoted by A^β and is defined by $A^\beta = \{x \in X : \nu_{\tilde{A}^I}(x) \leq \beta\}$, such that $\alpha, \beta \in [0, 1]$ and $\alpha + \beta \leq 1$. The (α, β)-cut of an IF set $\tilde{A}^I$ is denoted by $A_{(\alpha,\beta)}$ and is defined by $A_{(\alpha,\beta)} = \{x \in X : \mu_{\tilde{A}^I}(x) \geq \alpha$ and $\nu_{\tilde{A}^I}(x) \leq \beta\}$ such that $\alpha, \beta \in [0, 1]$ and $\alpha + \beta \leq 1$.

α-cut and β-cut for a TIFN:
Let $\tilde{A}^I = (\gamma_1, \gamma_2, \gamma_3; \gamma_1', \gamma_2, \gamma_3')$ be a TIFN, then its α-cut is defined to be a crisp set having all those x whose membership degree is greater than or equal to α, that is,

$$\mu_{\tilde{A}^I}(x) \geq \alpha \implies \frac{x - \gamma_1}{\gamma_2 - \gamma_1} \geq \alpha \text{ and } \frac{\gamma_3 - x}{\gamma_3 - \gamma_2} \geq \alpha$$

which further gives

$$\gamma_1 + \alpha(\gamma_2 - \gamma_1) \leq x \leq \gamma_3 - \alpha(\gamma_3 - \gamma_2).$$

Therefore, taking $(A_\alpha)^L = \gamma_1 + \alpha(\gamma_2 - \gamma_1)$ and $(A_\alpha)^U = \gamma_3 - \alpha(\gamma_3 - \gamma_2)$, we get $[(A_\alpha)^L, (A_\alpha)^U]$ to be an interval, that is,

$$A_\alpha = [(A_\alpha)^L, (A_\alpha)^U] = [\gamma_1 + \alpha(\gamma_2 - \gamma_1), \gamma_3 - \alpha(\gamma_3 - \gamma_2)].$$

Further, β-cut is given by a crisp set having all those x whose non-membership degree is less than or equal to β, such that

$$\nu_{\tilde{A}^I}(x) \leq \beta \implies \frac{\gamma_2 - x}{\gamma_2 - \gamma_1'} \leq \beta \text{ and } \frac{x - \gamma_2}{\gamma_3' - \gamma_2} \leq \beta.$$

This yields

$$\gamma_2 - \beta(\gamma_2 - \gamma_1') \leq x \leq \gamma_2 + \beta(\gamma_3' - \gamma_2).$$

Thus, choosing $(A^\beta)^L = \gamma_2 - \beta(\gamma_2 - \gamma_1')$ and $(A^\beta)^U = \gamma_2 + \beta(\gamma_3' - \gamma_2)$ gives $[(A^\beta)^L, (A^\beta)^U]$ an interval, that is,

$$A^\beta = [(A^\beta)^L, (A^\beta)^U] = [\gamma_2 - \beta(\gamma_2 - \gamma_1'), \gamma_2 + \beta(\gamma_3' - \gamma_2)].$$

Definition 11.2.9. A matrix $\tilde{A} = [\tilde{a}_{ij}]_{n\times n}$ is said to be a symmetric IF matrix if $\tilde{a}_{ij}$ and $\tilde{a}_{ji}$ are all IFNs and $\tilde{a}_{ij} = \tilde{a}_{ji}$, $\forall\, i, j = 1, 2, \ldots, n$.

Definition 11.2.10. Let $\tilde{A} = [\tilde{a}_{ij}]_{n\times n}$ be a symmetric IF matrix and A denotes the corresponding crisp matrix such that $A = [H(\tilde{a}_{ij})]_{n\times n}$, then $\tilde{A}$ is positive semi-definite/negative semi-definite according as A is positive semi-definite/negative semi-definite.

Arithmetic operations on TIFNs:

Let $\tilde{A}^I = (\gamma_1, \gamma_2, \gamma_3; \gamma_1', \gamma_2, \gamma_3')$ and $\tilde{B}^I = (\delta_1, \delta_2, \delta_3; \delta_1', \delta_2, \delta_3')$ be two TIFNs and λ be a real number. Then,

(i) $\tilde{A}^I \oplus \tilde{B}^I = (\gamma_1 + \delta_1, \gamma_2 + \delta_2, \gamma_3 + \delta_3; \gamma_1' + \delta_1', \gamma_2 + \delta_2, \gamma_3' + \delta_3')$.

(ii) $\tilde{A}^I \ominus \tilde{B}^I = (\gamma_1 - \delta_3, \gamma_2 - \delta_2, \gamma_3 - \delta_1; \gamma_1' - \delta_3', \gamma_2 - \delta_2, \gamma_3' - \delta_1')$.

(iii) $\lambda\tilde{A}^I = \begin{cases} (\lambda\gamma_1, \lambda\gamma_2, \lambda\gamma_3; \lambda\gamma_1', \lambda\gamma_2, \lambda\gamma_3'), & \text{if } \lambda \geq 0, \\ (\lambda\gamma_3, \lambda\gamma_2, \lambda\gamma_1; \lambda\gamma_3', \lambda\gamma_2, \lambda\gamma_1'), & \text{if } \lambda < 0. \end{cases}$

(iv) $\tilde{A}^I \otimes \tilde{B}^I \simeq (\kappa_1, \kappa_2, \kappa_3; \kappa_1', \kappa_2, \kappa_3')$

where

$$\kappa_1 = \min\{\gamma_1\delta_1, \gamma_1\delta_3, \gamma_3\delta_1, \gamma_3\delta_3\},\quad \kappa_3 = \max\{\gamma_1\delta_1, \gamma_1\delta_3, \gamma_3\delta_1, \gamma_3\delta_3\},\quad \kappa_2 = \gamma_2\delta_2,$$
$$\kappa_1' = \min\{\gamma_1'\delta_1', \gamma_1'\delta_3', \gamma_3'\delta_1', \gamma_3'\delta_3'\},\quad \kappa_3' = \max\{\gamma_1'\delta_1', \gamma_1'\delta_3', \gamma_3'\delta_1', \gamma_3'\delta_3'\}.$$

11.3 INTUITIONISTIC FUZZY QUADRATIC PROGRAMMING PROBLEM

11.3.1 Model Formulation

An IF quadratic programming problem (IF-QPP) having m constraints and n variables can be formulated as follows:

$$\textbf{(P1)}\ \min\ \tilde{Z}^I = (\tilde{C}^I)^T X + \frac{1}{2} X^T \tilde{Q}^I X$$
$$\text{s.t. } \tilde{A}^I X \preceq \tilde{b}^I,\ \ X \geq 0.$$

Equivalently, the problem (P1) can be expressed as

$$\textbf{(P2)}\ \min\ \tilde{Z}^I = \sum_{t=1}^{n} \tilde{c}_t^I x_t + \frac{1}{2} \sum_{r=1}^{n} \sum_{t=1}^{n} \tilde{q}_{rt}^I x_r x_t$$
$$\text{s.t. } \sum_{t=1}^{n} \tilde{a}_{st}^I x_t \preceq \tilde{b}_s^I,\ \ x_t \geq 0,$$
$$s = 1, 2, \ldots, m;\ \ t = 1, 2, \ldots, n$$

where $\tilde{C}^I = [\tilde{c}_t^I]_{n\times 1}$, $X = [x_t]_{n\times 1}$, $\tilde{A}^I = [\tilde{a}_{st}^I]_{m\times n}$, $\tilde{b}^I = [\tilde{b}_s^I]_{m\times 1}$, $\tilde{Q}^I = [\tilde{q}_{rt}^I]_{n\times n}$ such that $\tilde{Q}^I$ is a symmetric positive semi-definite IF matrix and $\tilde{c}_t^I$, $\tilde{a}_{st}^I$, $\tilde{b}_s^I$, $\tilde{q}_{rt}^I$ are TIFNs.

11.3.2 Proposed Solution Methodology

Applying the α-cut (w.r.t. membership) and β-cut (w.r.t. non-membership) for the objective function and constraints, the model (P2) is re-cast to

$$
\begin{aligned}
\textbf{(P3)}\ Z_{(\alpha,\beta)} = \min & \left\{ \left(\sum_{t=1}^{n} [((c_t)_\alpha)^L, ((c_t)_\alpha)^U] x_t + \frac{1}{2} \sum_{r=1}^{n} \sum_{t=1}^{n} [((q_{rt})_\alpha)^L, ((q_{rt})_\alpha)^U] x_r x_t \right), \right. \\
& \left. \left(\sum_{t=1}^{n} [((c_t)^\beta)^L, ((c_t)^\beta)^U] x_t + \frac{1}{2} \sum_{r=1}^{n} \sum_{t=1}^{n} [((q_{rt})^\beta)^L, ((q_{rt})^\beta)^U] x_r x_t \right) \right\} \\
\text{s.t.}\ & \sum_{t=1}^{n} [((a_{st})_\alpha)^L, ((a_{st})_\alpha)^U] x_t \le [((b_s)_\alpha)^L, ((b_s)_\alpha)^U], \\
& \sum_{t=1}^{n} [((a_{st})^\beta)^L, ((a_{st})^\beta)^U] x_t \le [((b_s)^\beta)^L, ((b_s)^\beta)^U], \\
& \alpha + \beta \le 1,\ \alpha \ge \beta,\ \alpha \ge 0,\ \beta \ge 0, \\
& x_t \ge 0,\ s = 1, 2, \ldots, m;\ t = 1, 2, \ldots, n.
\end{aligned}
$$

Taking the weights w and $1 - w$ for the membership (α-cut) and non-membership (β-cut) parts, the above bi-objective problem (P3) reduces to the following single-objective optimization problem:

$$
\begin{aligned}
\textbf{(P4)}\ & Z_{(\alpha,\beta)} \\
& = \min \left\{ w \left(\sum_{t=1}^{n} [((c_t)_\alpha)^L, ((c_t)_\alpha)^U] x_t + \frac{1}{2} \sum_{r=1}^{n} \sum_{t=1}^{n} [((q_{rt})_\alpha)^L, ((q_{rt})_\alpha)^U] x_r x_t \right) + \right. \\
& \quad \left. (1 - w) \left(\sum_{t=1}^{n} [((c_t)^\beta)^L, ((c_t)^\beta)^U] x_t + \frac{1}{2} \sum_{r=1}^{n} \sum_{t=1}^{n} [((q_{rt})^\beta)^L, ((q_{rt})^\beta)^U] x_r x_t \right) \right\} \\
\text{s.t.}\ & \sum_{t=1}^{n} [((a_{st})_\alpha)^L, ((a_{st})_\alpha)^U] x_t \le [((b_s)_\alpha)^L, ((b_s)_\alpha)^U], \\
& \sum_{t=1}^{n} [((a_{st})^\beta)^L, ((a_{st})^\beta)^U] x_t \le [((b_s)^\beta)^L, ((b_s)^\beta)^U], \\
& \alpha + \beta \le 1,\ \alpha \ge \beta,\ \alpha \ge 0,\ \beta \ge 0,\ 0 \le w \le 1, \\
& x_t \ge 0,\ s = 1, 2, \ldots, m;\ t = 1, 2, \ldots, n.
\end{aligned}
$$

Since, problem (P4) is an interval programming problem, therefore, the objective function is of the type $Z_{(\alpha,\beta)} = [Z^L_{(\alpha,\beta)}, Z^U_{(\alpha,\beta)}]$, where $Z^L_{(\alpha,\beta)}$ and $Z^U_{(\alpha,\beta)}$ are the lower and upper bounds of the objective function interval, respectively. Consequently, the problem (P4) further split into the two succeeding min-min and max-min type QPP models (P4-L) and (P4-U) so as to obtain the respective lower and upper bound values for $Z_{(\alpha,\beta)}$, given by:

$$\textbf{(P4-L)}\ Z^L_{(\alpha,\beta)} = \min_{\Delta}\left(\min_{x}\ Z = w\left(\sum_{t=1}^{n}(c_t)_\alpha x_t + \frac{1}{2}\sum_{r=1}^{n}\sum_{t=1}^{n}(q_{rt})_\alpha x_r x_t\right) + (1-w)\left(\sum_{t=1}^{n}(c_t)^\beta x_t + \frac{1}{2}\sum_{r=1}^{n}\sum_{t=1}^{n}(q_{rt})^\beta x_r x_t\right)\right)$$

$$\text{s.t. } \sum_{t=1}^{n}(a_{st})_\alpha x_t \le (b_s)_\alpha,$$

$$\sum_{t=1}^{n}(a_{st})^\beta x_t \le (b_s)^\beta,$$

$$\alpha+\beta \le 1,\ \ \alpha \ge \beta,\ \ \alpha \ge 0,\ \ \beta \ge 0,\ \ 0 \le w \le 1,$$

$$x_t \ge 0,\ \ s = 1,2,\ldots,m;\ \ t = 1,2,\ldots,n$$

and,

$$\textbf{(P4-U)}\ Z^U_{(\alpha,\beta)} = \max_{\Delta}\left(\min_{x}\ Z = w\left(\sum_{t=1}^{n}(c_t)_\alpha x_t + \frac{1}{2}\sum_{r=1}^{n}\sum_{t=1}^{n}(q_{rt})_\alpha x_r x_t\right) + (1-w)\left(\sum_{t=1}^{n}(c_t)^\beta x_t + \frac{1}{2}\sum_{r=1}^{n}\sum_{t=1}^{n}(q_{rt})^\beta x_r x_t\right)\right)$$

$$\text{s.t. } \sum_{t=1}^{n}(a_{st})_\alpha x_t \le (b_s)_\alpha,$$

$$\sum_{t=1}^{n}(a_{st})^\beta x_t \le (b_s)^\beta,$$

$$\alpha+\beta \le 1,\ \ \alpha \ge \beta,\ \ \alpha \ge 0,\ \ \beta \ge 0,\ \ 0 \le w \le 1,$$

$$x_t \ge 0,\ \ s = 1,2,\ldots,m;\ \ t = 1,2,\ldots,n$$

$$\begin{aligned}\text{where } \Delta = \ & \{(c_t)_\alpha \in [((c_t)_\alpha)^L, ((c_t)_\alpha)^U],\ (q_{rt})_\alpha \in [((q_{rt})_\alpha)^L, ((q_{rt})_\alpha)^U],\\ & (c_t)^\beta \in [((c_t)^\beta)^L, ((c_t)^\beta)^U],\ (q_{rt})^\beta \in [((q_{rt})^\beta)^L, ((q_{rt})^\beta)^U],\\ & (a_{st})_\alpha \in [((a_{st})_\alpha)^L, ((a_{st})_\alpha)^U], (b_s)_\alpha \in [((b_s)_\alpha)^L, ((b_s)_\alpha)^U],\\ & (a_{st})^\beta \in [((a_{st})^\beta)^L, ((a_{st})^\beta)^U],\ (b_s)^\beta \in [((b_s)^\beta)^L, ((b_s)^\beta)^U]\}.\end{aligned}$$

Further, we can write

$$\textbf{(P4-L1)}\ Z^L_{(\alpha,\beta)} = \min_{\Delta'}\left\{\min_{x}\left(w\left(\sum_{t=1}^{n}((c_t)_\alpha)^L x_t + \frac{1}{2}\sum_{r=1}^{n}\sum_{t=1}^{n}((q_{rt})_\alpha)^L x_r x_t\right) + (1-w)\left(\sum_{t=1}^{n}((c_t)^\beta)^L x_t + \frac{1}{2}\sum_{r=1}^{n}\sum_{t=1}^{n}((q_{rt})^\beta)^L x_r x_t\right)\right)\right\}$$

$$\text{s.t.} \sum_{t=1}^{n} (a_{st})_\alpha x_t \leq (b_s)_\alpha,$$
$$\sum_{t=1}^{n} (a_{st})^\beta x_t \leq (b_s)^\beta,$$
$$\alpha + \beta \leq 1, \;\; \alpha \geq \beta, \;\; \alpha \geq 0, \;\; \beta \geq 0, \;\; 0 \leq w \leq 1,$$
$$x_t \geq 0, \;\; s = 1, 2, \ldots, m; \;\; t = 1, 2, \ldots, n$$

and,

$$\textbf{(P4-U1)}\; Z^U_{(\alpha,\beta)} = \max_{\Delta'} \left\{ \min_x \left(w \left(\sum_{t=1}^{n} ((c_t)_\alpha)^U x_t + \frac{1}{2} \sum_{r=1}^{n} \sum_{t=1}^{n} ((q_{rt})_\alpha)^U x_r x_t \right) + \right.\right.$$
$$\left.\left. (1-w) \left(\sum_{t=1}^{n} ((c_t)^\beta)^U x_t + \frac{1}{2} \sum_{r=1}^{n} \sum_{t=1}^{n} ((q_{rt})^\beta)^U x_r x_t \right)\right)\right\}$$
$$\text{s.t.} \sum_{t=1}^{n} (a_{st})_\alpha x_t \leq (b_s)_\alpha,$$
$$\sum_{t=1}^{n} (a_{st})^\beta x_t \leq (b_s)^\beta,$$
$$\alpha + \beta \leq 1, \;\; \alpha \geq \beta, \;\; \alpha \geq 0, \;\; \beta \geq 0, \;\; 0 \leq w \leq 1,$$
$$x_t \geq 0, \;\; s = 1, 2, \ldots, m; \;\; t = 1, 2, \ldots, n$$

$$\text{where } \Delta' = \{(a_{st})_\alpha \in [((a_{st})_\alpha)^L, ((a_{st})_\alpha)^U],\; (b_s)_\alpha \in [((b_s)_\alpha)^L, ((b_s)_\alpha)^U],$$
$$(a_{st})^\beta \in [((a_{st})^\beta)^L, ((a_{st})^\beta)^U],\; (b_s)^\beta \in [((b_s)^\beta)^L, ((b_s)^\beta)^U]\}.$$

Now, our objective is to specify relevant values to the set Δ' for finding $Z^L_{(\alpha,\beta)}$ and $Z^U_{(\alpha,\beta)}$. The methodology to further evaluate the lower (P4-L1) and upper (P4-U1) bounds of the problem (P4) is described below:

Lower bound problem:

The problem (P4-L1) gives lower bound of the objective function of model (P2). Moreover, since model (P4-L1) has the same minimization operation on inner and outer programs, therefore, both of these can be integrated into a single optimization model as follows:

$$\textbf{(P5-L1)}\; Z^L_{(\alpha,\beta)} = \min_{\Delta',\, x} \left\{ w \left(\sum_{t=1}^{n} ((c_t)_\alpha)^L x_t + \frac{1}{2} \sum_{r=1}^{n} \sum_{t=1}^{n} ((q_{rt})_\alpha)^L x_r x_t \right) + \right.$$
$$\left. (1-w) \left(\sum_{t=1}^{n} ((c_t)^\beta)^L x_t + \frac{1}{2} \sum_{r=1}^{n} \sum_{t=1}^{n} ((q_{rt})^\beta)^L x_r x_t \right)\right\}$$

$$\text{s.t.} \sum_{t=1}^{n}(a_{st})_\alpha x_t \le (b_s)_\alpha,$$
$$\sum_{t=1}^{n}(a_{st})^\beta x_t \le (b_s)^\beta,$$
$$\alpha+\beta \le 1,\ \ \alpha \ge \beta,\ \ \alpha \ge 0,\ \ \beta \ge 0,\ \ 0 \le w \le 1,$$
$$x_t \ge 0,\ \ s=1,2,\ldots,m;\ \ t=1,2,\ldots,n$$

where $\Delta' = \{(a_{st})_\alpha \in [((a_{st})_\alpha)^L, ((a_{st})_\alpha)^U],\ (b_s)_\alpha \in [((b_s)_\alpha)^L, ((b_s)_\alpha)^U],$
$(a_{st})^\beta \in [((a_{st})^\beta)^L, ((a_{st})^\beta)^U],\ (b_s)^\beta \in [((b_s)^\beta)^L, ((b_s)^\beta)^U]\}$

or,

$$\textbf{(P5-L2)}\ Z^L_{(\alpha,\beta)} = \min_x \left\{ w\left(\sum_{t=1}^{n}((c_t)_\alpha)^L x_t + \frac{1}{2}\sum_{r=1}^{n}\sum_{t=1}^{n}((q_{rt})_\alpha)^L x_r x_t\right) + (1-w)\left(\sum_{t=1}^{n}((c_t)^\beta)^L x_t + \frac{1}{2}\sum_{r=1}^{n}\sum_{t=1}^{n}((q_{rt})^\beta)^L x_r x_t\right)\right\}$$

$$\text{s.t.} \sum_{t=1}^{n}((a_{st})_\alpha)^L x_t \le ((b_s)_\alpha)^U,$$
$$\sum_{t=1}^{n}((a_{st})^\beta)^L x_t \le ((b_s)^\beta)^U,$$
$$\alpha+\beta \le 1,\ \ \alpha \ge \beta,\ \ \alpha \ge 0,\ \ \beta \ge 0,\ \ 0 \le w \le 1,$$
$$x_t \ge 0,\ \ s=1,2,\ldots,m;\ \ t=1,2,\ldots,n$$

where $((a_{st})_\alpha)^L$, $((b_s)_\alpha)^U$ and $((a_{st})^\beta)^L$, $((b_s)^\beta)^U$ gives the maximum possible feasible region.

Upper bound problem:

As the model (P4-U1) has inner and outer operations in different directions, therefore, to combine these two operations, we consider the following two approaches:

(i) Duality approach

The Lagrangian dual of the inner problem of model (P4-U1) is obtained by maximizing the function $h(\mu, \mu', \nu)$, which is given as follows:

$$h(\mu,\mu',\nu) = \inf\left\{ w\left(\sum_{t=1}^{n}((c_t)_\alpha)^U x_t + \frac{1}{2}\sum_{r=1}^{n}\sum_{t=1}^{n}((q_{rt})_\alpha)^U x_r x_t\right) + (1-w)\left(\sum_{t=1}^{n}((c_t)^\beta)^U x_t + \frac{1}{2}\sum_{r=1}^{n}\sum_{t=1}^{n}((q_{rt})^\beta)^U x_r x_t\right) + \sum_{s=1}^{m}\mu_s\left(\sum_{t=1}^{n}(a_{st})_\alpha x_t - (b_s)_\alpha\right) + \sum_{s=1}^{m}\mu'_s\left(\sum_{t=1}^{n}(a_{st})^\beta x_t - (b_s)^\beta\right) - \sum_{t=1}^{n}\nu_t x_t\right\}$$

where μ_s, μ'_s, ν_t are Lagrange's multipliers such that $\mu_s, \mu'_s, \nu_t, x_t \geq 0$ and $(a_{st})_\alpha \in [((a_{st})_\alpha)^L, ((a_{st})_\alpha)^U]$, $(b_s)_\alpha \in [((b_s)_\alpha)^L, ((b_s)_\alpha)^U]$, $(a_{st})^\beta \in [((a_{st})^\beta)^L, ((a_{st})^\beta)^U]$, $(b_s)^\beta \in [((b_s)^\beta)^L, ((b_s)^\beta)^U]$, $\forall\, s, t$.

Since $[\tilde{q}^I_{rt}]_{n\times n}$ is a symmetric positive semi-definite matrix and convex linear combination of two positive semi-definite matrices is also a positive semi-definite matrix, therefore, the function $h(\mu, \mu', \nu)$ is a convex function. Thus, the necessary and sufficient condition to obtain maxima of $h(\mu, \mu', \nu)$ is that the gradient of $h(\mu, \mu', \nu)$ should vanish.

Hence, the inner problem of (P4-U1) gets converted to the following problem:

$$\max_{x,\mu,\mu',\nu} \Bigg\{ w\left(\sum_{t=1}^{n}((c_t)_\alpha)^U x_t + \frac{1}{2}\sum_{r=1}^{n}\sum_{t=1}^{n}((q_{rt})_\alpha)^U x_r x_t\right) + (1-w)\left(\sum_{t=1}^{n}((c_t)^\beta)^U x_t\right.$$
$$\left. + \frac{1}{2}\sum_{r=1}^{n}\sum_{t=1}^{n}((q_{rt})^\beta)^U x_r x_t\right) + \sum_{s=1}^{m}\mu_s\left(\sum_{t=1}^{n}(a_{st})_\alpha x_t - (b_s)_\alpha\right) + \sum_{s=1}^{m}\mu'_s$$
$$\left(\sum_{t=1}^{n}(a_{st})^\beta x_t - (b_s)^\beta\right) - \sum_{t=1}^{n}\nu_t x_t \Bigg\}$$

$$\text{s.t. } w\left(((c_t)_\alpha)^U + \sum_{r=1}^{n}((q_{rt})_\alpha)^U x_r\right) + (1-w)\left(((c_t)^\beta)^U + \sum_{r=1}^{n}((q_{rt})^\beta)^U x_r\right)$$
$$+ \sum_{s=1}^{m}\mu_s (a_{st})_\alpha + \sum_{s=1}^{m}\mu'_s (a_{st})^\beta - \nu_t = 0, \quad t = 1, 2, \dots, n$$

where $(a_{st})_\alpha \in [((a_{st})_\alpha)^L, ((a_{st})_\alpha)^U]$, $(b_s)_\alpha \in [((b_s)_\alpha)^L, ((b_s)_\alpha)^U]$, $(a_{st})^\beta \in [((a_{st})^\beta)^L, ((a_{st})^\beta)^U]$, $(b_s)^\beta \in [((b_s)^\beta)^L, ((b_s)^\beta)^U]$, $\mu_s, \mu'_s, \nu_t, x_t \geq 0$, $s = 1, 2, \dots, m$, $t = 1, 2, \dots, n$.

Therefore, taking into consideration the duality theory, the problem (P4-U1) to find the max value becomes

$$Z^U_{(\alpha,\beta)} = \max_{\Delta', x,\mu,\mu',\nu} \Bigg\{ w\left(\sum_{t=1}^{n}((c_t)_\alpha)^U x_t + \frac{1}{2}\sum_{r=1}^{n}\sum_{t=1}^{n}((q_{rt})_\alpha)^U x_r x_t\right) + (1-w)$$
$$\left(\sum_{t=1}^{n}((c_t)^\beta)^U x_t + \frac{1}{2}\sum_{r=1}^{n}\sum_{t=1}^{n}((q_{rt})^\beta)^U x_r x_t\right)$$
$$+ \sum_{s=1}^{m}\mu_s\left(\sum_{t=1}^{n}(a_{st})_\alpha x_t - (b_s)_\alpha\right) + \sum_{s=1}^{m}\mu'_s$$
$$\left(\sum_{t=1}^{n}(a_{st})^\beta x_t - (b_s)^\beta\right) - \sum_{t=1}^{n}\nu_t x_t \Bigg\}$$

$$\text{s.t. } w\left(((c_t)_\alpha)^U + \sum_{r=1}^{n}((q_{rt})_\alpha)^U x_r\right) + (1-w)\big(((c_t)^\beta)^U$$

$$+\sum_{r=1}^{n}((q_{rt})^{\beta})^{U}x_r\Bigg)+\sum_{s=1}^{m}\mu_s(a_{st})_{\alpha}+\sum_{s=1}^{m}\mu'_s(a_{st})^{\beta}-\nu_t=0,$$
$$\mu_s,\mu'_s,\nu_t,x_t\geq 0,\quad s=1,2,\dots,m,\quad t=1,2,\dots,n$$

where $\Delta' = \{(a_{st})_{\alpha}\in[((a_{st})_{\alpha})^{L},((a_{st})_{\alpha})^{U}],\ (b_s)_{\alpha}\in[((b_s)_{\alpha})^{L},((b_s)_{\alpha})^{U}],$
$(a_{st})^{\beta}\in[((a_{st})^{\beta})^{L},((a_{st})^{\beta})^{U}],\ (b_s)^{\beta}\in[((b_s)^{\beta})^{L},$
$((b_s)^{\beta})^{U}],\ \forall\, s,t\}.$

Since $w\left(((c_t)_{\alpha})^{U}+\sum_{r=1}^{n}((q_{rt})_{\alpha})^{U}x_r\right)+(1-w)\left(((c_t)^{\beta})^{U}+\sum_{r=1}^{n}((q_{rt})^{\beta})^{U}x_r\right)$ $+\sum_{s=1}^{m}\mu_s(a_{st})_{\alpha}+\sum_{s=1}^{m}\mu'_s(a_{st})^{\beta}-\nu_t=0,\quad t=1,2,\dots,n,$ therefore, we have $w\left(\sum_{t=1}^{n}((c_t)_{\alpha})^{U}x_t\right)+(1-w)\left(\sum_{t=1}^{n}((c_t)^{\beta})^{U}x_t\right)+\sum_{s=1}^{m}\sum_{t=1}^{n}\mu_s(a_{st})_{\alpha}x_t+\sum_{s=1}^{m}\sum_{t=1}^{n}\mu'_s(a_{st})^{\beta}x_t-\sum_{t=1}^{n}\nu_t x_t=-w\left(\sum_{r=1}^{n}\sum_{t=1}^{n}((q_{rt})_{\alpha})^{U}x_r x_t\right)-(1-w)\left(\sum_{r=1}^{n}\sum_{t=1}^{n}((q_{rt})^{\beta})^{U}x_r x_t\right).$
Thus, the above model can be re-written as:

$$Z^{U}_{(\alpha,\beta)}=\max_{\Delta',x,\mu,\mu',\nu}\left\{-\frac{w}{2}\left(\sum_{r=1}^{n}\sum_{t=1}^{n}((q_{rt})_{\alpha})^{U}x_r x_t\right)-\frac{(1-w)}{2}\right.$$
$$\left.\times\left(\sum_{r=1}^{n}\sum_{t=1}^{n}((q_{rt})^{\beta})^{U}x_r x_t\right)-\sum_{s=1}^{m}\mu_s(b_s)_{\alpha}-\sum_{s=1}^{m}\mu'_s(b_s)^{\beta}\right\}$$
$$\text{s.t.}\quad w\left(\sum_{r=1}^{n}((q_{rt})_{\alpha})^{U}x_r\right)+(1-w)\left(\sum_{r=1}^{n}((q^{\beta}_{rt})^{U}x_r\right)+\sum_{s=1}^{m}\mu_s(a_{st})_{\alpha}$$
$$+\sum_{s=1}^{m}\mu'_s(a_{st})^{\beta}-\nu_t=-w((c_t)_{\alpha})^{U}-(1-w)((c_t)^{\beta})^{U},$$
$$\mu_s,\mu'_s,\nu_t,x_t\geq 0,\quad s=1,2,\dots,m,\quad t=1,2,\dots,n$$

where $\Delta' = \{(a_{st})_{\alpha}\in[((a_{st})_{\alpha})^{L},((a_{st})_{\alpha})^{U}],\ (b_s)_{\alpha}\in[((b_s)_{\alpha})^{L},((b_s)_{\alpha})^{U}],$
$(a_{st})^{\beta}\in[((a_{st})^{\beta})^{L},((a_{st})^{\beta})^{U}],\ (b_s)^{\beta}\in[((b_s)^{\beta})^{L},((b_s)^{\beta})^{U}],\ \forall\, s,t\}.$

Moreover, since $((b_s)_{\alpha})^{L}\leq(b_s)_{\alpha}\leq((b_s)_{\alpha})^{U}$ and $((b_s)^{\beta})^{L}\leq(b_s)^{\beta}\leq((b_s)^{\beta})^{U}$, μ_s, $\mu'_s\geq 0$, $\forall\, s$, thus we have

$$Z^U_{(\alpha,\beta)} = \max_{\Delta',x,\mu,\mu',\nu}\left\{-\frac{w}{2}\left(\sum_{r=1}^{n}\sum_{t=1}^{n}((q_{rt})_\alpha)^U x_r x_t\right) - \frac{(1-w)}{2}\right.$$
$$\left.\times\left(\sum_{r=1}^{n}\sum_{t=1}^{n}((q_{rt})^\beta)^U x_r x_t\right) - \sum_{s=1}^{m}\mu_s((b_s)_\alpha)^L - \sum_{s=1}^{m}\mu'_s((b_s)^\beta)^L\right\}$$
$$\text{s.t. } w\left(\sum_{r=1}^{n}((q_{rt})_\alpha)^U x_r\right) + (1-w)\left(\sum_{r=1}^{n}((q_{rt})^\beta)^U x_r\right) + \sum_{s=1}^{m}\mu_s(a_{st})_\alpha$$
$$+\sum_{s=1}^{m}\mu'_s(a_{st})^\beta - \nu_t = -w((c_t)_\alpha)^U - (1-w)((c_t)^\beta)^U,$$
$$\mu_s, \mu'_s, \nu_t, x_t \geq 0, \quad s = 1,2,\ldots,m, \quad t = 1,2,\ldots,n.$$

Finally, since $((a_{st})_\alpha)^L \leq (a_{st})_\alpha \leq ((a_{st})_\alpha)^U$ and $((a_{st})^\beta)^L \leq (a_{st})^\beta \leq ((a_{st})^\beta)^U$, $\forall\, s,t$, it follows that

$$\textbf{(P6)}\ Z^U_{(\alpha,\beta)} = \max_{x,\mu,\mu',\nu}\left\{-\frac{w}{2}\left(\sum_{r=1}^{n}\sum_{t=1}^{n}((q_{rt})_\alpha)^U x_r x_t\right) - \frac{(1-w)}{2}\right.$$
$$\left.\times\left(\sum_{r=1}^{n}\sum_{t=1}^{n}((q_{rt})^\beta)^U x_r x_t\right) - \sum_{s=1}^{m}\mu_s((b_s)_\alpha)^L - \sum_{s=1}^{m}\mu'_s((b_s)^\beta)^L\right\}$$
$$\text{s.t. } w\left(\sum_{r=1}^{n}((q_{rt})_\alpha)^U x_r\right) + (1-w)\left(\sum_{r=1}^{n}((q_{rt})^\beta)^U x_r\right) + \sum_{s=1}^{m}\mu_s((a_{st})_\alpha)^L$$
$$+\sum_{s=1}^{m}\mu'_s((a_{st})^\beta)^L - \nu_t \leq -w((c_t)_\alpha)^U - (1-w)((c_t)^\beta)^U,$$
$$w\left(\sum_{r=1}^{n}((q_{rt})_\alpha)^U x_r\right) + (1-w)\left(\sum_{r=1}^{n}((q_{rt})^\beta)^U x_r\right) + \sum_{s=1}^{m}\mu_s((a_{st})_\alpha)^U$$
$$+\sum_{s=1}^{m}\mu'_s((a_{st})^\beta)^U - \nu_t \geq -w((c_t)_\alpha)^U - (1-w)((c_t)^\beta)^U,$$
$$\alpha+\beta \leq 1, \ \alpha \geq \beta, \ \alpha \geq 0, \ \beta \geq 0, \ 0 \leq w \leq 1,$$
$$\mu_s, \mu'_s, \nu_t, x_t \geq 0, \quad s = 1,2,\ldots,m, \quad t = 1,2,\ldots,n.$$

Hence, the problem (P6) can be used to find the upper bound of the problem (P4).

(ii) Direct approach

Here, we propose a modified and simplified model for (P4-U1) to obtain the upper bound value $Z^U_{(\alpha,\beta)}$. It is asserted that the following optimization problem (P7-U1) defined as

$$\textbf{(P7-U1)}\ Z^U_{(\alpha,\beta)} = \min_{x}\left\{w\left(\sum_{t=1}^{n}((c_t)_\alpha)^U x_t + \frac{1}{2}\sum_{r=1}^{n}\sum_{t=1}^{n}((q_{rt})_\alpha)^U x_r x_t\right) + \right.$$
$$\left.(1-w)\left(\sum_{t=1}^{n}((c_t)^\beta)^U x_t + \frac{1}{2}\sum_{r=1}^{n}\sum_{t=1}^{n}((q_{rt})^\beta)^U x_r x_t\right)\right\}$$

$$\text{s.t.} \sum_{t=1}^{n}((a_{st})_\alpha)^U x_t \leq ((b_s)_\alpha)^L,$$

$$\sum_{t=1}^{n}((a_{st})^\beta)^U x_t \leq ((b_s)^\beta)^L,$$

$$\alpha + \beta \leq 1, \ \ \alpha \geq \beta, \ \ \alpha \geq 0, \ \ \beta \geq 0, \ \ 0 \leq w \leq 1,$$

$$x_t \geq 0, \ \ s = 1, 2, \ldots, m, \ \ t = 1, 2, \ldots, n$$

yields the same optimal solution as obtained from model (P6).

Proof. Since $((a_{st})_\alpha)^L \leq ((a_{st})_\alpha)^U$ and $((a_{st})^\beta)^L \leq ((a_{st})^\beta)^U$, $\forall\ s, t$, therefore, model (P7-U1) is equivalent to

$$\textbf{(P7-U2)} \quad Z^U_{(\alpha,\beta)} = \min_x \left\{ w\left(\sum_{t=1}^{n}((c_t)_\alpha)^U x_t + \frac{1}{2}\sum_{r=1}^{n}\sum_{t=1}^{n}((q_{rt})_\alpha)^U x_r x_t\right) + (1-w)\left(\sum_{t=1}^{n}((c_t)^\beta)^U x_t + \frac{1}{2}\sum_{r=1}^{n}\sum_{t=1}^{n}((q_{rt})^\beta)^U x_r x_t\right)\right\}$$

$$\text{s.t.} \sum_{t=1}^{n}((a_{st})_\alpha)^U x_t \leq ((b_s)_\alpha)^L,$$

$$\sum_{t=1}^{n}((a_{st})_\alpha)^L x_t \leq ((b_s)_\alpha)^L,$$

$$\sum_{t=1}^{n}((a_{st})^\beta)^U x_t \leq ((b_s)^\beta)^L,$$

$$\sum_{t=1}^{n}((a_{st})^\beta)^L x_t \leq ((b_s)^\beta)^L,$$

$$\alpha + \beta \leq 1, \ \ \alpha \geq \beta, \ \ \alpha \geq 0, \ \ \beta \geq 0, \ \ 0 \leq w \leq 1,$$

$$x_t \geq 0, \ \ s = 1, 2, \ldots, m, \ \ t = 1, 2, \ldots, n.$$

Now, we shall prove that the dual problem of (P7-U2) is identical to the problem (P6). Assuming $\mu_s, \mu'_s, \nu_s, \nu'_s, \delta_t$ to be the Lagrange's multipliers corresponding to the constraints of the above problem, then the dual of the problem (P7-U2) is given by:

$$\max_{x,\mu,\nu,\delta} \left\{ w\left(\sum_{t=1}^{n}((c_t)_\alpha)^U x_t + \frac{1}{2}\sum_{r=1}^{n}\sum_{t=1}^{n}((q_{rt})_\alpha)^U x_r x_t\right) + (1-w)\left(\sum_{t=1}^{n}((c_t)^\beta)^U x_t + \frac{1}{2}\sum_{r=1}^{n}\sum_{t=1}^{n}((q_{rt})^\beta)^U x_r x_t\right) + \sum_{s=1}^{m}\mu_s\left(\sum_{t=1}^{n}((a_{st})_\alpha)^U x_t - ((b_s)_\alpha)^L\right) + \sum_{s=1}^{m}\mu'_s \left(\sum_{t=1}^{n}((a_{st})_\alpha)^L x_t - ((b_s)_\alpha)^L\right) + \sum_{s=1}^{m}\nu_s\left(\sum_{t=1}^{n}((a_{st})^\beta)^U x_t - ((b_s)^\beta)^L\right) + \right.$$

$$\left.\sum_{s=1}^{m}\nu_s'\left(\sum_{t=1}^{n}((a_{st})^{\beta})^{L}x_t-((b_s)^{\beta})^{L}\right)-\sum_{t=1}^{n}\delta_t x_t\right\}$$

$$\text{s.t. } w\left(((c_t)_{\alpha})^{U}+\sum_{r=1}^{n}((q_{rt})_{\alpha})^{U}x_r\right)+(1-w)\left(((c_t)^{\beta})^{U}+\sum_{r=1}^{n}((q_{rt})^{\beta})^{U}x_r\right)+$$

$$\sum_{s=1}^{m}\mu_s((a_{st})_{\alpha})^{U}+\sum_{s=1}^{m}\mu_s'((a_{st})_{\alpha})^{L}+\sum_{s=1}^{m}\nu_s((a_{st})^{\beta})^{U}+\sum_{s=1}^{m}\nu_s'((a_{st})^{\beta})^{L}-\delta_t=0,$$

$$\alpha+\beta\le 1,\ \ \alpha\ge\beta,\ \ \alpha\ge 0,\ \ \beta\ge 0,\ \ 0\le w\le 1,$$

$$\mu_s,\mu_s',\nu_s,\nu_s',\delta_t,x_t\ge 0,\ \ s=1,2,\dots,m,\ \ t=1,2,\dots,n.$$

Now, the equation $w\left(((c_t)_{\alpha})^{U}+\sum_{r=1}^{n}((q_{rt})_{\alpha})^{U}x_r\right)+(1-w)\left(((c_t)^{\beta})^{U}+\sum_{r=1}^{n}((q_{rt})^{\beta})^{U}x_r\right)+\sum_{s=1}^{m}\mu_s((a_{st})_{\alpha})^{U}+\sum_{s=1}^{m}\mu_s'((a_{st})_{\alpha})^{L}+\sum_{s=1}^{m}\nu_s((a_{st})^{\beta})^{U}+\sum_{s=1}^{m}\nu_s'((a_{st})^{\beta})^{L}-\delta_t=0,\ t=1,2,\dots,n$ gives

$$w\left(\sum_{t=1}^{n}((c_t)_{\alpha})^{U}x_t\right)+(1-w)\left(\sum_{t=1}^{n}((c_t)^{\beta})^{U}x_t\right)+\sum_{s=1}^{m}\sum_{t=1}^{n}\mu_s((a_{st})_{\alpha})^{U}x_t+$$
$$\sum_{s=1}^{m}\sum_{t=1}^{n}\mu_s'((a_{st})_{\alpha})^{L}x_t+\sum_{s=1}^{m}\sum_{t=1}^{n}\nu_s((a_{st})^{\beta})^{U}x_t+\sum_{s=1}^{m}\sum_{t=1}^{n}\nu_s'((a_{st})^{\beta})^{L}x_t-\sum_{t=1}^{n}\delta_t x_t=$$
$$-w\left(\sum_{r=1}^{n}\sum_{t=1}^{n}((q_{rt})_{\alpha})^{U}x_r x_t\right)-(1-w)\left(\sum_{r=1}^{n}\sum_{t=1}^{n}((q_{rt})^{\beta})^{U}x_r x_t\right).$$

Consequently, we have

$$Z^{U}_{(\alpha,\beta)}=\max_{x,\mu,\nu,\delta}\left\{-\frac{w}{2}\left(\sum_{r=1}^{n}\sum_{t=1}^{n}((q_{rt})_{\alpha})^{U}x_r x_t\right)-\frac{(1-w)}{2}\left(\sum_{r=1}^{n}\sum_{t=1}^{n}((q_{rt})^{\beta})^{U}x_r x_t\right)\right.$$
$$-\sum_{s=1}^{m}\mu_s((b_s)_{\alpha})^{L}-\sum_{s=1}^{m}\mu_s'((b_s)_{\alpha})^{L}$$
$$\left.-\sum_{s=1}^{m}\nu_s((b_s)^{\beta})^{L}-\sum_{s=1}^{m}\nu_s'((b_s)^{\beta})^{L}\right\}$$

$$\text{s.t. } w\left(\sum_{r=1}^{n}((q_{rt})_{\alpha})^{U}x_r\right)+(1-w)\left(\sum_{r=1}^{n}((q_{rt})^{\beta})^{U}x_r\right)+\sum_{s=1}^{m}\mu_s((a_{st})_{\alpha})^{U}$$
$$+\sum_{s=1}^{m}\nu_s((a_{st})^{\beta})^{U}+\sum_{s=1}^{m}\nu_s'((a_{st})^{\beta})^{L}-\delta_t=-w((c_t)_{\alpha})^{U}$$
$$-(1-w)((c_t)^{\beta})^{U},$$
$$\alpha+\beta\le 1,\ \ \alpha\ge\beta,\ \ \alpha\ge 0,\ \ \beta\ge 0,\ \ 0\le w\le 1,$$
$$\mu_s,\mu_s',\nu_s,\nu_s',\delta_t,x_t\ge 0,\ \ s=1,2,\dots,m,\ \ t=1,2,\dots,n.$$

Further, $((a_{st})_\alpha)^L \le ((a_{st})_\alpha)^U$ and $((a_{st})^\beta)^L \le ((a_{st})^\beta)^U$, $\forall\, s,t$ yields

$$Z^U_{(\alpha,\beta)} = \max_{x,\mu,\nu,\delta}\left\{-\frac{w}{2}\left(\sum_{r=1}^{n}\sum_{t=1}^{n}((q_{rt})_\alpha)^U x_r x_t\right) - \frac{(1-w)}{2}\right.$$
$$\times\left(\sum_{r=1}^{n}\sum_{t=1}^{n}((q_{rt})^\beta)^U x_r x_t\right) - \sum_{s=1}^{m}\mu_s((b_s)_\alpha)^L - \sum_{s=1}^{m}\mu'_s((b_s)_\alpha)^L$$
$$\left.-\sum_{s=1}^{m}\nu_s((b_s)^\beta)^L - \sum_{s=1}^{m}\nu'_s((b_s)^\beta)^L\right\}$$

$$\text{s.t. } w\left(\sum_{r=1}^{n}((q_{rt})_\alpha)^U x_r\right) + (1-w)\left(\sum_{r=1}^{n}((q_{rt})^\beta)^U x_r\right) + \sum_{s=1}^{m}\mu_s((a_{st})_\alpha)^L$$
$$+\sum_{s=1}^{m}\mu'_s((a_{st})_\alpha)^L + \sum_{s=1}^{m}\nu_s((a_{st})^\beta)^L + \sum_{s=1}^{m}\nu'_s((a_{st})^\beta)^L - \delta_t \le -w((c_t)_\alpha)^U$$
$$(1-w)((c_t)^\beta)^U,$$
$$w\left(\sum_{r=1}^{n}((q_{rt})_\alpha)^U x_r\right) + (1-w)\left(\sum_{r=1}^{n}((q_{rt})^\beta)^U x_r\right) + \sum_{s=1}^{m}\mu_s((a_{st})_\alpha)^U$$
$$\sum_{s=1}^{m}\mu'_s((a_{st})_\alpha)^U + \sum_{s=1}^{m}\nu_s((a_{st})^\beta)^U + \sum_{s=1}^{m}\nu'_s((a_{st})^\beta)^U - \delta_t \ge -w((c_t)_\alpha)^U$$
$$-(1-w)((c_t)^\beta)^U,$$
$$\alpha+\beta \le 1,\ \ \alpha \ge \beta,\ \ \alpha \ge 0,\ \ \beta \ge 0,\ \ 0 \le w \le 1,$$
$$\mu_s, \mu'_s, \nu_s, \nu'_s, \delta_t, x_t \ge 0,\ \ s = 1,2,\ldots,m,\ \ t = 1,2,\ldots,n.$$

Lastly, replacing $\mu_s + \mu'_s$ and $\nu_s + \nu'_s$ by λ_s and λ'_s, respectively, we obtain the model (P6). This establishes our claim. □

Remark 11.3.2.1. Moving onto the similar lines, we get the parallel results for the maximization type problem. Consider the following optimization problem:

$$\textbf{(M1)}\quad \max\quad \tilde{Z}^I = \sum_{t=1}^{n}\tilde{c}^I_t x_t + \frac{1}{2}\sum_{r=1}^{n}\sum_{t=1}^{n}\tilde{q}^I_{rt}x_r x_t$$
$$\text{s.t.}\quad \sum_{t=1}^{n}\tilde{a}^I_{st}x_t \ge \tilde{b}^I_s,\ \ x_t \ge 0,$$
$$s = 1,2,\ldots,m,\ \ t = 1,2,\ldots,n$$

where $\tilde{c}^I_t$, $\tilde{a}^I_{st}$, $\tilde{b}^I_s$, $\tilde{q}^I_{rt}$ are taken to be TIFNs and matrix $[\tilde{q}^I_{rt}]_{n\times n}$ is a symmetric negative semi-definite IF matrix. Then, the lower and upper bound problems are, respectively, given by

$$\textbf{(M2)}\ Z^L_{(\alpha,\beta)} = \max_{x}\left\{w\left(\sum_{t=1}^{n}((c_t)_\alpha)^L x_t + \frac{1}{2}\sum_{r=1}^{n}\sum_{t=1}^{n}((q_{rt})_\alpha)^L x_r x_t\right)+\right.$$

$$(1-w)\left(\sum_{t=1}^{n}((c_t)^{\beta})^{L}x_t + \frac{1}{2}\sum_{r=1}^{n}\sum_{t=1}^{n}((q_{rt})^{\beta})^{L}x_r x_t\right)\Bigg\}$$

$$\text{s.t.}\ \sum_{t=1}^{n}((a_{st})_{\alpha})^{L}x_t \geq ((b_s)_{\alpha})^{U},$$

$$\sum_{t=1}^{n}((a_{st})^{\beta})^{L}x_t \geq ((b_s)^{\beta})^{U},$$

$$\alpha+\beta \leq 1,\ \ \alpha \geq \beta,\ \ \alpha \geq 0,\ \ \beta \geq 0,\ \ 0 \leq w \leq 1,$$

$$x_t \geq 0,\ \ s = 1,2,\ldots,m;\ \ t = 1,2,\ldots,n$$

$$\textbf{(M3)}\ Z^{U}_{(\alpha,\beta)} = \max_{x}\Bigg\{w\left(\sum_{t=1}^{n}((c_t)_{\alpha})^{U}x_t + \frac{1}{2}\sum_{r=1}^{n}\sum_{t=1}^{n}((q_{rt})_{\alpha})^{U}x_r x_t\right) +$$

$$(1-w)\left(\sum_{t=1}^{n}((c_t)^{\beta})^{U}x_t + \frac{1}{2}\sum_{r=1}^{n}\sum_{t=1}^{n}((q_{rt})^{\beta})^{U}x_r x_t\right)\Bigg\}$$

$$\text{s.t.}\ \sum_{t=1}^{n}((a_{st})_{\alpha})^{U}x_t \geq ((b_s)_{\alpha})^{L},$$

$$\sum_{t=1}^{n}((a_{st})^{\beta})^{U}x_t \geq ((b_s)^{\beta})^{L},$$

$$\alpha+\beta \leq 1,\ \ \alpha \geq \beta,\ \ \alpha \geq 0,\ \ \beta \geq 0,\ \ 0 \leq w \leq 1,$$

$$x_t \geq 0,\ \ s = 1,2,\ldots,m,\ \ t = 1,2,\ldots,n.$$

Finally, the membership and non-membership functions of the optimal objective function can be obtained by varying α and β so as to get the various distinct values of the bounds $Z^{L}_{(\alpha,\beta)}$ and $Z^{U}_{(\alpha,\beta)}$.

11.4 NUMERICAL ILLUSTRATION

Solve the following IF-QPP:

$$\textbf{(N1)}\ \text{Minimize}\ \tilde{Z}^{I} = \Big[(-3,-2,-1;-4,-2,0)x_1 \oplus (1,1.5,2;0,1.5,4)x_2$$

$$\oplus\ \frac{1}{2}\Big((3,5,7;1,5,9)x_1^2$$

$$\oplus\ (-6,-4,-2;-8,-4,0)x_1x_2 \oplus (2,4,6;1,4,7)x_2^2\Big)\Big]$$

$$\text{subject to}\ (0,1,3;-2,1,4)x_1 \oplus (-3,-2,-1;-4,-2,1)x_2 \preceq (1,4,5;0,4,8),$$

$$(1,2,3;0,2,4)x_1 \oplus (0.5,1,1.5;-1,1,2)x_2 \preceq (5,7,9;4,7,10),$$

$$x_1, x_2 \geq 0.$$

Solution: Here, for the IF-QPP (N1), we have matrix

$$\tilde{Q} = \begin{bmatrix} (3,5,7;1,5,9) & (-3,-2,-1;-4,-2,0) \\ (-3,-2,-1;-4,-2,0) & (2,4,6;1,4,7) \end{bmatrix}$$

so that after applying the accuracy function, the corresponding crisp matrix is given by

$$Q = \begin{bmatrix} 5 & -2 \\ -2 & 4 \end{bmatrix}$$

which is a positive definite matrix.

Now, using the proposed methodology for $\alpha, \beta \in [0, 1]$ such that $\alpha + \beta \leq 1$, the model (P5-L2) yields

$$\textbf{(N2)}\ Z^L_{\alpha,\beta} = \min_{x_1,x_2} \Big[w\Big((-3+\alpha)x_1 + (1+0.5\alpha)x_2 + \frac{1}{2}\big((3+2\alpha)x_1^2 + (-6+2\alpha)x_1x_2 + (2+2\alpha)x_2^2\big)\Big) + (1-w)\Big((-2-2\beta)x_1 + (1.5-1.5\beta)x_2 + \frac{1}{2}\big((5-4\beta)x_1^2 + (-4-4\beta)x_1x_2 + (4-3\beta)x_2^2\big)\Big)\Big]$$

$$\begin{aligned} \text{s.t.}\ & \alpha x_1 + (-3+\alpha)x_2 \leq 5 - \alpha, \\ & (1+\alpha)x_1 + (0.5+0.5\alpha)x_2 \leq 9 - 2\alpha, \\ & (1-3\beta)x_1 + (-2-2\beta)x_2 \leq 4 + 4\beta, \\ & (2-2\beta)x_1 + (1-2\beta)x_2 \leq 7 + 3\beta, \\ & \alpha + \beta \leq 1,\ \alpha \geq \beta, \\ & \alpha, \beta, x_1, x_2 \geq 0, \end{aligned}$$

and, the model (P7-U1) results into

$$\textbf{(N3)}\ Z^U_{\alpha,\beta} = \min_{x_1,x_2} \Big[w\Big((-1-\alpha)x_1 + (2-0.5\alpha)x_2 + \frac{1}{2}\big((7-2\alpha)x_1^2 + (-2-2\alpha)x_1x_2 + (6-2\alpha)x_2^2\big)\Big) + (1-w)\Big((-2+2\beta)x_1 + (1.5+2.5\beta)x_2 + \frac{1}{2}\big((5+4\beta)x_1^2 + (-4+4\beta)x_1x_2 + (4+3\beta)x_2^2\big)\Big)\Big]$$

$$\begin{aligned} \text{s.t.}\ & (3-2\alpha)x_1 + (-1-\alpha)x_2 \leq 1 + 3\alpha, \\ & (3-\alpha)x_1 + (1.5-0.5\alpha)x_2 \leq 5 + 2\alpha, \\ & (1+3\beta)x_1 + (-2+3\beta)x_2 \leq 4 - 4\beta, \\ & (2+2\beta)x_1 + (1+\beta)x_2 \leq 7 - 3\beta, \\ & \alpha + \beta \leq 1,\ \alpha \geq \beta, \\ & \alpha, \beta, x_1, x_2 \geq 0 \end{aligned}$$

where w and $1-w$ are the weights attached with the membership and non-membership counterparts, respectively, such that $0 \leq w \leq 1$. Moreover, the crisp optimization models (N2) and (N3) are solved using the "LINGO-17.0" software and the optimal objective function values obtained from models (N2) and (N3) for distinct values of w, α and β are presented in Tables 11.1 and 11.2.

TABLE 11.1
Value of $Z = [Z^L_{\alpha,\beta}, Z^U_{\alpha,\beta}]$ at Different α, β-cuts for $w = 0.6$

$\alpha \backslash \beta$	0.0	0.1	0.2	0.3	0.4	0.5
0.2	[−0.7839, −0.1938]	[−0.9417, −0.1694]	[−1.2272, −0.1473]	Infeasible	Infeasible	Infeasible
0.4	[−0.6506, −0.2351]	[−0.7325, −0.2069]	[−0.8585, −0.1813]	[−1.0684, −0.1581]	[−1.4691, −0.137]	Infeasible
0.6	[−0.5551, −0.2826]	[−0.6172, −0.2502]	[−0.689, −0.2207]	[−0.7929, −0.1938]	[−0.9561, −0.1694]	Infeasible
0.8	[−0.4721, −0.3372]	[−0.5261, −0.3]	[−0.5854, −0.2661]	Infeasible	Infeasible	Infeasible
1.0	[−0.4,−0.4]	Infeasible	Infeasible	Infeasible	Infeasible	Infeasible

TABLE 11.2
Value of $Z = [Z^L_{\alpha,\beta}, Z^U_{\alpha,\beta}]$ at Different α, β-cuts for $w = 0.4$

$\alpha \backslash \beta$	0.0	0.1	0.2	0.3	0.4	0.5
0.2	[−0.6172, −0.2502]	[−0.7335, −0.2069]	[−0.9513, −0.1694]	Infeasible	Infeasible	Infeasible
0.4	[−0.551, −0.2826]	[−0.6507, −0.2351]	[−0.7929, −0.1938]	[−1.0758, −0.1581]	[−1.8148, −0.1272]	Infeasible
0.6	[−0.4985, −0.3182]	[−0.5854, −0.2661]	[−0.6901, −0.2207]	[−0.8688, −0.1813]	[−1.2414, −0.1473]	Infeasible
0.8	[−0.4469, −0.3572]	[−0.5261, −0.3]	[−0.6172, −0.2502]	Infeasible	Infeasible	Infeasible
1.0	[−0.4,−0.4]	Infeasible	Infeasible	Infeasible	Infeasible	Infeasible

11.5 CONCLUSION

In the present study, we have discussed the construction and solution methodology of quadratic programming problems under an intuitionistic fuzzy environment. The proposed algorithm is of great significance since it efficiently deals with the ambiguous data inherent in a realistic problem along with its possible reduction to solve the fuzzy and the crisp quadratic programming problems. Here, by employing the technique of α-cut, the original IF-QPP is transformed to the lower and upper bound optimization sub-problems. The lower bound problem can be solved directly, however, we have investigated two different approaches to deal with the max-min type upper bound problem. First, a duality theory based method is suggested to obtain the optimal upper bound value and later on, a direct solution approach is introduced which yields the same optimal solution as that of the dual approach. The latter technique to find the upper bound value is much more computationally efficient than the former, since it involves a lesser number of constraints and variables in the model. Moreover, the presented direct approach has substantially less computing time, allowing it to be used to handle big-data problems of different sectors. Although we have successfully handled an IF-QPP, the problem of solving a fully IF-QPP still remains a point of concern. Therefore, in the future, the methodology can be modified/extended to solve the fully single/multi-objective IF-QPPs.

ACKNOWLEDGEMENTS

The authors would like to express their gratitude to the anonymous reviewers for their insightful comments and recommendations, which have significantly improved the quality and presentation of the paper. The first author is also grateful to the MHRD, Govt. of India, for financial support to carry out this research work.

REFERENCES

Abdel-Malek, L.L., and Areeratchakul, N. (2007). "A quadratic programming approach to the multi-product newsvendor problem with side constraints." *European Journal of Operational Research*, 176(3), 1607–1619.

Ammar, E.E. (2008). "On solutions of fuzzy random multiobjective quadratic programming with applications in portfolio problem." *Information Sciences*, 178, 468–84.

Angelov, P.P. (1997). "Optimization in an intuitionistic fuzzy environment." *Fuzzy Sets and Systems*, 86, 299–306.

Cruz, C., Silva, R.C., and Verdegay, J.L. (2011). "Extending and relating different approaches for solving fuzzy quadratic problems." *Fuzzy Optimization and Decision Making*, 10, 193–210.

Dorn, W.S. (1960). "Duality in quadratic programming." *Quarterly of Applied Mathematics*, 18(2), 155–162.

Dubois, D., and Prade, H. (1980). "Fuzzy sets and systems: Theory and application." San Diego, USA: Academic Press.

Huang, G.H. (1996). "IPWM: an interval parameter water quality management model." *Engineering Optimization*, 26(2), 79–103.

Khalifa, A.E.W. (2020). "A new approach for solving fully fuzzy quadratic programming problems." *International Journal of Optimization in Civil Engineering*, 10(2), 333–344.

Khalifa, H.A., Alharbi, M., and Kumar, P. (2021). "A new method for solving quadratic fractional programming problem in neutrosophic environment." *Open Engineering*, 11(1), 880–886.

Kheirfam, B., and Verdegay, J.L. (2012). "Strict sensitivity analysis in fuzzy quadratic programming." *Fuzzy Sets and Systems*, 198, 99–111.

Kumar, C.A., and Jeyalakshmi, R.R. (2017). "The maximum index method for solving the fuzzy quadratic programming problem with interval numbers using different levels of α-cuts." *Journal of Engineering Technology*, 6, 240–247.

Liu, S.T. (2009). "A revisit to quadratic programming with fuzzy parameters." *Chaos, Solitons and Fractals*, 41, 1401–1407.

Mahajan, S., and Gupta, S.K. (2019). "Methods to solve QPPs with fuzzy parameters and fuzzy variables." *Journal of Intelligent & Fuzzy Systems*, 37(2), 2757–2767.

Mahajan, S., and Gupta, S.K. (2020). "Development and analysis of a sustainable garbage disposal model for environmental management under uncertainty." *Science of the Total Environment*, 709, 135037.

Mahajan, S., and Gupta, S.K. (2021). "On inexact quadratic programming problems involving mixed terms with an application to tea industry." *Computers & Industrial Engineering*, 156, 107264.

Mahajan, S., Gupta, S.K., Ahmad, I., and Al-Homidan, S. (2021). "Using concave optimization methods for inexact quadratic programming problems with an application to waste management." *Journal of Inequalities and Applications*, 2021(1), 1–19.

Maheswari, U. (2019). "A new approach for the solution of fuzzy quadratic programming problems." *Journal of Advanced Research in Dynamical and Control Systems*, 11(1), 342–349.

Malik, M., Gupta, S.K., and Ahmad, I. (2021). "A new approach to solve fully intuitionistic fuzzy linear programming problem with unrestricted decision variables." *Journal of Intelligent & Fuzzy Systems*, 41, 6053–6066.

Mirmohseni, S.M., and Nasseri, S.H. (2017). "A Quadratic programming with Triangular Fuzzy numbers." *Journal of Applied Mathematics and Physics*, 5, 2218–2227.

Silva, R.C., Cruz, C., and Verdegay, J.L. (2013). "Fuzzy costs in quadratic programming problems." *Fuzzy Optimization and Decision Making*, 12, 231–248.

Silva, R.C., Verdegay, J.L., and Yamakami, A. (2007). "Two-phase method to solve fuzzy quadratic programming problems." *In 2007 IEEE International Fuzzy Systems Conference* (pp. 1–6). IEEE.

Singh, S.K., and Yadav, S.P. (2015). "Modeling and optimization of multi objective nonlinear programming problem in intuitionistic fuzzy environment." *Applied Mathematical Modelling*, 39, 4617–4629.

Zadeh, L.A. (1965). "Fuzzy Sets." *Information and Control*, 8, 338–353.

Zhou, X.G., Cao, B.Y., and Nasseri, S.H. (2014). "Optimality conditions for fuzzy number quadratic programming with fuzzy coefficients." *Journal of Applied Mathematics*, 2014, 1–8.

Index